Mathematical Classification and Clustering

Nonconvex Optimization and Its Applications

Volume 11

The titles published in this series are listed at the end of this volume.

Mathematical Classification and Clustering

by

Boris Mirkin
DIMACS, Rutgers University

KLUWER ACADEMIC PUBLISHERS
DORDRECHT / BOSTON / LONDON

A C.I.P. Catalogue record for this book is available from the Library of Congress

ISBN-13: 978-1-4613-8057-3   e-ISBN-13: 978-1-4613-0457-9
DOI: 10.1007/978-1-4613-0457-9

Published by Kluwer Academic Publishers,
P.O. Box 17, 3300 AA Dordrecht, The Netherlands.

Kluwer Academic Publishers incorporates
the publishing programmes of
D. Reidel, Martinus Nijhoff, Dr W. Junk and MTP Press.

Sold and distributed in the U.S.A. and Canada
by Kluwer Academic Publishers,
101 Philip Drive, Norwell, MA 02061, U.S.A.

In all other countries, sold and distributed
by Kluwer Academic Publishers Group,
P.O. Box 322, 3300 AH Dordrecht, The Netherlands.

Printed on acid-free paper

Table of Contents

Variable Weights. Approximate Conjunctive Concepts. Selecting the Variables. Transforming the Variable Space. Knowledge Discovery. Discussion

Row/Column Partitioning Bipartitioning Discussion

Rooted Labeled Tree. Indexed Tree and Ultrametric. Hierarchy and Additive Structure. Nest Indicator Function. Edge-Weighted Tree and Tree Metric. T-Splits. Neighbors Relation. Character Rooted Trees. Comparing Hierarchies. Discussion

Monotone Equivariance and Threshold Graphs. Isotone Cluster Methods. Classes of Isotone Methods. Discussion

Ultrametric and Minimum Spanning Trees. Tree Metric and Its Adjustment. Discussion

Split Metrics and Canonical Decomposition. Mathematical Properties. Weak Clusters and Weak Hierarchy. Discussion

Pyramids. Least-Squares Fitting. Discussion

Binary Hierarchy Decomposition of a Data Matrix. Cluster Value Strategy for Divisive Clustering. Approximation of Square Tables. Discussion

Foreword

I am very happy to have this opportunity to present the work of Boris Mirkin, a distinguished Russian scholar in the areas of data analysis and decision making methodologies.

The monograph is devoted entirely to clustering, a discipline dispersed through many theoretical and application areas, from mathematical statistics and combinatorial optimization to biology, sociology and organizational structures. It compiles an immense amount of research done to date, including many original Russian developments never presented to the international community before (for instance, cluster-by-cluster versions of the K-Means method in Chapter 4 or uniform partitioning in Chapter 5). The author's approach, approximation clustering, allows him both to systematize a great part of the discipline and to develop many innovative methods in the framework of optimization problems. The optimization methods considered are proved to be meaningful in the contexts of data analysis and clustering.

The material presented in this book is quite interesting and stimulating in paradigms, clustering and optimization. On the other hand, it has a substantial application appeal. The book will be useful both to specialists and students in the fields of data analysis and clustering as well as in biology, psychology, economics, marketing research, artificial intelligence, and other scientific disciplines.

Panos Pardalos, Series Editor.

Preface

The world is organized via classification: elements in physics, compounds in chemistry, species in biology, enterprises in industries, illnesses in medicine, standards in technology, firms in economics, countries in geography, parties in politics — all these are witnesses to that. The science of classification, which deals with the problems of how classifications emerge, function and interact, is still unborn. What we have in hand currently is clustering, the discipline aimed at revealing classifications in observed real-world data. Though we can trace the existence of clustering activities back a hundred years, the real outburst of the discipline occurred in the sixties, with the computer era coming to handle the real-world data.

Within just a few years, a number of books appeared describing the great opportunities opened in many areas of human activity by algorithms for finding "coherent" clusters in a data "cloud" put in a geometrical space (see, for example, Benzécri 1973, Bock 1974, Clifford and Stephenson 1975, Duda and Hart 1973, Duran and Odell 1974, Everitt 1974, Hartigan 1975, Sneath and Sokal 1973, Sonquist, Baker, and Morgan 1973, Van Ryzin 1977, Zagoruyko 1972).

The strict computer eye was supposed to substitute for imprecise human vision and transform the art of classification into a scientific exercise (for instance, numerical taxonomy was to replace handmade and controversial taxonomy in biology). The good news in that was that the algorithms did find clusters. The bad news was that there was no rigorous theoretical foundation underlying the algorithms. Moreover, for a typical case in which no clear cluster structure prevailed in the data, different algorithms produced different clusters. More bad news was the lack of any rigorous tool for interpreting the clusters found, which yielded eventually to the emergence of the so-called conceptual clustering as a counterpart to the traditional one.

The pessimism generated by these obstacles can be felt in popular sayings like these: "There are more clustering techniques suggested than the number of real-world problems resolved with them", and "Clustering algorithms are worth a dime a dozen." However, the situation is improving, in the long run. More and more real-world problems, such as early diagnostics in medicine, knowledge discovery and message understanding in artificial intelligence, machine vision and robot planning in engineering, require developing a sound theory for clustering.

In the last two decades, beyond the traditional activity of inventing new clustering concepts and algorithms, we can distinguish two overlapping mainstreams potentially leading to bridging the gaps within the clustering discipline. One is related to modeling cluster structures in terms of observed data, and the other is connected with analyzing particular kinds of phenomena, such as image processing or biomolecular-data-based phylogeny reconstructing – even though in the latter kind of analyses, clustering is only a part, however important, of the entire problem.

Within the former movements, initially, the effort was concentrated on developing probabilistic models in a statistical framework (see, for example, monographs by Breiman et al. 1984, Jain and Dubes 1988, McLachlan and Basford 1988), leaning more to testing rather than to revealing the cluster structures. However, all along, work was being done on modeling of clusters in the data just as it is, without any connection to a possible probabilistic mechanism of data generation. In this paradigm, probabilistic clusters are just a particular clustering structure, and the clustering discipline seems more related to mathematics and artificial intelligence than to statistics. The present book offers an account of clustering in the framework of this wider paradigm.

Actually, the book's goal is threefold. First, it is supposed to be a reference book for the enormous amount of existing clustering concepts and methods; second, it can be utilized as a clustering text-book; and, third, it is a presentation of the author's and his Russian colleagues' results, put in the perspective of the current development.

As a reference book, it features:

(a) a review of classification as a scientific notion;

(b) an updated review of clustering algorithms based on a systematic typology of input-data/output-cluster-structures (the set of cluster structures considered is quite extensive and includes such structures as neural networks);

(c) a detailed description of the approaches in single cluster clustering, partitioning, and hierarchical clustering, including most recent developments made in various countries (Canada, France, Germany, Russia, USA);

(d) development of a unifying approximation approach;

(e) an extensive bibliography, and

(f) an index.

To serve in the text-book capacity, the monograph includes:

(a) a dozen illustrative and small, though real-world, data sets, along with clustering problems quite similar to those for larger real data sets;

(b) detailed description and discussion of the major algorithms and underlying theories;

(c) solutions to the illustrative problems found with the algorithms described (which can be utilized as a stock of exercises).

It should be pointed out that the data sets, mostly, are taken from published sources and have been discussed in the literature extensively, which provides the reader with opportunity to look at them from various perspectives. The examples are printed with a somewhat smaller font, like this.

The present author's results are based on a different approach to cluster analysis, which can be referred to as *approximation clustering*, developed by him and his collaborators starting in the early seventies. Some similar work is being done in the USA and in the other countries. In this approach, clustering is considered to approximate data by a simpler, cluster-wise structure rather than to reveal the geometrically explicit "coherent clusters" in a data point-set. The results found within the approximation approach amount to a mathematical theory for clustering involving the following directions of development: (a) unifying a considerable part of the clustering techniques, (b) developing new techniques, (c) finding relations among various notions and algorithms both within the clustering discipline and outside – especially in statistics, machine learning and combinatorial optimization.

The unifying capability of approximation clustering is grounded on convenient relations which exist between approximation problems and geometrically explicit clustering. Based on this, the major clustering techniques have been reformulated as locally optimal approximation algorithms and extended to many situations untreatable with explicit approaches such as mixed-data clustering. Firm mathematical relations have been found between traditional and conceptual clustering; moreover, unexpectedly, some classical statistical concepts such as contingency measures have been found to have meaning in the approximation framework. These yield a set of simple but efficient interpretation tools. Several new methods have been developed in the framework, such as additive and principal cluster analyses, uniform partitioning, box clustering, and fuzzy additive type clustering. In a few cases, approximation clustering goes into substantive phenomena modeling, as in the case of aggregating mobility tables.

The unifying features of the approximation approach fit quite well into some general issues raised about clustering goals (defined here in the general classification context) and the kinds of data tables treated. Three data types – column-conditional, comparable and aggregable table – defined with regard to extent of comparability among the data entries, are considered here through all the material in terms of different approximation clustering models.

Though all the mathematical notions used are defined in the book, the reader is assumed to have an introductory background in calculus, linear algebra, graph theory, combinatorial optimization, elementary set theory and logic, and statistics and multivariate statistics.

The contents of the book are as follows. In Chapter 1, the classification forms and functions are discussed, especially as involved in the sciences. Such an analysis is considered a prerequisite to properly defining the scope and goals of clustering; probably, it has never been undertaken before, which explains why the discussion takes more than two dozen pages. The basic data formats are discussed, and a set of illustrative clustering problems is presented based on small real-world data sets. In Chapter 2, the data table notions are put in a geometrical perspective. The major low-rank approximation model is considered as related to data analysis techniques such as the principal component and correspondence analyses, and its extension to arbitrary additive approximation problems is provided. In Chapter 3, a systematic review of the clustering concepts and techniques is given, sometimes accompanied by examples. Chapters 4 through 7, the core of the book, are devoted to a detailed account of the mathematical theories, including the most current ones, on clustering, with three kinds of discrete clustering structures: single cluster (Chapter 4), partition (Chapters 5 and 6), and hierarchy and its extensions (Chapter 7). There are not too many connections between the latter Chapters,

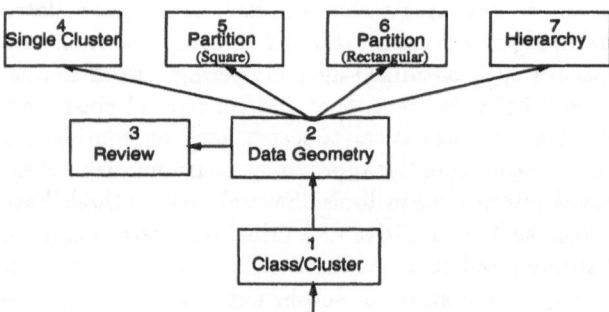

Figure 0.1: Basic dependence structure.

which allows us to present the structure of the book in the following fan-shaped format (see Fig.0.1).

The Sections are accompanied by reviewing discussions while the Chapters' main features are listed as their preambles.

The following are suggested as subjects for a college course/seminar, based on the material presented: a review of clustering (Chapters 1 through 3), clustering algorithms (any subset of algorithms presented in Chapters 3 through 6 along with the illustrative examples from these chapters and corresponding data descriptions from Chapter 1), and combinatorial clustering (Chapters 4 through 7).

Last, but not least, the author would like to acknowledge the role of some researchers and organizations in preparing of this volume: my collaborators in Russia, who participated in developing the approximation approach, especially Dr. V. Kupershtoh, Dr. V. Trofimov and Dr. P. Rostovtsev (Novosibirsk); Dr. S. Aivazian (Moscow), who made possible the development of a program, ClassMaster, implementing (and, thus, testing) many of the approximation clustering algorithms in the late eighties; Ecole Nationale Supérieure des Télécommunications (ENST, Paris), which provided a visiting position for me at 1991-1992, and Dr. L. Lebart and Dr. B. Burtschy from ENST, who helped me in understanding and extending the contingency data analysis techniques developed in France; Dr. F.S. Roberts, Director of the Center for Discrete Mathematics and Theoretical Computer Science (DIMACS, a NSF Science and Technology Center), in the friendly atmosphere of which I did most of my research in 1993-1996; support from the Office of Naval Research (under a grant to Rutgers University) that provided me with opportunities for further developing the approach as reflected in my most recent papers and talks, the contents of which form the core of the monograph presented; discussions with Dr. I. Muchnik (Rutgers University) and Dr. T. Krauze (Hofstra University) have been most influential for my writing; the Editor of the series, Dr. P. Pardalos, has encouraged me to undertake this task; and Mr. R. Settergren, a PhD student, has helped me in language editing. I am grateful to all of them.

Chapter 1

Classes and Clusters

FEATURES

• The concept of classification, along with its forms and purposes, is discussed.

• A review of classification in the sciences is provided emphasizing the current extension-driven phase of its development.

• Clustering is considered as data-based classification.

• Three kinds of table data, column-conditional, comparable and aggregable, are defined.

• A set of illustrative data sets are introduced, along with corresponding clustering problems.

1.1   Classification: a Review

It is a common opinion that narrative becomes science when it involves classification.

A definition of classification, going back to J.S. Mill (1806-1873) is this:

"Classification is the actual or ideal arrangement together of those which are like, and the separation of those which are unlike; the purpose of this arrangement being primarily

(a) to facilitate the operations of the mind in clearly conceiving and retaining in the memory the characters of the objects in question,

(b) to disclose the correlations or laws of union of properties and circumstances, and

(c) to enable the recording of them that they may be referred to conveniently."
(Sayers 1955, p.38-39).

Agreeing with the definition in principle, I suggest a different set of purposes:

Classification is the actual or ideal arrangement together of those which are like, and the separation of those which are unlike; the purpose of this arrangement being primarily

(1) to shape and keep knowledge;

(2) to analyze the structure of phenomena; and

(3) to relate different aspects of a phenomenon in question to each other.

In the definition framed, item (b) from the former definition has been split into items (2) and (3), while items (a) and (c) have been merged into (1). This change reflects the present author's opinion on the relative importance of the purposes.

Aristotle (384-322 B.C.) has been recognized as the first scientist to introduce a scientific meaning to the concept of classification. In particular, he proposed using the the so-called *Five Predicables* to describe the logic of classifying: genus, species, difference, property, and accident.

A *genus* is a class of entities called *species* serving as divisions to the genus. For example, "Sciences" (genus) consists of "Mathematics," "Physics," "Biology," and so on (species). Any genus can be presented in two different ways. The first, *extension*, concerns all the things covered by the genus as they are. The second, *intension*, refers to the meaning of the genus concept as it is expressed by sequence description. Using a modernized terminology, we could say that extension of the genus refers to enumeration of the set of all related objects, while its intension is just its description written in semantically loaded language. Roughly, the intension can be presented as a set of descriptors, that is, values (or gradations, or grades, or categories) of some variables (or attributes or features).

Five Predicables:

1. *Genus*: a set of species.

2. *Species*: an element of a genus.

3. *Difference*: an attribute added to the genus name to specify a species.

4. *Property*: a species modality which is characteristic to the genus, although not involved in the genus definition.

5. *Accident*: a species attribute, modalities of which differ for different species.

Aristotle's concepts should not be considered as being of a historic interest only. The five predicables still seem valid, though their meaning should be adjusted slightly. The notion of the phenomenon/process in question stands for the concept of *genus*, though the meaning of the former one is much more indefinite than that of genus. Distinguishing between *intension* and *extension* has become quite essential. Due to tremendous progress in observational facilities and computer techniques, a necessity has emerged to deal with (huge) empirical data about phenomena lacking clear theoretical concepts and definitions, to say more about the regularities unknown. In such a case, classification becomes a principal tool and aim of the analysis. This is the case when a marketing researcher investigates consumer behavior using data on factual or intentional purchases, or a sociologist studies the life-style of a social group, or a geologist wants to predict the mineral stock of a territory based on a data set of some other territories already investigated. Important, still mainly unresolved, questions arise in such a framework: What are the criteria for classification? How can the entire set of the data available be taken into account? How can one judge the importance of a particular variable? How can one produce an extension-driven classification when there is no theory to provide the Aristotelian differences? How can one make a clear interpretation of the classes found?

Previously, such questions would arise quite rarely. The traditional intensional classification completely depends on substantive theories of the phenomena classified following those in every single operation. The only logical concept of *difference* is involved as the only way for producing the classes from the genera: specifying them by their differences. No specific theories of the classification process itself is necessary, in this framework. This is why, in the present author's opinion, for the two millenia after Aristotle no general classification theory has been developed.

In Aristotelian terms, the genus extension can be available for studying even when no clear definition of the genus (intension) has been provided. It must be a theory of classification developed to meet the challenge.

Some extended versions of the concepts of *property* and *accident* seem quite relevant to at least certain of the extensional classification problems. The accident should be understood as a variable having no strong association to the classification while the property should be considered a strongly associated variable. Thus, the accident represents a stock of the variables to be used for the subsequent divisions of the classes while the property is a stock of the variables to be used for the interpretation purposes.

1.1.1   Classification in the Sciences

Let us discuss, in brief, some of the classification ideas developed and employed in the sciences.

Library Science and Information Retrieval

This is the only field where authors allow themselves to title their monographs with the name of "Classification", as they feel they are the only people dealing with "classification in general" because they classify the knowledge universe. To do that, two problems must be solved. The first is to express relationships between different subjects (as in the book title "Behavior of Animals") with a classification of knowledge; the second, to relate that classification to printed matter (especially, in view of syntactic interrelations between subjects in the documents such as "Statistics of State" and "State of Statistics"). The first contemporary classification made was what can be called conceptual one.

> Conceptual classification is a hierarchy of classes, each subdivided according to the hierarchical structure of the corresponding concept, as in Melvin Dewey's (1851– 1932) or the Universal Decimal Classification (UDC) system.

Completeness, simplicity, extendibility and other advantages of UDC and similar classification systems are at odds with their shortcomings implied by the lack of their reflecting many important kinds of relationships between subjects and/or documents such as the form of presentation, the process involved, comparison, time/space, and so on. Any aspect of this kind, (called a *facet*), can be added to classification code as a "parallel" characteristic to be applied to all classes, which is complementary to the hierarchical structure of conceptual classification.

Modern key-word or reference-cited descriptions can be considered as faceted ones. Using these along with modern computing facilities suggests a new perspective for providing and maintaining specialized classifications. For example, Classification itself, as a subject, is spread over all subdivisions of the library classifications in use. To cope with such a situation, the Editorial Board of *Classification Literature Automated Search Service* (Day 1993) maintains a list of relevant publications (currently, of 82 items), called a *profile*, as well as a list of relevant key-words. A journal paper is considered relevant to the field if it meets the following two criteria: 1) it cites at least one item from the profile; 2) its title contains at least one key-word from the list. Obviously, such an idea could be extended to create an update classification of a discipline based on cross-citations and key-word associations in such a way that any sub-discipline can be singled out as presented by a set of papers with high internal and low external key-word and citation associations.

Current problems of the information storing and retrieving in computers much resemble those in library science for both seek making comprehensive search and retrieval of information in either format, files or printed matter.

There is a great activity in many industries and institutions to create and maintain large data bases to keep records of specific things, like bank accounts or chemical compounds, in computers. Computer scientists have analyzed problems related to maintaining data bases; they have developed rather universal principles and concepts to describe relations between the records of any structure (like those reflected in the relational data base model); and, currently, various software tools based on those concepts and principles (see, for instance, Dutka and Hanson 1989).

The data base usually is organized as a set of files concerning different items or their characteristics to provide an easy interface with the user activities concerning various inquiries and easy insertion/deletion of data and/or attributes. Such an organization very much resembles our intuitive meaning of the notion of "classification": a subject field is divided into partly overlapping divisions somehow associated with each other providing the user with facilities for all the classification purposes mentioned. This is why the knowledge base discipline has emerged as a development of the data base field to serve, basically, as a classification tool (see, for example, Clancey 1985, 1992).

The practice of developing computer programming languages leads step by step to the inclusion of more and more classification structures: the richer the structures, the more powerful the language; the latest version is C++ and the like (see, for instance, Andrews 1993).

Mathematics

Mathematics comprises two kinds of activities: computations (finding exact or approximate solutions to various equations and optimization problems) and deductions on the properties of mathematical concepts (which are frequently related to computational purposes).

> The deductive part of mathematics can be considered as *an art to construct, analyze and connect classifications of mathematical objects by means of logical tools.*

Let us consider, for example, mathematical problems associated with the square algebraic equation

$$x^2 + px + q = 0$$

where p, q are given reals, and x stands for an unknown value ("root") which satisfies the equation. Figure 1.1 represents a great part of the equation theory arranged as a classification made in terms of the coefficient-based variables related to classes described in terms of the properties of the roots.

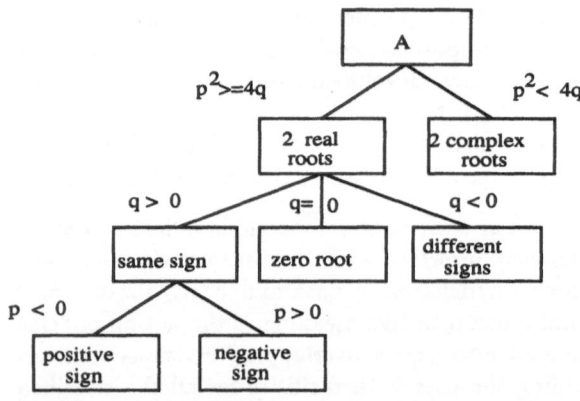

Figure 1.1: Set of all the equations (**A**) hierarchically classified with regard to the number of real/complex roots and pattern of signs of the real roots; the classes being exactly described in terms of the equation coefficients.

This example should be considered as a representative model of mathematical theorems regarding existence (no-solution subclass is empty), necessary conditions (a class includes another one), necessary and sufficient conditions (a class coincides with another one), and classification. It should be pointed out that, actually, there are two coinciding classifications presented in Figure 1.1: one is described in terms of the properties of the roots (number of roots, their nature, signs of the roots), and the other, in terms of the coefficients (parameters $p, q, p^2 - 4q$).

It is not always so; sometimes a result could concern one classification only (like, for example, the theorem stating that there exists only one isomorphic class of n-dimensional linear spaces [in linear algebra], or "Classification Theorem" in the theory of finite groups [Gorenstein 1994]).

The observation above leads to clarifying the possible uses of the classification nature of mathematics: artificial intelligence research in mathematics should be based upon the extensional issues, as it has been done by K. Appel and W. Haken 1977 for solution of the well known "Four Color Problem" in graph theory. The intensional approach related to automatical deduction was unsuccessfully tried many times in recent decades.

Classification in Physics and Chemistry

After Thales, who claimed existence of a constant matter underlying all the changeable things in nature and called it "water", it was Empedocles who distinguished between four basic elements: earth, water, air, and fire – classification which was strongly supported by Plato and his school. Currently, these "elements" are related to the states of matter: solid, liquid, gas and plasma. Yet, we have accumulated a deeper knowledge fixed in more refined classifications.

> The Periodic Law belongs among the most profound achievements in the discipline, as it links four aspects of the elements: the internal structure of the atoms, their bondage into molecules, their chemical interaction properties, and their physical features.

The Periodic Chart (Fig. 1.2) was designed (D.I. Mendeleev 1869) as a purely empirical observation (involving 56 elements available at that time). Beyond its use as a form of knowledge maintenance, " Chart provides a stimulus and a guide in chemical research, constantly suggesting as it does new experiments to be tried and providing a basis for critically evaluating and checking information already obtained. ... The very existence of the Periodic Law as an empirical principle provided a tremendous stimulus to the development of our knowledge of atomic structure and greatly accelerated the growth of our understanding of the relationship of the structure and the properties of matter." (Sisler 1963, p.34.)

Currently, scientists face even a greater challenge: what is the association between the structure of a molecule and its physical and chemical properties? One of the most impressive cases when such an association has been established based on a classification is the theory of symmetric crystal forms. Though the crystals

1a	2a	3b	4b	5b	6b	7b	8b	8b	8b	1b	2b	3a	4a	5a	6a	7a	0
1 H																1 H	2 He
3 Li	4 Be											5 B	6 C	7 N	8 O	9 F	10 Ne
11 Na	12 Mg											13 Al	14 Si	15 P	16 S	17 Cl	18 Ar
19 K	20 Ca	21 Sc	22 Ti	23 V	24 Cr	25 Mn	26 Fe	27 Co	28 Ni	29 Cu	30 Zn	31 Ga	32 Ge	33 As	34 Se	35 Br	36 Kr
37 Rb	38 Sr	39 Y	40 Zr	41 Nb	42 Mo	43 Tc	44 Ru	45 Rh	46 Pd	47 Ag	48 Cd	49 In	50 Sn	51 Sb	52 Te	53 I	54 Xe
55 Cs	56 Ba	57-71	72 Hf	73 Ta	74 W	75 Re	76 Os	77 Ir	78 Pt	79 Au	80 Hg	81 Tl	82 Pb	83 Bi	84 Po	85 At	86 Rn
87 Fr	88 Ra	89-103															

Figure 1.2: A version of the Periodic Chart from Sisler 1963, p.32.

may have different shapes, their symmetries are considered only in terms of the *isometries*, that is, the symmetrical transformations of a sphere. There exist only 32 different (non-isomorphic) finite groups of isometries in the three-dimensional space: 5 of them are rotations about an axis through an angle $n\alpha$ where n is an integer and α is equal to $360°, 180°, 120°, 90°$, or $60°$, respectively; 6 groups are formed by combining rotations (around different axes) as the base elements; and 21 groups are obtained by adding (rotary) reflections about planes to the rotations. These *point* groups form the basis of the other classifications of the crystals, based either on their refinement (with permitted transformations of translation added) to the 230 so-called *space (or, Fedorov) groups*, or on their aggregation in 6 (or sometimes 7) so-called *crystal systems* (see Senechal 1990).

Geology

The science of earth deals with all earth structures: minerals, rocks, soils, glaciers, water, mountains, etc, as well as with many associated phenomena: the moon and planets, magnetic field, earthquakes, climate, etc. Each of these can be observed in such varied forms, that neither purposeful use nor understanding of their nature can be accomplished without classification representations, which have been made for the latest hundred years quite extensively.

> The most impressive achievements of classification in geology are in the line of relationship between structure and history (origin and evolution) rather than between structure and function (properties), as it has been in physics and chemistry.

The following principles underlie that. The rocks, ordinarily, are organized in layers (strata) that are rather clearly distinguished; any layer is considered as older than the layer just above ("law of superposition"); when originally formed, the strata were laterally continuous; the fossils found in the rock are remains of the organisms living in the time of the formation of the rock.

A general method, both based on these principles and supporting them, has been developed, called "geological correlation", leading to finding many oil or coal deposits in practice, as well as to some theoretical breakthroughs, like periodization of the geological time (see Table 1.1).

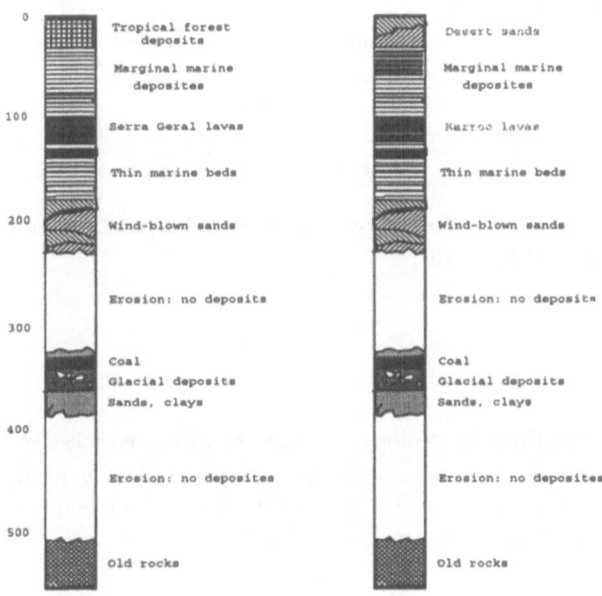

Figure 1.3: Simplified columnar section of rocks from SE Brazil (left) and SW Africa (right): great similarity before 100 mln years ago (based on figure 19-6 from Putnam 1978, p. 610).

The principle of correlation allows for concluding that, for any given well (considered as a vertical column), the sequence of the strata in a well nearby will be (almost) the same; moreover, the fact that the sequence of the strata in a well is almost the same, yields that its place belongs in the same formation as the first one (see Fig. 1.3).

ERA	PERIOD	LIFE FORMS
Cenozoic (Recent life)	Quaternary Tertiary	Rise of mammals and appearance of modern marine animals
Mesozoic (Middle life)	Cretaceous Jurassic Triassic	Abundant reptiles (including dinosaurs) more advanced marine invertebrates
Paleozoic (Ancient life)	Permian Pennsylvanian Mississippian Devonian Silurian Ordovician Cambrian	First reptiles First land animals (amphibia) First fish Primitive invertebrate fossils
Precambrian		Meager evidence of life

Table 1.1: Relative geological time scale, based primarily on superposition and character of fossils; after Putnam 1978, p. 16.

There are many other classifications in geology. In classification of rocks, instead of an exact description of the class, the so–called stratotype method is used, involving "case-based" comparison of a rock with a typical standard, which is much easier when the system of the categories is not well described.

Biology

Classification in biology can be employed to analyze the relationship between all three kinds of aspects noted above: structure, function, and history.

Although natural languages distinguish rather clearly between many of the living organisms and taxa (for example, "the bird flies" and "the fish swims"), the first systematic effort in biological taxonomy was done by C. Linnaeus (1707–1778) who completed his descriptive catalogues for plants and animals at 1758.

> Biological taxonomy consists of the following four parts:
> 1) a *hierarchical classification* of the organisms arranged in distinct classes — taxa (plural of *taxon*);
> 2) the *descriptions* of the taxa;
> 3) *nomenclature*; that is, a list of the names of those taxa;
> 4) *identification* keys to relate the particular organisms to the classification (see Abbott, Bisby, and Rogers 1985).

The biological taxonomy is a live, changing system; scientists are eager to reconsider extensions of many taxa, sometimes in rather high levels. For instance, no convenient classification of viruses has been created yet.

From the theoretical point of view, the most important feature of taxonomy is that, although it is created by the observable character resemblance (called *phenetic*), the classification hierarchy reflects evolutionary, *phyletic* relations between taxa. However, the *phenetic* classification (based on phenetic similarities) and *phyletic* classification (based on evolutionary considerations, see Fig. 1.4) are not coinciding. For example, salmon and lungfish are much more phenetically similar to each other than to cows, which contradicts the generally accepted opinion that it was salmon diverged from the common ancestor of the cow and lungfish (see Abbott, Bisby, and Rogers 1985, p. 228-229).

It appears, for many taxa in phyletic classification, that it has been impossible to find any specific set of characters to single out a particular taxon. To deal with that, a new concept of class as a "polythetic", not a "monothetic" one, has been developed (Sokal and Sneath 1973).

> A *polythetic* class, defined with a set of attributes, consists of the objects, each of which holds a majority of attributes from the set, while any of the attributes occurs at majority of the objects.

Such a concept immediately led to developing various machineries for finding polythetic classes in real data sets. The methods developed belong to the core of this book and will be discussed further. What is important here is that the concept involves a preassumption of the "equal weight" of all the attributes considered, which contradicts the traditional view that some of the attributes are essential and others are not, expressed in the concept of monothetical class.

In contrast to the initial expectations, a sound progress achieved in the mathematical classification and clustering methodology has not led to a corresponding effect in the biological classification. The cause, perhaps, is that no adequate progress in development of the theory of the biological characters (variables) has been achieved yet.

The term *character* refers to a variable related to a part of a living organism under consideration (like the structure or the length of leaf-blades in a plant, or color of eye of an animal). To compare two taxa, one needs to compare the *states* (that is, categories or values) of the characters related to *homologous* parts (organs) of the taxa members. But how one can decide which organs are homologous, that is, "alike"? Are the wings of fly homologous to the wings of bird or to the arms of man? To date, the homology concept is considered as just the structural correspondence (Sneath and Sokal 1973, p. 77), having no theoretical support. It may be that the support desired can be based on representing the living organism as a system consisting of subsystems that are responsible for providing some specific supporting goals (functions): nutrient procurement, gas exchange, internal transport, regulations of body fluids, coordination of regulatory activity, motion, reproduction, control, etc. The subsystems may be divided into sub-subsystems corresponding to their subgoals (subfunctions, or tasks), etc. Homologous subsystems should be defined in terms of similarity of their functions/goals, which could eventually lead to a threefold organization of the biology taxonomy instead of the singular current structure.

> The threefold taxonomy should consist of the three hierarchies corresponding to each other: hierarchy of the goals (functions), hierarchy of the subsystems responsible for those functions, hierarchy of taxa defined in terms of the characters based on the subsystems. This would make associations between structure and function in living organisms explicit.

Linguistics

Everybody who has read an impressive overview of the work undertaken to transform English into "Newspeak", described by George Orwell in his novel *1984* (see also Atkinson 1988, p. 29), will understand that language can be considered as a natural classification machinery created for shaping and keeping a portrayal of the world via human thoughts and communications. The language is an excellent model for investigating and learning how classification works in the human world. In its potential capability toward the future mathematical classification developments, language could be likened to mechanical motion as the major natural phenomenon studied for developing the modern mathematical theories for analyzing real variable functions. All of the various of the elementary functions (exponent, logarithm, sine, tangent, etc) as well as the most important operations (derivatives and integrals) were found to be useful in the analysis of motion phenomena. Likely, the most interesting classification forms and operations could be found through and for analyzing language phenomena.

Among linguistic classifications, we must distinguish between those involved in the language phenomena as they are and those created by the scientists. The word meanings in a language, the sentence parts (reflecting classification of the world phenomena by variables: who (what), what does (did), when, why, where,

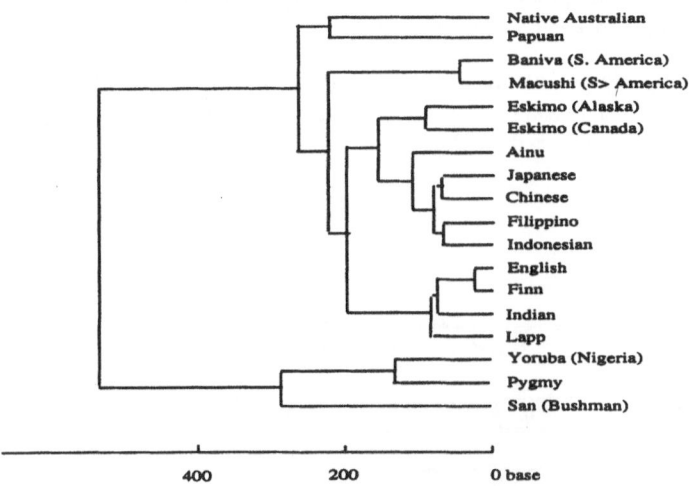

Figure 1.4: A phylogeny for eighteen human populations based on genetic distances measured in number of bases (after M. Nei and T. Ota from Osawa and Honjo 1991, p.421).

etc.), the speech parts (nouns for things, verbs for actions,...), etc. are examples of the natural language classifications. Classification of the word constituents such as phonemes and morphemes (for a kind of "Phonetic Chart" see Atkinson, Kilby, and Roca 1988, pp. xiv-xv, 71-86, 105-123) and evolutionary classification of the languages are examples of the scientific classifications.

Some work combines these two kinds of classification. For example, there are so-called "implicational universals" found in some language families (Sherzer 1976, Ch. 15). An implicational universal is a predicate of the form "attribute X implies attribute Y for all of the languages from a given group".

Psychology

> Living organisms classify and generalize real world phenomena with simplifying categories to perceive, learn, predict, and to behave (see, for instance, Ornstein 1985).

All psychological systems depend very heavily on their physiological construction, which can be interpreted in terms of the structure (physiology) underlying the function (psychology). One of those devices, neuron and neural network, gave rise to an interesting concept in the pattern recognition theory, and was later extended into many other disciplines in the computer sciences.

Several para-scientific classifications emerged on the premise of a close relationship between anatomical features and the character (an example of such a classification, nail forms, is presented in Fig. 1.5).

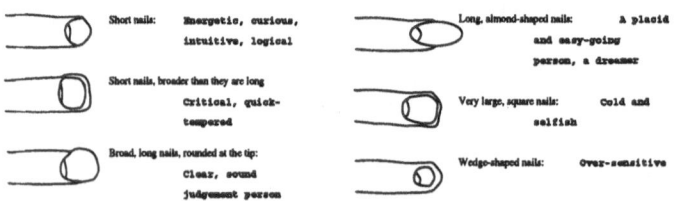

Figure 1.5: Nail forms and corresponding personality traits as claimed in Bosanko 1983, p. 11.

There have been made several ground-breaking discoveries on the nature of the categorization and classification in the human mind. Currently, the psychologists seem to prefer analyzing the processes of categorization with mathematical modeling. Initially, they developed models based on the similarity concept: any pattern is considered as a point of an "internal feature space" in such a way that any exemplar could be related to the most similar "prototype" (Shepard 1988). Then, models based on logical rules ("production systems") became popular: each rule is an expression like "if the exemplar pattern is A, the category must be B" (see Estes 1994).

A lot of work has been done to analyze the differences among the personalities. The best known psychological classification has come out from ancient times, assigning people by their temper to four types: choleric, sanguinic, phlegmatic, and melancholic. These types were derived from a theory which dominated in medieval Europe that the human body contained four kind of fluids: bile, blood, phlegm, and black bile; the dominance of one of them supposedly predetermined the temper and character of a person. In contrast to the classifications involving quite definitely bounded classes (discussed for the other sciences above), this concept

Temper	Humor	Reaction	
		Strength	Speed
Choleric	Bile	High	High
Sanguinic	Blood	Low	High
Phlegmatic	Phlegm	High	Low
Melancholic	BlBile	Low	Low

Table 1.2: Medieval and modernized presentation of the common character types.

suggests patterns, like the concept of stratotype in geology or the nail form in Fig 1.5, rather than partitions the people. Since the theory of four fluids failed, a modernized explanation of the typology appeared relating it to the strength and speed of nervous reactions (see Table 1.2) (Nebylitsyn 1972). The principal feature of this presentation is the same: patterns, not subsets, are fixed; yet technically it is different: all combinations of the unidimensional patterns are considered here (in 2-variable space) while only unidimensional patterns uncombined have been presented in the medieval typology (in 4-variable space).

Social Theme	**Investigative Theme**
Typical Traits:	
Ethical, responsible, kind, generous, friendly, understanding, concerned for the welfare of others	Analytical, curious, independent, rational, original, creative
Interests and Preferences:	
Training, teaching, curing, helping others	Science, gathering and analyzing information, working on their own
Particular Skills Developed:	
Interpersonal skills, verbal ability, listening, empathy	Writing and mathematical skills, critical thinking
Typical Work Activities:	
Teaching, training, coaching, leading discussions, group projects	Solving problems through thinking, scientific or laboratory work, collecting and organizing data
Preferred Lifestyles and Work Situations:	
Religious organizations, family life, helping professions, working in groups, personnel offices, volunteering	Computer-related industries, achievement-oriented organizations, unstructured organizations that allow freedom in the work styles

In the box above, an inventory of investigative type versus social type from Brew 1987 is presented in a similar pattern-wise fashion, which is quite characteristic for the modern empirically-driven theories of personality (see, for example, Ornstein 1985, Good and Branther 1974). We can see how the classification presented connects three different aspects of the personality: function (Typical traits), attitude (Interests and Preferences), and action (Skills, Work Activities and Preferred Lifestyles and Work Situations). This adds two more aspects to the "history/structure/function" framework above, thus yielding a fivefold system: "history/structure/function/attitude/action".

There exists a growing area related to developing technical sensory and data processing systems dealing with classification-wise problems: systems of pattern recognition (to read and process printed or hand-written letters, to recognize sounds, to perceive pictures, to analyze and compare cardiograms, etc), machine learning, robot vision, and so forth.

Social and Political Sciences

In social and political sciences the classification paradigm involves four basic dimensions: history/structure/attitude/action:

Structure

The concept of social class underlies much of the social and political theory and practice (Edgell 1993). Strata differentiating people by income, power, prestige and perhaps other dimensions is another classification concept involved. Yet one more classification concept developed in sociology is of the ideal type (M. Weber [1864-1920]). The *ideal type* is such a combination of characteristics that no real entity can satisfy all of them, though the entities can be compared by their proximity to the ideal type.

The class structure is influenced by an external factor — economy and technology. There should be several other basic social structures considered as also determined mainly by the outer factors: states and nations (by geography), race, kinship, gender and age (by biology).

Attitude and Action

Society functions through institutions and organizations (family, education, polity, economy, religion, law, etc) that are heavily associated with classification (see, for example, an account of organization systematics in McKelvey 1982). Moreover, a great part of societal control is made through classifications. It is especially clear in the case of the so-called socialist countries, like the USSR, where all aspects of social life were arranged via classifications mixing the party/administrative hierarchy with the industry/organization/location ranking used as the basis of a priority system for distributing limited goods and services (from meat and cars, to housing, to medicine, recreation and education). However, this phenomenon can be seen in any other society.

History

Amazingly, societal classifications themselves are an important part of the evolutionary description. For instance, aboriginal Australian tribes were found to classify the universe according to a simple classification of their society in marriage groups; that classification "extends to all facts of life; its impress is seen in all the principal rites" (Durkheim and Mauss 1958, p. 14-15). The social borders and their "sparseness" is another important classification parameter of the evolution (Indian castes, medieval aristocracy, etc.).

A review of societal taxonomies is presented by Lenski 1994.

1.1.2 Discussion

1. A definition of classification is given; it is an arrangement of the entities in question, which is instrumental in analyzing the structure, relating different aspects to each other, and keeping the knowledge.

2. The ancient concepts of the five predicables are discussed and their modern meanings are suggested.

3. The former developments in classification theory were intension-driven: it was not much to develop in the classification context; everything had to be considered in the framework of substantive sciences. Currently, the sciences face a new, extension-driven phase of their developments, which raises corresponding problems of classification theory.

4. Some classificational ideas in sciences may lead to an impact both in the substantive disciplines and classification theory. It concerns, primarily, classificational interpretation of certain mathematical results, the history/structure/function associations in biology, and the fivefold system history/structure/function/attitude/action in human sciences. The classifications are designed within these aspects in such a way that they are closely related or interact across the aspects.

5. Several extension-driven classification concepts which emerged in the sciences are: polythetic class and the "equal-weight-of-the-attributes", in biology; factor score, cluster, and type in psychology; ideal type, in sociology; extensionally found production rule "A includes (implies) B", in mathematics, linguistics and psychology.

1.2 Forms and Purposes of Classification

1.2.1 Forms of Classification

Based on the material above, we can quite clearly distinguish between the class and type concepts.

> The unity of its intensional and extensional descriptions underlies the concept of a "classical" class, while there is no such unity in the concept of type: type can be represented as a combination of the attribute values or a particular "prototype" entity, and it may have no attachment to the empirical entities presumed.

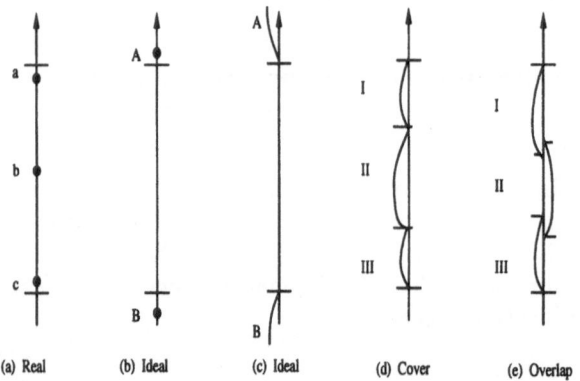

Figure 1.6: Unidimensional typology forms.

Let us take a closer look at some of the simplest forms of type-based classification, *typology*. Let us consider a quantitative variable with respect to a set of individuals, for instance, "the education level", to distinguish among various forms of the typology of the individuals made by this variable. In Fig. 1.6, five of the unidimensional typology forms are presented on the vertical axis representing the variable, with its range limits shown by the horizontal lines. The forms (a) and

(b) correspond to the situation when the types are represented by some particular "typical" values either taken either from theoretical considerations or characterizing some particular entities, real (a) or ideal (b). The other three drawings correspond to the case when the types are represented by the variable intervals: ideal (c) or real, (d) and (e). In both of the (d) and (e) pictures, the intervals cover all the range of the variable. The nonoverlapping pattern in (d) causes the set of type intervals, I, II, and III, to form a partition of the range as, for instance, "less than one year of studies" (type III), "from one to eight years of studies" (type II), and "more than eight years of studies" (type I). With this particular form, the concept of typology overlaps the classical classification concept since here the types are exactly described both intensionally and extensionally, and thus they are classical classes, in this case. Some other unidimensional forms of typology can be considered employing the concepts of probabilistic distribution or fuzzy membership functions, as well as nonquantitative variables.

Two major kinds of the classification structures are *hierarchical* and *non-hierarchical.*

Classification is hierarchical if it is nested like the conceptual library classifications or taxonomy in biology. The examples mentioned present two important types of classification. A typical library classification, such as Dewey's or Universal Decimal Classification, can be considered as arranged in the Aristotelian style: the part of the universe in question is divided in certain subparts that, in their turn, are subdivided in sub-subparts, etc., in such a way that the divisions are made based on some logical concepts. The notion of the logical concept can be expressed more or less formally using the so-called nominal variables. A *nominal variable* maps the entities of a class into its categories (values) in such a way that any entity corresponds to one and only one category; no relations among the categories are assumed as, for example, among different occupations in sociology. The concept of nominal variable is an abstract one since, in the real world, the variables rarely satisfy the definition exactly: some individuals could have several occupations or no occupation at all, some occupations admit comparisons, etc.

Another type of hierarchical classification, called *systematics*, is based on an opposite process of combining smaller classes into larger ones due to their similarity with regard to various and/or different attributes. The biology taxonomy has been constructed in such a way, which is reflected in that fact that the major classes, such as Chordata and Mollusca, are described in distinctive but rather indefinite terms ("largely marine invertebrates" [Mollusca], "all vertebrates and certain marine animals having notochord" [Chordata]) because there are too many special cases to be involved in a general definition. The conceptual classification may have empty classes as defined by logical combinations of the categories some of which can never be met in the context considered; the systematics class may not be empty since it is based on the generalization of the empirical facts.

There are two major kinds of the hierarchical classifications: conceptual classification and systematics. The conceptual classification corresponds to that considered by Aristotle and defined with the top-bottom sequential divisions by nominal variables. The systematics is defined bottom-up and, usually, lacks the unambiguously dividing variables.

Two types of non-hierarchical classifications can be distinguished rather clearly: *typology* and *structural classification*. The concept of typology is defined through one or several of the essential variables for the domain classified. Some explicit models of the unidimensional typology have been considered just above (Fig.1.6). Multidimensional typologies usually are created from the unidimensional ones using, basically, either of the two approaches met in the temper classifications, medieval and modernized (see Table 1.2). When all the combinations of the categories/intervals of the variables participating are considered as multidimensional types, the typology can be called *faceted*. The other extreme is when every multidimensional type is defined by a corresponding single prevalent variable (so, the number of the types here equals the number of the variables), as was done in the medieval typology. Such a typology is called *characterological* in psychology. The difference between faceted and characterological typologies can be easily seen in Fig. 1.7 where two variables are taken to define a characterological typology (a) based on the corresponding prevalent variable, A or B, and a faceted typology (b) based on all the four combinations, AB, Ab, aB, and ab, of the unidimensional types A, a and B, b.

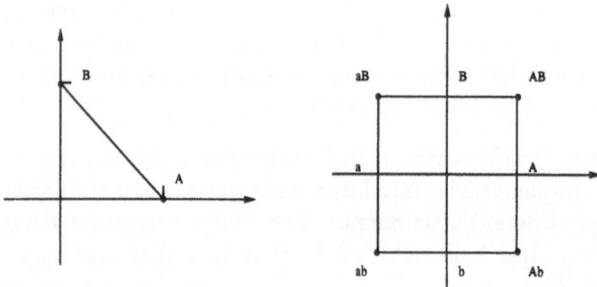

Figure 1.7: Patterns of twovariate typology.

The *structural classification* concept relates to situations when classes correspond to subsystems of the domain classified, which is considered as a complex system in such a way that subsystem-to-subsystem interactions must be included in the classification as its class-to-class interrelations. For example, the set of the national economy industries is structurally classified when the sectors of energy, primary sources, processing industries, consumer goods, service sectors are considered.

A particular type of non-hierarchical classification, which can be considered both as a typology and as a structural one, is the *stratification* introduced to reflect inequality between entities rather than similarity. Stratification in society based on correlated variables as income, power and prestige, can be defined through corresponding range intervals (see Fig. 1.8 where four strata are presented within a cone representing the feasible domain).

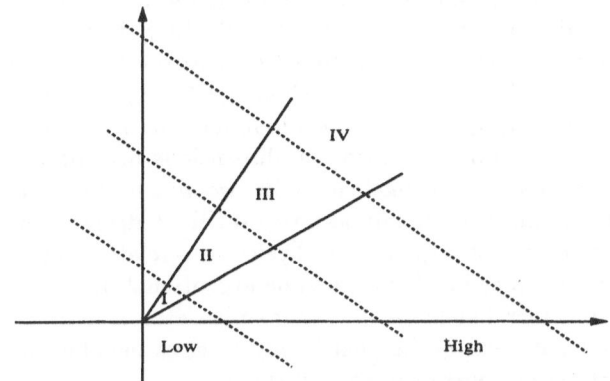

Figure 1.8: Pattern of strata in a stratification.

There are three major kinds of non-hierarchical classifications: typology, structural classification, and stratification. Unidimensional typology can be defined either with standard point (prototype) indication or with variable intervals (categories). Three important kinds of typologies are: real type, ideal type and partition typology.

1.2.2   Purposes of the Classification

In the beginning, the purposes of classification have been listed as follows:

1. to analyze the structure of phenomena;

2. to relate different aspects of a phenomenon in question to each other; and

3. to shape and keep knowledge.

Now, we can update these items in more detail.

To clear the first two purposes, let us refer to the quintet revealed: history/structure/function/attitude/action. Classification of the domain considered by any of these dimensions helps in understanding the structure of the domain as well as in revealing its relationship to the other dimensions. While moving from the "dead" nature to living organisms to human society, the relative importance of the particular dimensions moves from the left to the right part of the quintet. In physics and chemistry (and in mathematics), the most important dimensions are structure/function; in geology, structure/history; in biology, structure/history/function; in psychology, structure/function/attitude/action; in sociology, history/structure/attitude/action; and in technology (though omitted from the current review), attitude/action. Moreover, while moving from the right to the left of the quintet, the associations between the dimensions become less rigid, and also change their causal one-way dependencies for two-way feedback relations. The most rigid relations are in mathematics: two different descriptions of a class must relate to the same set of objects. In physics, the relationship is still quite hard, though it can be rather indefinite in some exceptional boundary cases. Those relations become rather indefinite in a human society: a human's motives and behavior only partly depend on her/his family or class membership (structure): a complicated motivation sphere controls all these.

Relating different dimensions of a phenomenon can be done in either way. For example, the classification by structure, the place in the Periodic Chart (currently, the atom number), determines the functional, physical and chemical properties of an element. On the other hand, classification of the square equations by their roots (function) is used to determine the corresponding coefficient-based (structural) variables describing the classes.

Actually, it is the classes and their descriptions (when possible, in different variable spaces) that represent the knowledge kept and shaped in the classifications. As is known, knowledge can be presented, somewhat simplistically, as a set of concepts (notions) and associations between them, expressed, mostly, through logical statements (rules) $A \rightarrow B$ involving the concepts A, B. Any concept corresponds to a class of phenomena; moreover, combining like phenomena into the same class, we facilitate perceiving, recording and analysis of the phenomena as a whole. In this respect, we can say that knowledge is nothing but a set of interrelated classifications.

In practice, classification, as any kind of knowledge, is used for prediction and control. Prediction of behavior of a member of class is based on description of this class imprinted in interrelations among the variables. Control actions can be undertaken in a reasonable way when the class reactions can be predicted.

1.2.3 Content of the Classification

Classification, as a natural phenomenon, has rather indefinite contents. Still, we can distinguish the following seven aspects of classification as its rather separated parts: (a) *domain* of the universe classified; (b) underlying *theory*, (c) *nomenclature*: the names of the classes, (d) *structure*: the structure (and names) of relations between classes; (e) *description*: the definitions of the classes; (f) *key*: operational rules for identification of the entities as belonging to corresponding classes, (g) *membership*: the lists of the entities which are class members or representatives, for any class.

> Seven components of a classification: (a) domain, (b) underlying theory, (c) nomenclature, (d) structure, (e) description, (f) key, (g) membership.

In particular types of classifications, different items can be presented differently. The biological taxonomy, for example, has all the sections nonempty, while an ideal type typology may have the sections (d), (e), and (f) empty.

Let us indicate the items in the librarian and biology classifications, respectively: (a) domain: set of the printed matters, manuscripts and documents, or, set of the living organisms; (b) theory: presentation of the universe according to the structure of the sciences, culture, technologies, and activities in the society, or an understanding of the reproduction and other biology processes (from rather primitive at the time of Lamarck to much more sophisticated contemporary theories); (c) nomenclature: see the volumes of the library catalogues, or the lists in biological manuals; (d) structure: in both classifications, hierarchy, although in the library classification there are some "horizontal", faceted relations; (e) and (f): description and key: the description is an intensional form of the definition as expressed in a narrative style to be understood by public at large, and the key is a rigid form of the description defining the class in an algorithmic style; biological descriptions (see, for example, Abbott, Bisby, and Rogers 1985, p. 23 - 31) are more strict than the rather vague indications in the library classifications; (g) membership: no common membership lists are provided for the library classifications, and in biology, any species is arranged as a top-bottom series of the including classes.

1.2.4   What is Clustering?

> *Clustering* is a mathematical technique designed for revealing classification structures in the data collected on real-world phenomena.

A *cluster* is a piece of data (usually, a subset of the objects considered, or a subset of the variables, or both) consisting of the entities which are much "alike", in terms of the data, versus the other part of the data.

Earlier developments of clustering techniques should be credited, primarily, to three areas of research: numerical taxonomy in biology (Sneath and Sokal 1973), factor analysis in psychology (Holzinger and Harman 1941), and unsupervised learning in pattern recognition (Duda and Hart 1973). In the seventies, a number of monographs were published demonstrating emergence of the discipline (see Jardine and Sibson 1971, Duran and Odell 1974, Everitt 1974, Clifford and Stephenson 1975, Hartigan 1975, Van Ryzin 1977); the most important methods of clustering — moving centers (K-Means) and agglomerative techniques — were included in major statistical packages as BMDP, SAS and SPSS. The principal idea of the various clustering techniques developed is this: measure somehow similarity value between any two of the entities classified, and then design clusters in such a way that the entities within the clusters are similar to each other while those in different clusters are dissimilar. In the eighties, the research continued, concentrating primarily on the questions of substantiation of the techniques developed both experimentally and theoretically (Jain and Dubes 1988, McLachlan and Basford 1988).

The purposes of clustering are primarily the same as of classification in general, although applied to data, not to real-world phenomena.

> Purposes of clustering:
> 1) to analyze the structure of the data;
> 2) to relate different aspects of the data to each other; and
> 3) to assist in classification designing.

The last item here has substituted the first item ("shaping and keeping knowledge") in the list of classification purposes, since clustering is an empirical, data-based tool for classification developing through performing the other two tasks.

So far, the mainstream of work in clustering has been concentrated on the first of the purposes, which is much reflected in the material of this monograph. The second purpose will be considered also, though quite moderately. As to the third, it has not been formalized yet since we have no clear understanding what "classification designing" means.

1.2.5 Discussion

1. Though the concept of classification as a way to structurize and understand the real-world phenomena and relationships could be treated in a very wide meaning (Clancey 1985, Zacklad and Fontaine 1993), here a much narrower class of classificational structures is considered as the genuine classifications: the hierarchical and partition-wise non-hierarchical ones.

2. Among the hierarchical classifications, the conceptual one and systematics are picked out, while typology, structural classification, and stratification are distinguished from the non-hierarchical classifications.

3. The concept of typology comprises a wide set of classifications pertaining to such seemingly different items as a set of prototype individual entities, or a partition of the domain along with the intensional descriptions of the classes as the variable categories (intervals) combined.

4. The classical concept of class, involving both intensional and extensional descriptions, pertains to quite distinctive and well understood natural entities, like the chemical elements or the biological species. Some relaxed versions of that concept, the type included, are quite useful when an understudied domain is analyzed.

5. The knowledge kept in classifications is the relations among the classes, most common forms of which can be expressed as logical implication or equivalence statements (rule base).

6. Since a classification is a real-world object, its contents cannot be characterized quite definitely, although the seven components listed above can be distinguished.

7. Clustering is a part of the classification process pertaining to analysis of a set of data; the clustering goals are just those of the classification applied in a limited area; the clustering forms should match those in the general classification.

1.3 Table Data and Its Types

1.3.1 Kinds of Data

The following kinds of the real-life data are the main interest for mathematical and computational processing: pictures (images), graphics (signals), texts (letter and word series), chemical formulas (graph structures), maps (spatial structures), and tables. Although images are of the most importance in medicine and machine vision, graphics and signals, in engineering and seismography, chemical formulas, in pharmacy and chemistry, and spatial structures, in meteorology and navigation; only tables will be considered in this book. First, tables are the most universal form of information storage in numerical computational devices (the only kind which is currently available): any other kind of information can be presented in a table form. Second, the specifics of the former kinds of data require specific theories for their analysis, which cannot be covered in this, quite general, presentation. Third, the diversity of table data forms is so extensive that even for this particular kind of data, too much remains to be done.

A quite suitable classification of data tables has been suggested by Tucker 1964 based on two numerical variables: the number of ways and the number of modes. A formal account of the classification is this. Let $I_1, I_2, ..., I_n$ be sets of some entities: kinds of plant, kinds of bird, geographical sites, periods of times, etc. It is permitted that some of the sets be coinciding, with k the number of different sets among the given n sets. An n-dimensional array of, usually numeric, code values $a(i_1, ..., i_n)$ given for any combination of $i_1 \in I_1, i_2 \in I_n, ..., i_n \in I_n$, is referred to as a n-way k-mode table. For example, a similarity matrix a_{ij} between the entities $i, j \in I$ is a two-way one-mode table, while a rectangular entity-to-variable matrix x_{ik} where $i \in I$, $k \in K$, and I is set of the entities and K is set of the variables, is a two-way two-mode table.

In this book, mostly two-way, two-mode and one-mode, data tables are considered. We will also lean upon the following subdivision of data tables concerning comparability of the entries:

1. Column-Conditional table.

2. Comparable table.

3. Aggregable table.

The first two are well-known (see, for instance, in Arabie, Carroll, and De Sarbo 1987 where comparable data are referred to as unconditional); the latter seems to have never been singled out before. We discuss them in the following three subsections.

1.3.2   Column-Conditional Data Table

Primarily, such a two-way table is an entity-to-variable table, that is, a rectangular array like that one presented in Table 1.3, having the rows corresponding to the entities and columns corresponding to the variables, with the entries coding the values of the variables at the entities. The variables are called also attributes, features, characteristics, parameters, etc. Such terms as case, object, observation, or instance are in use as synonymous to the "entities".

Planet	Distance kilomile	Diameter mile	Period year	Day	Moons amount	Matter	EBalance
Mercury	36	3000	0.24	59	0	Solid	Negative
Venus	67	7500	0.62	243	0	Solid	Negative
Earth	93	7900	1	1	1	Solid	Negative
Mars	142	4200	1.88	1	2	Solid	Negative
Jupiter	483	89000	12	0.42	17	Liquid	Positive
Saturn	885	74600	30	0.42	22	Liquid	Positive
Uranus	1800	32200	84	0.67	15	Mixed	Positive
Neptune	2800	30800	165	0.75	8	Liquid	Positive
Pluto	3660	1620	248	6.40	1	Solid	Negative

Table 1.3: **Planets:** Planets of the Solar system along with some of their characteristics; EBalance is the difference between the received and emitted energies.

Table 1.3 is an update of a table cited by W.S. Jevons (1835-1882) in his account
of the classification subject (Jevons 1958). This format of data often arises directly
from experiments or observations, from surveys, from industrial or governmental
statistics, and so on. This is a most conventional form for presenting data base
records.

It must be noted that, usually, no evident data structure can be seen in the
table directly, in contrast to the case of Table 1.3 which shows an obvious two-class
pattern: the first four planets have each of the variables presented quite differently
from the following four planets. For instance, the planets of the first group have
just a few moons while there are at least 8 moons at the planets of the second group.
This poses two challenging problems still unresolved: (1) What is the regularity
underlying such a huge difference? (2) How can the deviant behavior of the most
recently discovered Pluto be explained?

Often the entity-to-variable data table is considered as a raw data for transforming it into the other table formats.

The entity-to-variable data table can be denoted as $X = (x_{ik}), i \in I, k \in K$,
(where I is the set of entities, K is the set of variables, and x_{ik} is the value of the
variable $k \in K$ for the entity $i \in I$).

Basically, any variable $k \in K$ can be considered as a mapping of the set of
the entities into its value set, $x_k : I \rightarrow X_k$, with $x_k(i) = x_{ik}$. For example, in
Table 1.3, the value set of the first variable, Distance (the average distance from
Sun to the planet), is $X_1 = [36, 3600]$, the interval between (and including) 36 and
3600 thousand miles. The value set for the variable Matter (kind of the surface) is
$X_6 = \{Solid, Mixed, Liquid\}$. In both cases, only data-related values are included

since the data as they are do not provide any information about whether there exists any planet with its distance from Sun beyond the range [36, 3600], or with its surface consisting of a different kind of matter. The question might be asked: why is the interval [36, 3600] considered the value set, and not just particular values placed in the table (36, 67, 95,...)? The answer is that it is only a matter of convenience since the quantitative variable values are supposed to be averaged or/and compared with some other values.

> A distinctive feature of the variable-to-entity table is that its values are compared only within the columns (variables).

Having in mind that a cluster corresponds to a subset of the entities, it is quite useful to have its intensional description in terms of the variables. For instance, the group of planets $S = \{Mercury, Venus, Earth, Mars\}$ can be characterized as the set of planets whose distance from Sun is less than 150 kilomiles. More vague description is presented with a type-wise indication that the distance is approximately 85 kilomiles (which is the average distance). These are the most popular descriptions currently in use in clustering. Obviously, any intensional cluster description, while looking quite theoretical, has an empirical nature since it is based only on the data table under consideration. Any intensional description to an extensional cluster obtained somehow is referred to as its *interpretation*. Consistent interpretations become parts of the theory of the corresponding phenomena.

There exist also two-way one-mode column-conditional data, as, for instance, a table $A = (a_{ij})$ of inter-industrial supply, where a_{ij} is the supply of the industry i product to industry j $(i, j \in I)$ measured in natural units such as coal supply in tons, electricity supply in kilowatts.

1.3.3   Comparable Data Tables

Comparable Rectangular Table

Table 1.4 represents an extract from the results of the following sorting experiment: each out of 50 respondents partitioned 20 terms related to the human body by intuitive similarity, and, for any two terms, the number of subjects who did not put them in the same category, was considered their dissimilarity (the experiment was carried out by G. Miller (1968) as reported in Rosenberg 1982). The columns relate to the "larger" body parts, "Head", "Arm", "Chest", and "Leg", respectively, while the rows represent the other 16 body terms as presented in Table 1.4.

Again, this is a matrix $X = (x_{ik}), i \in I, k \in K$, where I and K do not overlap. But, this time, all the values x_{ik} across the table are comparable. This means that operation of averaging the values, within a part of the table corresponding to or

No	Term	Symbolic	Head	Arm	Chest	Leg
1	Body	Bo	45	50	37	50
2	Cheek	Ch	19	50	49	50
3	Ear	Ea	18	49	50	49
4	Elbow	El	49	8	50	47
5	Face	Fa	14	48	47	48
6	Hand	Ha	48	14	50	46
7	Knee	Kn	49	47	50	8
8	Lip	Li	18	49	50	49
9	Lung	Lu	48	49	17	49
10	Mouth	Mo	19	49	50	49
11	Neck	Ne	31	45	38	45
12	Palm	Pa	50	16	49	48
13	Thigh	Th	47	45	48	5
14	Toe	To	49	47	50	13
15	Trunk	Tr	42	46	19	45
16	Waist	Wa	44	45	26	46

Table 1.4: **Body:** An extract from Miller's sorting data (1968): number of subjects (out of 50) who did not put any given row-terms into the same category with the four column terms. (Treated: pp. 391 - 395.)

considered as a cluster, may be considered meaningful, at least for the close values. For instance, the dissimilarities between the four face parts and Arm and Chest, in the following subtable,

	Arm	Chest
Cheek	50	49
Ear	49	50
Lip	49	50
Mouth	49	50

have 49.5 as their average, which can be considered an aggregate characteristic of the subtable.

A great source of comparable data tables is rating the entities by various features (see Table 1.11 as an example).

Comparable data can be treated also just as a general entity-to-variable table, especially if such a treatment can help generating intensional descriptions of the

No	Genus	Human	Chimpanzee	Gorilla	Orangutan
2	Chimpanzee	1.45			
3	Gorilla	1.51	1.57		
4	Orangutan	2.98	2.94	3.04	
5	Rhesus monkey	7.51	7.55	7.39	7.10

Table 1.5: **Primates:** The mean number of nucleotide substitutions per 100 sites of 5.3 kb of noncoding DNA globin regions from Li and Grauer 1991, p.122. (Treated: pp. 129, 141, 357, 361, 366, 367.)

clusters in terms of the columns considered as the variables.

Proximity/Dissimilarity Data

In Table 1.5, the data on genetic distances between Human and four ape genera from Li and Grauer 1991 are presented. The data relate to a long discussed issue of the Human's origin: one view, by Darwin, claimed that the African apes, Chimpanzee and Gorilla, are man's closest relatives; another view was in favor of the Asian Orangutan as of the Homo clade; and yet another, homocentric, view gave the man a family of its own.

The data is a typical square matrix $D = (d_{ij})$ of the dissimilarity values between entities $i, j \in I$; both the rows and columns relate to the same entity set I while all the values d_{ij} are measured in the same scale and thus comparable across the table. (Only subdiagonal distances, d_{ij} with $i > j$, are presented in Table 1.5.)

Such a table can be found as a result of transforming an entity-to-variable table into entity-to-entity (or variable-to-variable) proximity matrix.

Sometimes, square association matrices can be observed directly, for instance, in sociometry, when individuals somehow rate their feelings about the other group members, or in labor statistics (occupational mobility tables), or in genetics (results of recombination testing) or in national economy statistics (inter-industrial input-output flows), etc. Graphs, binary relations, and weighted graphs can be considered as special cases of this type of data.

To interpret extensional clusters $S \subset I$, supplementary information on the relevant variables of the entities is necessary. For the human origin problem considered, this kind of data can be expected from paleontology or physiology research.

Sometimes, an interpretation can be provided in terms of the structure of the data revealed (especially in structural classification problems when data reflect interactions between the subsystems sought). Sometimes, an interpretation can be found considering the columns and the rows to be different entities, thus doubling I into two nonoverlapping copies, I_r, for the rows, and I_c, for the columns, and referring to the data as just a comparable rectangular table.

Yet another peculiarity of the proximity data comes from its size, which is just the number of the entities squared (or a half of it when the proximities are symmetrical). This makes it difficult to process large proximity data tables involving thousands or more entities.

1.3.4 Aggregable Data Tables

Rectangular Aggregable Data

Let us take a look at Table 1.6 (from L. Guttman (1971), cited by Greenacre 1988) which cross-tabulates 1554 Israeli adults according to their living places as well as, in some cases, that of their fathers (column items) and "principal worries" (row items). There are 5 column items considered: EUAM - living in Europe or America, IFEA - living in Israel, father living in Europe or America, ASAF - living in Asia or Africa, IFAA- living in Israel, father living in Asia or Africa, IFI - living in Israel, father also living in Israel. The principal worries are: POL, MIL, ECO - political, military and economical situation, respectively; ENR - enlisted relative, SAB - sabotage, MTO - more than one worry, PER - personal economics, OTH - other worries.

	EUAM	IFEA	ASAF	IFAA	IFI
POL	118	28	32	6	7
MIL	218	28	97	12	14
ECO	11	2	4	1	1
ENR	104	22	61	8	5
SAB	117	24	70	9	7
MTO	42	6	20	2	0
PER	48	16	104	14	9
OTH	128	52	81	14	12

Table 1.6: **Worries:** The original data on cross-classification of 1554 individuals by their worries and origin places. (Treated: pp. 54, 84, 100, 224, 325-326.)

The columns and the rows of such a matrix correspond to qualitative categories, and its entries represent counts or proportions of the cases fitting both the column and the row categories. In clustering constructions this kind of matrix still has been used rather rarely, though it has obvious advantage of using a quite homogeneous (counting) scale only! Both of the drawbacks of the proximity data are neatly avoided: 1) the size of the data matrix is relatively small because it is determined by number of the categories, not entities themselves; 2) the row and column items here are quite important aids to the interpretation purposes, being connected-to-data categories (in the area the data are from), not just ordinary cases.

The cross–classification data are called usually *contingency* data (tables). An extraordinary feature of this kind of data, distinguishing it from the general homogeneous data tables, is its *aggregability* property: the row or/and column items can be aggregated, according to their meaning, in such a way that the corresponding rows and columns are just summed together. For instance, let us aggregate the columns in Table 1.6 according to person's living places, thus summing up the columns IFEA, IFAA, and IFI into the aggregate column I (living in Israel) while aggregating their worries into two basic kinds: the worries coming outside their families (OUT=POL+MIL+ECO+SAB+MTO) and inside the families (FAM=ENR+PER); the other worries row OTH remains non-aggregated. The resulting data set:

	EUAM	I	ASAF
OUT	506	147	223
FAM	152	74	165
OTH	128	78	81

The table still counts 1554 individuals. Such an aggregability seems quite important in cluster analysis problems since it makes a natural aggregated representation for any data part as related to a cluster.

It remains to say that the aggregable data can be of the one-way data format, too, as, for instance, the data in Table 1.7 related to inter-generational mobility in the USA in 1973 (Hout 1986).

This table, though too aggregate for any real-world use, can be employed for an illustrative discussion of the Weberian issue: whether the social classes can be determined as being self-cohesive in the inter-generational mobility process, or not? This will be done using the original 17 × 17 mobility table (see Table 1.21, p.55) which has been aggregated into Table 1.7.

Similar *interaction* matrices can be drawn on industrial interactions (industry-to-industry commodity flows) or on international trade (nation-to-nation trade flows), etc. Note that the latter two examples bring another scale of counting (money, not just pieces) which still has the aggregability property. Actually, any

Father's Occupation	Son's Occupation					Total
	Upper Nonma	Lower Nonma	Upper Manual	Lower Manual	Farm	
Upper Nonmanual	1,414	521	302	643	40	2,920
Lower Nonmanual	724	524	254	703	48	2,253
Upper Manual	798	648	856	1,676	108	4,086
Lower Manual	756	914	771	3,325	237	6,003
Farm	409	357	441	1,611	1,832	4,650
Total	4,101	2,964	2,624	7,958	2,265	19,912

Table 1.7: **Mobility 5**: Cross-classification of father's occupation by son's first full-time civilian occupation for U.S. men 20 to 64 years of age in 1973, from Hout 1986, p. 11. (Treated: pp. 129-132, 135, 136, 137, 138, 142, Section 5.6.)

parameter of mass or volume (not just of intensity) of interaction between the row/column items in a two-way table, creates an aggregable data table.

1.3.5 Discussion

In this section, a classification of the data tables is considered. One base of the classification involves the number of different entity sets involved (ways) and the number of those of them which are different (modes). Most attention will be given, in the remainder, to two-way data tables, of which one-mode tables are usually similarity/interaction/distance data while two-mode tables are usually entity-to-variable or contingency data.

The second base involves degree of comparability of the entries across the table permitted: (1) column-conditional data (within-column comparing and averaging), (2) comparable data (overall comparing and averaging), (3) aggregable data (overall comparing and adding up to the total volume). There are also the so-called Boolean, or binary data combining peculiarities of all the three kinds.

1.4 Column-Conditional Data and Clustering

1.4.1 Boolean Data Table: Tasks and Digits

In Table 1.8 (cited by Rosenberg, Van Mechelen and De Boeck 1995) seven persons are described in terms of their success/failure patterns on a set of six tasks.

Task	Voca-bulary	Syno-nims	Jigsaw puzzles	Mathem. problems	Planning routes	Geography problems
Olivia	0	0	1	0	0	0
Ann	0	0	1	0	0	0
Peter	1	1	0	1	0	0
Mark	1	1	1	0	1	1
John	1	1	1	0	1	1
Dave	1	1	1	1	1	1
Andrew	1	1	1	1	1	1

Table 1.8: **Tasks:** Hypothetical example of persons' success/failure (1/0) pattern on a set of tasks. (Treated: p. 147)

The columns of the table can be considered as the so-called Boolean, or binary variables.

A Boolean (binary) variable shows presence/absence of a quality or feature or an attribute.

A convenient way of perceiving a Boolean variable is to think of it as a yes-or-no question. Can Olivia solve the jigsaw puzzles? Yes. Can Peter? No. (See 1 (for Yes) and 0 (for No) in Table 1.8.)

Clustering the rows in Table 1.8 by similarity may reveal some "typical" task resolving patterns, thus assigning individuals to types, implying that there should be some basic strategies or skills involved in forming the patterns. Similarly, the column-clusters could reveal individuals performing similarly on the within-cluster tasks, again implying that there must be some basic solution strategies involved. This allows us to think of simultaneous clustering of both the rows and columns to reveal the task patterns and the individuals employing them, keeping in mind that the cluster structure found could lead to skill/strategy interpretation of the clusters.

Yet another example, in Table 1.9, shows presence/absence of the seven enumerated segments in the rectangle in Fig 1.9 used to draw numeral digits in a stylized manner (this example was considered by many authors, Breiman et al. 1984 and Corter and Tversky 1986 included). The seven binary variables correspond to the columns of the data matrix, and the digits, to the rows. No-answer is presented with missing entries in Table 1.9.

Two kinds of problems may involve row-clustering in Table 1.9. The first is finding a cluster structure (patterns of similar rows) in the table as it is; then, the

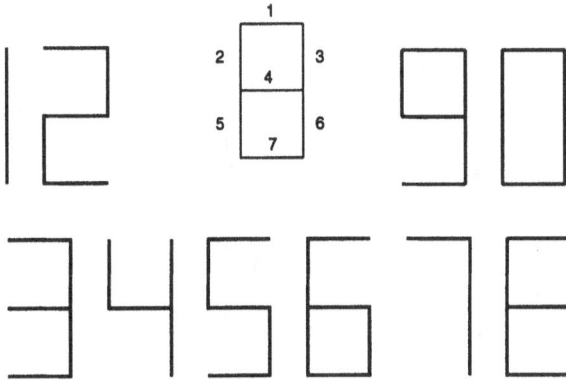

Figure 1.9: **Digits:** Integer digits presented by segments of the rectangle.

clusters found are to be interpreted in terms of some substantive properties of the numerals, like being odd/even or prime/composite. Although the digit character images seem quite arbitrary, trying such an approach (finding clusters with subsequent interpretation of them in terms of different variables) is a traditional use of clustering.

Problems of the second kind come out of the opposite direction: initially, clusters are found in terms of a different set of variables (say, by the digit confusion data in Table 1.16); then, they are to be interpreted in terms of the Boolean segment variables of Table 1.9.

Binary data emerge usually as a result of preliminary processing of some more complicated primary data like images (with their pixels dark or light) or questionnaires (with their categories of answers coded). Also, they are an important way for presenting specific structures and relations as simplified models for real-life situations.

1.4.2 Examples of Quantitative Entity-to-Variable Data: Iris, Disorders and Body

Quantitative (multivariate) entity-to variable data is a rectangular matrix having the rows corresponding to the entities (cases or objects or observations or items or elements, and so on), and the columns corresponding to the variables, with the

Digit	e1	e2	e3	e4	e5	e6	e7
1			1		1		
2	1		1	1	1		1
3	1		1	1		1	1
4		1	1	1		1	
5	1	1		1		1	1
6	1	1		1	1	1	1
7	1		1			1	
8	1	1	1	1	1	1	1
9	1	1	1	1		1	1
0	1	1	1		1	1	1

Table 1.9: **Digits:** Segmented numerals presented with seven binary variables corresponding to presence/absence of the corresponding segment in Figure 1.9. (Treated: p. 312.)

entries coding the values of the variables for the entities.

This kind of data often arises directly from experiments, from surveys, from industrial or government statistics, and so on. Often such data are considered as a primary source for transformation into other table formats.

Table 1.10 may be considered the most popular data set in classification, machine learning and data analysis research. The data is composed of 150 iris specimens, each measured on four morphological variables: sepal length v1 and width v2 and petal length v3 and width v4, as collected by a botanist, E. Anderson, and published by R. Fisher in his founding paper (Fisher 1936). There are three species, *Iris setosa* (diploid), *Iris versicolor* (tetraploid), and *Iris virginica* (hexaploid), each represented by 50 consecutive entities in the table.

> Treated: pp.   158,
> 200, 202, 202, 204,
> 305, 305, 312, 315,
> 316, 317

Two kinds of problems are associated with Anderson-Fisher data.

The first problem is developing clustering techniques that find a cluster structure which is in good accord with the division of the specimens in the species. It is well known from the previous studies that the two latter species are not well separable in the space of the four variables presented (look, for example, on spec-

Entity in a Class	Class 1 Iris setosa				Class 2 Iris versicolor				Class 3 Iris virginica			
	v1	v2	v3	v4	v1	v2	v3	v4	v1	v2	v3	v4
1	5.1	3.5	1.4	0.3	6.4	3.2	4.5	1.5	6.3	3.3	6.0	2.5
2	4.4	3.2	1.3	0.2	5.5	2.4	3.8	1.1	6.7	3.3	5.7	2.1
3	4.4	3.0	1.3	0.2	5.7	2.9	4.2	1.3	7.2	3.6	6.1	2.5
4	5.0	3.5	1.6	0.6	5.7	3.0	4.2	1.2	7.7	3.8	6.7	2.2
5	5.1	3.8	1.6	0.2	5.6	2.9	3.6	1.3	7.2	3.0	5.8	1.6
6	4.9	3.1	1.5	0.2	7.0	3.2	4.7	1.4	7.4	2.8	6.1	1.9
7	5.0	3.2	1.2	0.2	6.8	2.8	4.8	1.4	7.6	3.0	6.6	2.1
8	4.6	3.2	1.4	0.2	6.1	2.8	4.7	1.2	7.7	2.8	6.7	2.0
9	5.0	3.3	1.4	0.2	4.9	2.4	3.3	1.0	6.2	3.4	5.4	2.3
10	4.8	3.4	1.9	0.2	5.8	2.7	3.9	1.2	7.7	3.0	6.1	2.3
11	4.8	3.0	1.4	0.1	5.8	2.6	4.0	1.2	6.8	3.0	5.5	2.1
12	5.0	3.5	1.3	0.3	5.5	2.4	3.7	1.0	6.4	2.7	5.3	1.9
13	5.1	3.3	1.7	0.5	6.7	3.0	5.0	1.7	5.7	2.5	5.0	2.0
14	5.0	3.4	1.5	0.2	5.7	2.8	4.1	1.3	6.9	3.1	5.1	2.3
15	5.1	3.8	1.9	0.4	6.7	3.1	4.4	1.4	5.9	3.0	5.1	1.8
16	4.9	3.0	1.4	0.2	5.5	2.3	4.0	1.3	6.3	3.4	5.6	2.4
17	5.3	3.7	1.5	0.2	5.1	2.5	3.0	1.1	5.8	2.7	5.1	1.9
18	4.3	3.0	1.1	0.1	6.6	2.9	4.6	1.3	6.3	2.7	4.9	1.8
19	5.5	3.5	1.3	0.2	5.0	2.3	3.3	1.0	6.0	3.0	4.8	1.8
20	4.8	3.4	1.6	0.2	6.9	3.1	4.9	1.5	7.2	3.2	6.0	1.8
21	5.2	3.4	1.4	0.2	5.0	2.0	3.5	1.0	6.2	2.8	4.8	1.8
22	4.8	3.1	1.6	0.2	5.6	3.0	4.5	1.5	6.9	3.1	5.4	2.1
23	4.9	3.6	1.4	0.1	5.6	3.0	4.1	1.3	6.7	3.1	5.6	2.4
24	4.6	3.1	1.5	0.2	5.8	2.7	4.1	1.0	6.4	3.1	5.5	1.8
25	5.7	4.4	1.5	0.4	6.3	2.3	4.4	1.3	5.8	2.7	5.1	1.9
26	5.7	3.8	1.7	0.3	6.1	3.0	4.6	1.4	6.1	3.0	4.9	1.8
27	4.8	3.0	1.4	0.3	5.9	3.0	4.2	1.5	6.0	2.2	5.0	1.5
28	5.2	4.1	1.5	0.1	6.0	2.7	5.1	1.6	6.4	3.2	5.3	2.3
29	4.7	3.2	1.6	0.2	5.6	2.5	3.9	1.1	5.8	2.8	5.1	2.4
30	4.5	2.3	1.3	0.3	6.7	3.1	4.7	1.5	6.9	3.2	5.7	2.3
31	5.4	3.4	1.7	0.2	6.2	2.2	4.5	1.5	6.7	3.0	5.2	2.3
32	5.0	3.0	1.6	0.2	5.9	3.2	4.8	1.8	7.7	2.6	6.9	2.3
33	4.6	3.4	1.4	0.3	6.3	2.5	4.9	1.5	6.3	2.8	5.1	1.5
34	5.4	3.9	1.3	0.4	6.0	2.9	4.5	1.5	6.5	3.0	5.2	2.0
35	5.0	3.6	1.4	0.2	5.6	2.7	4.2	1.3	7.9	3.8	6.4	2.0
36	5.4	3.9	1.7	0.4	6.2	2.9	4.3	1.3	6.1	2.6	5.6	1.4
37	4.6	3.6	1.0	0.2	6.0	3.4	4.5	1.6	6.4	2.8	5.6	2.1
38	5.1	3.8	1.5	0.3	6.5	2.8	4.6	1.5	6.3	2.5	5.0	1.9
39	5.8	4.0	1.2	0.2	5.7	2.8	4.5	1.3	4.9	2.5	4.5	1.7
40	5.4	3.7	1.5	0.2	6.1	2.9	4.7	1.4	6.8	3.2	5.9	2.3
41	5.0	3.4	1.6	0.4	5.5	2.5	4.0	1.3	7.1	3.0	5.9	2.1
42	5.4	3.4	1.5	0.4	5.5	2.6	4.4	1.2	6.7	3.3	5.7	2.5
43	5.1	3.7	1.5	0.4	5.4	3.0	4.5	1.5	6.3	2.9	5.6	1.8
44	4.4	2.9	1.4	0.2	6.3	3.3	4.7	1.6	6.5	3.0	5.5	1.8
45	5.5	4.2	1.4	0.2	5.2	2.7	3.9	1.4	6.5	3.0	5.8	2.2
46	5.1	3.4	1.5	0.2	6.4	2.9	4.3	1.3	7.3	2.9	6.3	1.8
47	4.7	3.2	1.3	0.2	6.6	3.0	4.4	1.4	6.7	2.5	5.8	1.8
48	4.9	3.1	1.5	0.1	5.7	2.6	3.5	1.0	5.6	2.8	4.9	2.0
49	5.2	3.5	1.5	0.2	6.1	2.8	4.0	1.3	6.4	2.8	5.6	2.2
50	5.1	3.5	1.4	0.2	6.0	2.2	4.0	1.0	6.5	3.2	5.1	2.0

Table 1.10: **Iris:** Anderson-Fisher data on three classes (species) of Iris specimens.

imens 28, 33 and 44 from class 2 and 18, 26, and 33 from class 3 which are more similar to each other than to other specimens of the same species). This leads to yet another problem, of the so-called *constructive induction*: deriving some new variables from the given ones to allow a better discriminating between the classes.

The second kind of problems is easier: can we describe somehow the classes presented intensionally, in terms of the variables given? Here the classes are considered as they are given, and no development of any techniques for class finding is required.

Analogous problems, (1) adjusting the cluster structure and (2) intensional describing the classes, can be considered for the data presented in Table 1.11. The entities are "archetypal psychiatric patients" fabricated by experienced psychiatrists from Stanford University along with the list of seventeen variables, as follows:

w1. Somatic concern

w2. Anxiety

w3. Emotional withdrawal

w4. Conceptual disorganization

w5. Guilt feelings

w6. Tension

w7. Mannerisms and posturing

w8. Grandiosity

w9. Depressive mood

w10. Hostility

w11. Suspiciousness

w12. Hallucinatory behavior

w13. Motor retardation

w14. Uncooperativeness

w15. Unusual thought content

w16. Blunted effect

w17. Excitement

The values of the variables are severity ratings from 0 to 6. The patients are partitioned into four classes, each containing eleven consecutive individuals considered typical for each of the following four mental disorders: depressed (manic-depressive illness), manic (manic-depressive illness), simple schizophrenia, and paranoid schizophrenia, respectively. The table is published in Mezzich and Solomon 1980 along with a detailed description of the data on pp. 60-63.

Treated: pp.   201,
214, 309, 309, 311,
315, 319

Some may doubt whether the Disorders data really is column-conditional: all the variables are measured in the same scale, which makes all the values comparable with each other. Others may dispute this opinion, explaining that the severity ratings cannot be compared because the variables refer to different aspects.

Another example of the comparable data which will be treated here as a column-conditional one is Body (Table 1.4, p. 29). The natural hierarchy of the body parts should be reflected in the underlying cluster structure.

No	w1	w2	w3	w4	w5	w6	w7	w8	w9	w10	w11	w12	w13	w14	w15	w16	w17
1	4	3	3	0	4	3	0	0	6	3	2	0	5	2	2	2	1
2	5	5	6	2	6	1	0	0	6	1	0	1	6	4	1	4	0
3	6	5	6	5	6	3	2	0	6	0	5	3	6	5	5	0	0
4	5	5	1	0	6	1	0	0	6	0	1	2	6	0	3	0	2
5	6	6	5	0	6	0	0	0	6	0	4	3	5	3	2	0	0
6	3	3	5	1	4	2	1	0	6	2	1	1	5	2	2	1	1
7	5	5	5	2	5	4	1	1	6	2	3	0	6	3	5	2	3
8	4	5	5	1	6	1	1	0	6	1	1	0	5	2	1	1	0
9	5	3	5	1	6	3	1	0	6	2	1	1	6	2	5	5	0
10	3	5	5	3	2	4	2	0	6	3	2	0	6	1	4	5	1
11	5	6	6	4	6	3	1	0	6	2	0	0	6	4	4	6	0
12	2	2	1	2	0	3	1	6	2	3	3	2	1	4	4	0	6
13	0	0	0	4	1	5	0	6	0	5	4	4	0	5	5	0	6
14	0	3	0	5	0	6	0	6	0	3	2	0	0	3	4	0	6
15	0	0	0	3	0	6	0	6	1	3	1	1	0	2	3	0	6
16	3	4	0	0	0	5	0	6	0	6	0	0	0	5	0	0	6
17	2	4	0	3	1	5	1	6	2	5	3	0	0	5	3	0	6
18	1	2	0	2	1	4	1	5	1	5	1	1	0	4	1	0	6
19	0	2	0	2	1	5	1	5	0	2	1	1	0	3	1	0	6
20	0	0	0	6	0	5	1	6	0	5	5	4	0	5	6	0	6
21	5	5	1	4	0	5	5	6	0	4	4	3	0	5	5	0	6
22	1	3	0	4	1	4	2	6	3	3	2	0	0	4	3	0	6
23	3	2	5	2	0	2	2	1	2	1	2	0	1	2	2	4	0
24	4	4	5	4	3	3	1	0	4	2	3	0	3	2	4	5	0
25	2	0	6	3	0	0	5	0	0	3	3	2	3	5	3	6	0
26	1	1	6	2	0	0	1	0	0	3	0	1	0	1	1	6	0
27	3	3	5	6	3	2	5	0	3	0	2	5	3	3	5	6	2
28	3	0	5	4	0	0	3	0	2	1	1	1	2	3	3	6	0
29	3	3	5	4	2	4	2	1	3	1	1	1	4	2	2	5	2
30	3	2	5	2	2	2	2	1	2	2	3	1	2	2	3	5	0
31	3	3	6	6	1	3	5	1	3	2	2	5	3	3	6	6	1
32	1	1	5	3	1	1	3	0	1	1	1	0	5	1	2	6	0
33	2	3	5	4	2	3	0	0	3	2	2	0	0	2	4	5	0
34	2	4	3	5	0	3	1	4	2	5	6	5	0	5	6	3	3
35	2	4	1	1	0	3	1	6	0	6	6	4	0	6	5	0	4
36	5	5	5	6	0	5	5	6	2	5	6	6	0	5	6	0	2
37	1	4	2	1	1	1	0	5	1	5	6	5	0	6	6	0	1
38	4	5	6	3	1	6	3	5	2	6	6	4	0	5	6	0	5
39	4	5	4	6	2	4	2	4	1	5	6	5	1	5	6	2	4
40	3	4	3	4	1	5	2	5	2	5	5	3	1	5	5	1	5
41	2	5	4	3	1	4	3	4	2	5	5	4	0	5	4	1	4
42	3	3	4	4	1	5	5	5	0	5	6	5	1	5	5	3	4
43	4	4	2	6	1	4	1	5	3	5	6	5	1	5	6	2	4
44	3	5	5	5	2	5	4	5	2	4	6	5	0	5	6	5	5

Table 1.11: **Disorders:** Data on archetypal patients measured on 17 psychopatho-logical items from Mezzich and Solomon 1980, p.62.

1.4.3   Mixed Variable Tables: Planets and Russian Master-pieces

In Table 1.12, some illustrative data are presented on eight masterpieces of Russian literature along with the values of 5 variables, which are: 1) LenSent - Average length of sentences (number of words); 2) LenDial - Average length of dialogues (number of sentences); 3) NChar - Number of principal characters in the novel; 4) InMon - Does the author use internal monologues of the characters or not; 5) Presentat - Principal way of presentation of the subject by the author.

The variables 1 to 3 are *quantitative*, which means that, typically, statements involving quantitative comparisons of their values or quantitative transformations of those, are meaningful. Variable 4 is *Boolean (binary)*; its categories are Yes or No. Variable 5 is *nominal*: it has three *categories* (called also grades): Direct - meaning that the author prefers direct descriptions and comments, Behav - the author prefers expressing his ideas through behavior of the characters, and Thought - the subject is shown, mainly, through characters' thoughts. Actually, any nominal variable can be considered as a partition-wise classification. It is important from the technical point of view since it allows thinking of partitional classifications as nominal variables: a partition-wise clustering structure may be thought of as a nominal variable approximating the variables given.

Title	LenSent	LenDial	NChar	InMon	Presentat
Eug.Onegin	15.0	16.6	2	No	Direct
Doubrovski	12.0	9.8	1	No	Behav
Captain's Daughter	11.0	10.4	1	No	Behav
Crime and Punishment	20.2	202.8	2	Yes	Thought
Idiot	20.9	228.0	4	Yes	Thought
A Raw Youth	29.3	118.6	2	Yes	Thought
War & Peace	23.9	30.2	4	Yes	Direct
A.Karenina	27.2	58.0	5	Yes	Direct

Table 1.12: **Masterpieces:** Russian masterpieces of 19th century: the first three by A. Pushkin, the next three by F. Dostoevski, and the last two by L. Tolstoy. (Treated: pp. 76, 94, 106, 115, 152, 303, 312, 312, 316, 319.)

The following two problems could be related to the data in Table 1.12:

1. Classify the set of the masterpieces in groups (clusters) that would be homogeneous in terms of the variables presented and, simultaneously, describe the clusters in terms of the most "important" (for the classification) variables. In the particular case presented, the clusters should be connected to the authors, which could be interpreted as the variables measured indeed underlie the authors' writing manners.

2. Reveal interrelations between the following two aspects of writing style: (a) style features (as presented by two of the variables, LenSent and LenD), and (b) presentation features (the other three variables), via clustering. We expect, in this particular case, to have a high correlation between the aspects since both of them seem to be determined by the author personalities.

1.4.4 Discussion

Seven small data sets are presented to illustrate the following general classification problems:

(1) (extensional) finding a cluster structure (Tasks, Digits, Body, Masterpieces); the list of the cluster structures to seek includes partition (Digits and Masterpieces), hierarchy (Body), and bipartition/bihierarchy, that is, simultaneous clustering of both the row and column sets (Tasks).

(2) finding intensional description of clusters given extensionally (all),

(3) adjusting variables to a pre-given cluster structure (Digits, Iris, Disorders),

(4) finding interrelation between entity and variable sets (Tasks) or between some variable subsets (Masterpieces) via clustering.

1.5 Clustering Problems for Comparable Data

1.5.1 Entity-to-Entity Distance Data: Primates

This kind of data is represented by Table 1.3 Primates, p. 27. It is obtained usually from the entity-to-variable data tables by calculating the inter-entity distances in the variable space as discussed in Section 2.1.

The basic clustering problem related to this kind of data is revealing its cluster structure based on the following principle: within-cluster distances should be smaller than those between the clusters. Typical questions answered are: which objects belong to the same clusters and which to the different ones? The problem of human origin is of this kind.

Cluster structures of interest can be of the following kinds: 1) just one cluster containing the human, 2) a partition, and 3) hierarchical clusters (to speculate on the species evolution).

Function	e^x	lnx	$1/x$	$1/x^2$	x^2	x^3	$\sqrt{x}$	$\sqrt[3]{x}$
lnx	7							
$1/x$	1	1						
$1/x^2$	1	1	7					
x^2	2	2	2	2				
x^3	3	2	1	1	6			
$\sqrt{x}$	2	4	1	2	5	4		
$\sqrt[3]{x}$	2	4	1	1	5	3	5	
$|x|$	2	3	1	1	5	2	3	2

Table 1.13: **Functions:** Similarities between nine elementary functions rated by a high-school mathematics teacher. (Treated: pp. 259, 221.)

1.5.2   Entity-to-Entity Similarity Data: Functions

Similarity measures differ from distance in that they have opposite direction: the greater the similarity, the more alike are the entities, while the greater the distance, the less alike are the entities.

Psychologists measure similarity by asking the respondents to evaluate similarity between some stimuli, subject to the respondent's personal feeling.

For example, an educational research team in Russia has proposed a nontraditional methodology for knowledge control and testing based on the respondent's evaluation of the similarities between the basic concepts of the discipline tested rather than on traditional direct questioning on substantive topics. The basic idea is that there exists a (cognitive) structure of semantic relationship among the concepts, which must be acquired by learning; the discrepancy between a student's personal structure and that one to be acquired may be used for scoring the student's knowledge degree (Satarov 1991). The working tool is a concept-to-concept similarity matrix as produced by the respondent.

The Table 1.13 of similarities between 9 elementary algebraic functions has been produced by a high-school teacher of mathematics as supposedly related to the correct cognitive structure.

Usually, corresponding semantic structure is sought with quantitative methods of multidimensional scaling and ordination, aimed at finding such a conceptual variable space that the entities can be embedded there with the inter-entity distances conforming to the given similarities. However, the dimensions of the spaces found (along with corresponding concepts) usually correspond to some entity clusters, which means that the metric structure recovered is not informative, but only the clusters are meaningful. There are two problems to resolve, in the approach to knowledge control considered: (1) finding a cluster structure from a similarity matrix; (2) measuring the difference between the student's cluster structure and the standard one.

1.5.3 Three-way similarity matrix: Kinship

This is a list of fifteen kinship terms:

1. Aunt
2. Brother
3. Cousin
4. Daughter
5. Father

6. Granddaughter
7. Grandfather
8. Grandmother
9. Grandson
10. Mother

11. Nephew
12. Niece
13. Sister
14. Son
15. Uncle

Kim and Rosenberg (1975) asked the respondents to sort the kinship terms by as many classes as they wanted; the data on dissimilarities between the terms derived from six sorting experiments are taken from Arabie, Carroll, and DeSarbo 1987. The dissimilarity index is the number of the respondents who put the terms in question in different classes. There are six matrices presented by their under-diagonal elements, two digits each, in Tables 1.14 and 1.15. The first two matrices relate to groups of females and males, respectively, who made a single sorting; the other four matrices contain results of each of two sorting experiments conducted repeatedly in a female group (matrices three and four) and in a male group (matrices five and six). The data, actually, is a three-way table (a_{ijt}) where i and j relate to the kinship terms while t relates to the different respondent groups.

The problem here is to find out whether there exists any unique clustering structure of the kinship terms underlying the six matrices presented? If yes, what are the deviations of each of the six groups from that?

Similarities	Number
79	1
5670	2
366671	3
76227863	4
3473702577	5
763575763261	6
36787534761748	7
7727717131491763	8
336876155031763077	9
57335574387437783080	10
1375543279317738733945	11
384870206728773474217735	12
77207248167035772662327266	13
4738597932783777357513577938	14
76	1
5563	2
605673	3
70427852	4
5774724478	5
725876764557	6
54797852702929	7
7454706855283056	8
526480242954704577	9
51524872617159785380	10
2671495380527859706126	11
582365346553795873437152	12
77357128256853754651547155	13
2856547750775574596926527559	14
78	1
5362	2
546671	3
73337263	4
5274754974	5
755270734852	6
48777652732731	7
7848697250302552	8
456077363150744776	9
58484572547451764780	10
3075524977497552745330	11
513269426149775275347447	12
78416728367251744763477467	13
3151477746774875517429577853	14

Table 1.14: **Kinship I:** Dissimilarities between fifteen kinship terms as based on Kim and Rosenberg (1975) data taken from Arabie, Carroll, and DeSarbo 1987.

Similarities	Number
74	1
6056	2
545767	3
66366862	4
5169653478	5
685369743962	6
43787447693431	7
7445616750343562	8
406274333250683976	9
64414567526549753978	10
3867514180397549665330	11
523162276340775169366840	12
78326335346849773763406757	13
2850547940764469516639637753	14
83	1
3877	2
796183	3
79558443	4
7982826983	5
847684837348	6
78838474823811	7
8573818074133848	8
726384341374827383	9
49745381778078847383	10
4283537485728579797912	11
771078526373837682558176	12
85528214348074836943748161	13
1077398572857884797942498379	14
84	1
5573	2
716780	3
76428261	4
7177795582	5
846981816660	6
70828265804621	7
8560797366214361	8
635983441966816382	9
59695879727971856485	10
4582596485638472807220	11
722175476159836876468267	12
84498022447565815660658064	13
1771558463857084727646608470	14

Table 1.15: **Kinship II:** Dissimilarities between fifteen kinship terms based on Kim and Rosenberg (1975) data (continued). (Treated: pp. 223, 384.)

It should be noted also that in spite of the fact that the entries are just counts, the data is not aggregable since there is no meaning in summation of the entries across the table.

1.5.4   Entity-to-Entity Interaction Table: Confusion

Table 1.16 represents results of an experiment on confusion between segmented numerals (similar to those presented in Figure 1.9), reported in Keren and Baggen 1981. A digit appeared on a screen for a very short time (stimulus), and a subject was to report the digit name. For every stimulus, the frequencies of the digits claimed stand in the corresponding row of Table 1.16. The matrix has two formal features: (1) it is asymmetric (as every interaction table), (2) its principal diagonal entries cover the most part of its "mass", which is a feature of the steady-state interaction processes (most part of the observations falls within the states).

Stimulus	Response									
	1	2	3	4	5	6	7	8	9	0
1	877	7	7	22	4	15	60	0	4	4
2	14	782	47	4	36	47	14	29	7	18
3	29	29	681	7	18	0	40	29	152	15
4	149	22	4	732	4	11	30	7	41	0
5	14	26	43	14	669	79	7	7	126	14
6	25	14	7	11	97	633	4	155	11	43
7	269	4	21	21	7	0	667	0	4	7
8	11	28	28	18	18	70	11	577	67	172
9	25	29	111	46	82	11	21	82	550	43
0	18	4	7	11	7	18	25	71	21	818

Table 1.16: **Confusion:** The Keren and Baggen 1981 data on confusion of the segmented numeral digits. (Treated: pp. 87, 260, 267, 312.)

The problem here is to find out general patterns of confusion and, then, to interpret them using some other data as, for instance, segment-based Boolean variables in Digit data table 1.9, p. 36.

1.5.5 Variable-to-Variable Correlation Table: Activities

Analysis of the structure of interrelation between variables via cluster analysis in variable-to-variable correlation tables was an important stage in the evolution of clustering methodology; moreover, the term itself emerged just to denote sets of mutually interrelated variables to be interpreted as being manifestations of the same "interior personality factor", like ability or aggressiveness. Such a clustering was considered as a fast but not quite proper (quite suspicious) way to get a sense of the underlying factors when exact computation of the factor scores was an issue because of the lack of appropriate computing facilities (Tryon 1939). Though computing routines are currently readily available, there are still situations when analysis of the structure of a correlation matrix might be useful. Such a situation arises, for instance, when correlations between the variables are low. In such a case, no analytical factor analysis model could work; still clusters could show a qualitative picture of interrelations.

	Work	Eat	Cook	Clea	Laun	Serv	Gard	Farm	Nurs	Leis
Sex	59	266	326	175	97	77	4	49	40	44
Age	7	5	20	7	8	6	8	36	110	38
Job	45	88	125	91	60	57	15	32	37	40
Family	9	4	26	7	3	3	5	15	7	78
Kids	0	0	1	1	0	0	2	0	78	1
UFemale	5	24	26	16	6	7	3	4	1	2
Home	2	1	3	4	2	1	3	12	5	1
DTech	1	2	1	0	0	0	0	3	3	1
Income	6	10	6	7	6	14	9	14	37	9
College	34	40	74	34	32	6	8	64	31	22
School	3	10	9	23	15	18	3	54	63	19
Farm	27	14	9	9	2	37	6	78	14	10
SWeek	114	9	6	7	11	29	10	70	16	9

Table 1.17: **Activity:** Correlation data between socio-economic variables of the agriculture personnel and the modes of their time-spending, in 0-1000 scale range (from Mirkin 1985, p.149). (Treated: p. 144.)

A rectangular variable-to-variable matrix with rather low correlation entries (which is a feature of the sociological large-scale surveys, in the present author's experience) is presented in Table 1.17 (Mirkin 1985, p. 149). The set of row-variables consists of some socio-economical parameters taken from a survey of 1024 employees in rural settlements at Novosibirsk region (Russia, 1979); all 13 variables presented have been categorized and considered as qualitative variables. These are as follows:

1. Sex.

2. Age (by decades).

3. Job Status (6 ranks).

4. Marital Status (4 categories).

5. Children in House (yes, no).

6. Unworking Female in Household (yes, no).

7. House Quality (good, bad).

8. Domestic Devices Available (combinations of vacuum cleaner, washing machine and refrigerator).

9. Income (7 intervals).

10. College/Technical Education Level (four grades, "no at all" included).

11. Years Spent at School (four grades).

12. Farmyard at Home (yes, no).

13. Number of Shifts a Week (6 grades).

The set of column-variables reflects the time spent for the following aggregated activities covering more than 70% of all the activities of the respondents: 1. Work at an enterprise. 2. Eating. 3. Cooking. 4. House cleaning. 5. Laundry. 6. Getting service (like hair-cut, mail, etc). 7. Gardening at home farmyard. 8. Work at home farmyard (poultry, cattle, etc). 9. Nursing children. 10. Leisure.

The problem is to find which patterns of association exist between the row and column variables based on the pair-wise correlation coefficients presented in Table 1.17.

1.5.6   Category-to-Category Proximity Table: Behavior

Let us consider a 15 × 15 table of proximities between 15 kinds of situations and 15 kinds of human behavior, based on the appropriateness of the behavior to the

Situation	Behavior							
	Run	Talk	Kiss	Write	Eat	Sleep	Mumble	Read
Class	-1.99	1.70	-2.41	3.66	-0.28	-0.91	-0.89	2.76
Date	0.49	4.05	4.22	-0.89	3.28	-0.74	-1.39	-1.63
Bus	-3.07	3.57	-0.24	0.36	0.97	2.53	0.66	2.66
FDinner	-1.95	4.01	0.41	-1.93	3.93	-2.22	-1.97	-0.55
Park	3.43	3.91	3.20	2.49	3.62	1.12	0.89	3.26
Church	-3.13	-1.22	-2.13	-1.66	-3.13	-2.74	-0.99	-0.93
JInterv	-2.57	3.95	-3.43	0.34	-2.78	-3.76	-3.20	-2.03
Sidewalk	1.07	3.68	0.24	-1.13	0.32	-3.05	0.45	0.30
Movies	-2.05	0.47	1.70	-1.78	2.97	-0.43	-0.38	-2.78
Bar	-2.55	3.74	0.66	0.87	3.16	-1.61	1.70	0.20
Elevator	-2.88	2.89	0.28	-1.47	0.59	-3.20	0.61	-0.03
Restroom	-1.68	2.74	-1.70	-1.05	-2.16	-1.68	0.53	0.24
Own room	1.64	4.07	4.01	3.78	3.43	4.34	3.16	4.07
DLounge	-0.11	3.37	2.03	3.22	2.68	1.57	0.99	4.05
FBGame	-0.39	3.57	0.57	0.05	3.53	-1.53	0.72	-0.82
	Fight	Belch	Argue	Jump	Cry	Laugh	Shout	
Class	-3.30	-2.74	0.82	-2.72	-2.30	1.72	-2.57	
Date	-0.93	-2.28	-0.01	-0.09	-1.47	3.49	-0.72	
Bus	-2.99	-2.36	-0.34	-1.39	-1.43	2.59	-1.51	
FDinner	-2.84	-2.01	-1.26	-2.22	-1.30	2.62	-2.55	
Park	-1.45	0.49	0.55	2.91	0.70	3.59	2.41	
Church	-3.89	-3.09	-2.59	-2.80	-1.38	- 1.91	-3.18	
JInterv	-3.47	-3.30	-2.68	-3.03	-3.14	1.37	-2.86	
Sidewalk	-3.05	-1.70	-0.43	-0.97	-0.80	2.89	0.37	
Movies	-3.14	-1.93	-2.80	-2.20	2.64	3.43	-2.09	
Bar	-2.61	0.53	-0.20	-0.76	-1.07	3.72	-0.38	
Elevator	-2.93	-1.97	-1.93	-2.39	-1.03	2.26	-2.78	
Restroom	-2.74	0.61	-1.03	-0.86	0.28	1.39	-0.99	
Own room	-0.26	2.30	3.01	2.22	3.49	3.66	1.93	
DLounge	-2.11	-0.51	0.37	0.07	-0.63	3.24	-0.91	
FBGame	-2.47	-0.66	0.47	2.61	-0.20	3.39	3.43	

Table 1.18: **Behavior:** Situation-Behavior centered appropriateness rates by Price and Bouffard (1974) from Eckes and Orlik 1993. (Treated: pp. 222-224.)

situation (rated on a scale from 0 to 9 by fifty-two subjects in an experiment by Price and Bouffard 1974 reported in Eckes and Orlik 1993, p. 66). The data in Table 1.18 are deviations of the raw proximities from their grand mean.

The basic motivation is to find which classes of the columns correspond to specific classes of the rows. The problem is quite similar to that in the preceding section; the only difference is that the data here are taken from the subjects directly while, in the former table, the correlation values have been derived from a raw table of entity-to-variable format, which might be taken into account in mathematical modeling of the problem.

1.5.7   Graphs and Binary Relations

A graph is a mathematical concept modeling a set of interconnected elements. A graph G consists of a nonempty set $V(G)$ of its *vertices* and a set $E(G)$ of unordered pairs $e = uv$ of the vertices called *edges*; edge $e = uv$ is said to *join* its *ends* $u, v \in V(G)$. Graphs are so named because they are usually represented graphically: each vertex is indicated by a point, and each edge by a line joining the points which represent its ends.

Data tables are a natural vehicle for deriving graphs representing them, at least partly. Let us consider any similarity/dissimilarity matrix $A = (a_{ij})$, $i, j \in I$, for instance, the similarity matrix Functions (Table 1.13, p. 42), and specify a similarity threshold, say, $t = 4$. This gives rise to threshold graph G having the entity set I as its vertex set $V(G)$; edge ij joins those and only those entities i and j for which the similarity $a_{ij} \geq t$. In our example, the threshold graph is presented in Figure 1.10 (in terms of the functions).

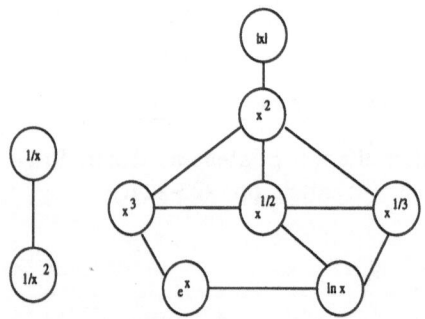

Figure 1.10: Threshold graph for Functions similarity data ($t = 4$).

Yet another way to represent graphs is through *adjacency* matrices. A square $|V(G)|$ by $|V(G)|$ symmetric matrix $r_G = (r_{uv})$ is the adjacency matrix if $r_{uv} = 1$ when $uv \in E(G)$ and $r_{uv} = 0$ when $uv \notin E(G)$. For instance, the threshold graph presented in Figure 1.10 has its adjacency matrix

$$
r_G = \begin{pmatrix}
1 & 1 & 0 & 0 & 0 & 1 & 0 & 0 & 0 \\
1 & 1 & 0 & 0 & 0 & 0 & 1 & 1 & 0 \\
0 & 0 & 1 & 1 & 0 & 0 & 0 & 0 & 0 \\
0 & 0 & 1 & 1 & 0 & 0 & 0 & 0 & 0 \\
0 & 0 & 0 & 0 & 1 & 1 & 1 & 1 & 1 \\
1 & 0 & 0 & 0 & 1 & 1 & 1 & 0 & 0 \\
0 & 1 & 0 & 0 & 1 & 1 & 1 & 1 & 0 \\
0 & 1 & 0 & 0 & 1 & 0 & 1 & 1 & 0 \\
0 & 0 & 0 & 0 & 1 & 0 & 0 & 0 & 1
\end{pmatrix}
$$

which can be written also in its "lower triangular" form:

$$
r_G =
\begin{array}{c|ccccccccc}
2 & 1 \\
3 & 0 & 0 \\
4 & 0 & 0 & 1 \\
5 & 0 & 0 & 0 & 0 \\
6 & 1 & 0 & 0 & 0 & 1 \\
7 & 0 & 1 & 0 & 0 & 1 & 1 \\
8 & 0 & 1 & 0 & 0 & 1 & 0 & 1 \\
9 & 0 & 0 & 0 & 0 & 1 & 0 & 0 & 0 \\
\hline
 & 1 & 2 & 3 & 4 & 5 & 6 & 7 & 8
\end{array}
$$

There is another kind of graph, the so–called *directed graphs* which have their vertices joined by directed *arcs* rather than undirected edges. An *arc* is an ordered pair $(u, v) \in V(G) \times V(G)$ of the vertices reflecting the direction of the arc. Directed graphs also can be represented by diagrams (the joining lines are provided with arrows to show direction of the arc) and square Boolean adjacency matrices which are not necessarily symmetric.

Directed graphs can arise as the threshold graphs for asymmetric data tables as the graph in Fig. 1.11, which represents the levels of confusion between the numeral digits larger than 40 in Table 1.16. The twofold role of the threshold graphs can be seen quite clearly: on one hand, they present information on interrelation between the entities; on the other hand, the information is cleared of unnecessary details to allow observation of its most important parts. We can see, for example, that 9 is the greatest source for confusion while 3 and 0 are important confusion targets. In this latter aspect, threshold graphs are a result of preprocessing the data, which sometimes gives a lot of information about the data structure.

Two-mode data tables lead to the so-called bipartite graphs. A graph $G = (V(G), E(G))$ is called a *bipartite* graph if the set of its vertices can be partitioned

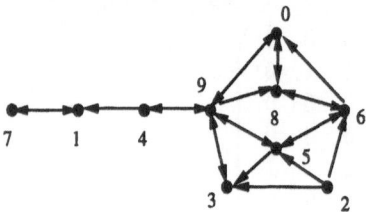

Figure 1.11: Directed threshold graph for digit Confusion data ($t = 40$).

in two subsets, V_1 and V_2, in such a way that no edge (or arc) joins the vertices within either of these subsets.

For example, data Task in Table 1.8 can be presented by the bipartite graph in Fig. 1.12 for which the data is its adjacency matrix. In contrast to the previous picture, this drawing does not give too many hints about the data structure. The reader is invited to draw the bipartite threshold graph for data Activities (Table 1.17), with threshold level, say, $t = 60$, to see what kind of structure can be found for that.

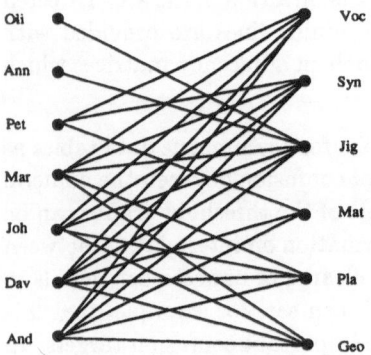

Figure 1.12: Bipartite graph for Task data.

Graphs can arise also as primary information from substantive research such as, for instance, in social psychology (sociometric matrices reporting results of mutual choices among the members of a small group) or in engineering (a scheme of connections between the elements of a device).

Considering a threshold graph is a simplest way to analyze the structure of the data if it is simple enough. This is why graphs could be considered sometimes as structures rather than data. We prefer thinking of arbitrary graphs as the data since, in our experience, the threshold graphs are too complicated to be considered a final result: this is a structure to be analyzed rather than described. Yet, in graph theory, there are several concepts (such as connected component, bicomponent, clique, coloring, tree) related very closely to clustering.

1.5.8 Discussion

Four out of six presented illustrative data sets are one-mode distance/similarity/interaction data (Primates, Functions, Kinship and Confusion) while the other two are two-mode similarity data (Activities and Behavior). The Kinship data set is special: it is the only three-way data set included.

One-mode data clustering is just to reveal the data structure in terms of a single cluster (Primates), or a partition (Primates and Confusion), or a set of possibly overlapping clusters (Functions and Kinship), or a partition along with associations between its classes (Confusion). Two-mode similarity data clustering is to reveal pairs of associated row/column subsets ("box clustering").

1.6 Clustering Problems for Aggregable Data

1.6.1 Category-to-Category Data: Worries

Let us consider contingency table for data Worries (Table 1.6 in Section 1.3.4) obtained from the original cross-classification by dividing each of its entries by the total number of the counted entities (1554 individuals). The results (which are the frequencies of the corresponding events, multiplied by 1000) are presented in Table 1.19. Its (i, j)-th entry can be denoted by p_{ij}, while the totals of the rows and columns called *marginals*, by p_{i+}, $i \in I$ (rows) and p_{+j}, $j \in J$, respectively.

It is supposed that living place influences the principal worry structure; the question is then how to analyze the influences based on the contingency table.

Traditionally, conditional probabilities $p(i/j) = p_{ij}/p_j$, not primary co-occurrence frequency values p_{ij}, are considered as reflecting the influences, though

	EUAM	IFEA	ASAF	IFAA	IFI	Total
POL	76	18	21	4	5	124
MIL	140	18	62	8	9	237
ECO	7	1	3	1	1	13
ENR	67	14	39	5	3	128
SAB	75	15	45	6	5	146
MTO	27	4	13	1	0	45
PER	31	10	67	9	6	123
OTH	82	33	52	9	8	184
Total	505	113	302	43	37	1000

Table 1.19: Cross-classification from Table 1.6 presented as a contingency table.

care should be taken to use the conditional probabilities in a proper way. For example, the main worry for the EUAM individuals, according to Table 1.6 or 1.19, is the military situation with $p(MIL/EUAM) = 140/505 = 27.7\%$; and the main worry for ASAF is the personal economy problems with $p(PER/ASAF) = 67/302 = 22.1\%$. Would this mean that the influence in the first case is higher than in the second? Not necessarily. To come to a reasonable conclusion, let us compare the conditional probability $p(i/j)$ to the average rate p_i of i for all 1544 observations.

To make the comparison, absolute change $p(i/j) - p_i$, or relative value $p(i/j)/p_i$, or relative change $q_{ij} = (p(i/j) - p_i)/p_i$ could be used. The relative value $p(i/j)/p_i = p_{ij}/(p_i p_j)$ called the (marginal) odds ratio, is a standard tool in contingency data analysis (see, for instance, Reynolds 1977).

Value	MIL/EUAM	PER/ASAF
$p(i)$	237	123
$p(i/j)$	277	221
$p(i/j) - p(i)$	40	98
$p(i/j)/p(i)$	1168	1804
q_{ij}	168	804

Table 1.20: Column-to-row interaction measured by different indices.

	Son's First Full-Time Occupation																
	1	2	3	4	5	6	7	8	9	10	11	12	13	14	15	16	17
1	25	107	20	11	3	27	8	8	8	8	5	12	11	4	7	2	1
2	8	395	64	42	9	116	40	34	59	15	40	61	75	20	30	4	16
3	14	317	116	89	6	144	68	40	56	31	33	82	89	16	60	5	7
4	7	120	34	45	2	73	27	13	25	7	11	31	33	5	21	1	4
5	19	187	69	52	33	112	82	34	50	36	35	69	103	24	63	11	14
6	4	203	41	26	4	145	42	23	49	19	34	74	95	28	42	3	13
7	5	77	20	20	2	57	47	13	23	8	12	48	46	11	19	3	4
8	7	208	49	33	2	174	52	151	67	29	59	104	262	81	75	8	16
9	6	215	54	38	5	172	54	65	195	53	57	195	175	47	110	6	34
10	6	132	29	20	6	131	51	54	71	170	43	142	158	39	130	4	40
11	8	122	47	24	3	170	49	56	65	36	126	130	174	51	110	6	37
12	5	142	33	18	6	184	71	55	102	49	81	391	239	67	149	11	58
13	9	160	37	28	1	188	75	108	93	36	89	171	529	118	123	10	36
14	2	33	5	5	0	40	13	26	22	13	25	55	97	92	38	1	26
15	4	54	11	8	6	86	24	37	42	30	53	101	126	47	142	6	45
16	13	252	58	34	10	188	94	86	145	121	102	323	399	150	259	457	981
17	2	39	8	3	3	39	23	27	42	21	40	96	114	46	83	18	376

Table 1.21: **Mobility 17**: Intergenerational mobility in the USA 1973 (males, aged 20 to 64) with seventeen status categories presented in Table 1.22; rows correspond to father's occupation (at son's sixteen birthday) and columns to son's first full-time civilian occupation. (Treated in Section 5.6.)

All three of these values (see Table 1.20) show that the conditional probability in the second case deviates from the average more than in the first case. This means that the influence of EUAM to MIL is less than the influence of ASAF to PER, in spite of the fact that the conditional probability in the first case is higher. This shows that the value $q_{ij} = (p(i/j) - p_i)/p_i$, referred to as Relative Change of Probability of i when j is taken into account, or RCP(i/j), is worth considering as a proper measure of influence.

Analysis of patterns of the column-to-rows influences via clustering can be done in both of the following directions: 1) analysis of within-row or -column set similarities, 2) analysis of between-row and -column set interrelations.

Similar problems can be addressed in many other areas of application: marketing research (for example, in analysis of association of a set of foods [rows] with their characteristics [columns] by the respondents' answers [counted observations]); attribution of texts' authorship (a set of texts, among them those to be attributed, as the row sets; a set of characteristic words/phrases as the column set, the entries are counts of occurrences of the words in the texts); ecology (sites as the rows, species as the columns, an entry is the count of the occurrences of a species in a site).

No	Occupation	Aggregate
1	Professionals, self-employed	
2	Professionals, salaried	Upper
3	Managers	nonmanual
4	Sales, other	
5	Proprietors	Lower
6	Clerks	nonmanual
7	Sales, retail	
8	Crafts, manufacturing	Upper
9	Crafts, other	manual
10	Crafts, construction	
11	Service	
12	Operatives, other	Lower
13	Operatives, manufacturing	manual
14	Laborers, manufacturing	
15	Laborers, other	
16	Farmers	Farm
17	Farm laborers	

Table 1.22: Occupations in Table 1.21 and aggregation of them in Table 1.7 Mobility 5.

1.6.2 Interaction Data: Mobility and Switching

The Mobility data in Table 1.7, as well as the original 17 by 17 data table on intergenerational occupational mobility in the USA in 1973 collected by Featherman and Hauser (1978) and reported by Breiger 1981 (Table 1.21), are interaction aggregable data tables.

This kind of data, primarily, can be used in the analysis of the patterns of interaction between the social groups represented by the occupational groups (see, for example, Breiger 1981, Goodman 1981, and Hout 1986). Clustering may be a tool for an aggregate analysis of the interactions and the social structure, although it should comply with the statistical considerations on the subject made previously (see, for instance, the citations above).

Similar kinds of data concerning consumer switching between various brands available in a market segment (such as cars, cereals, food snacks, CD-players, etc.) is quite popular in marketing research. Let us consider data on switching between soft drinks from an experiment conducted in 1972. Every entry in Table 1.23 represents the number of times a respondent used the column drink after drinking the row drink (once again, many of the occurrences are in the principal diagonal,

Soft drinks	Coke	7-Up	Tab	Like	Pepsi	Sprite	DPepsi	Fresca
Coke	188	33	3	10	41	17	4	11
7-Up	32	77	1	11	24	17	2	8
Tab	2	3	4	9	2	1	2	2
Like	4	7	4	7	11	2	6	5
Pepsi	47	35	2	8	137	20	7	10
Sprite	8	13	2	5	11	23	2	6
DPepsi	4	2	8	4	5	4	11	5
Fresca	17	7	4	8	11	8	5	15

Table 1.23: **Switching**: Switching (from row to column product) data on soft drinks from Bass, Pessemier, and Lehmann 1972 (DPepsi is Diet Pepsi). (Treated: pp. 208, 226.)

concerning loyal consumers).

The problem is to find general patterns of consumption, and clustering may be well used for that.

1.6.3 Discussion

A two-mode data set (Worries) is considered along with two classes of the problems: (1) row (or column) similarity clustering, (2) finding associated row-to-column subsets. One-mode interaction data tables are to be aggregated with least violation of the pattern of interaction (Mobility) or/and to have the patterns of interaction revealed (Switching).

Chapter 2

Geometry of Data Sets

FEATURES

• Entity-to-variable data table can be represented geometrically in three different settings of which one (row-points) pertains to conventional clustering, another (column-vectors), to conceptual clustering, and the third one (matrix space), to approximation clustering.

• Two principles for standardizing the conditional data tables are suggested as related to the data scatter.

• Standardizing the aggregable data is suggested based on the flow index concept introduced.

• Graph-theoretic concepts related to clustering are considered.

• Low-rank approximation of data, including the popular Principal component and Correspondence analysis techniques, are discussed and extended into a general Sequential fitting procedure, SEFIT, which will be employed for approximation clustering.

2.1   Column-Conditional Data

2.1.1   Three Data/Clustering Approaches

All the column-conditional data will be considered as presented in a quantitative
table format. A quantitative data table is a matrix $X = (x_{ik}), i \in I, k \in K$, where
I is the set of entities, K is the set of variables, and x_{ik} is the value of the variable
$k \in K$ at the entity $i \in I$. The number of the entities will be denoted as N, and
number of the variables, as n, which means that $N = |I|$ and $n = |K|$. Based on
the matrix X, the data can be considered in any of the following three geometrical
frameworks: (1) Space of the Entities, (2) Space of the Variables, and (3) Matrix
Space. Let us consider them in more detail.

(1) Space of the Entities.

The data is considered to be a set of the entities, $i \in I$, presented with corresponding row-vectors $x_i = (x_{ik}), k \in K$, as the elements of a space, usually
Euclidean space R^n (where $n = |K|$) consisting of all the n-dimensional vectors of
the form $x = (x_1, x_2, ..., x_n)$.

Euclidean space R^n involves the so-called *scalar product* operation: for any
$x = (x_k) \in R^n$ and $y = (y_k) \in R^n$, their scalar product (x, y) is defined as

$$(x, y) = \sum_{k \in K} x_k y_k.$$

The square root of the scalar product of a vector $x \in R^n$ with itself is called
its *(Euclidean) norm* and is denoted as $||x|| = (x, x)^{1/2} = \sqrt{\sum_k x_k^2}$. Geometri-
cally, $||x||$ is the distance between x and 0, the vector having all zero components
(called also the origin of the space). Such an interpretation corresponds to the
multidimensional analogue of the Pythagorean Theorem.

The *Euclidean distance* $d(x, y)$, $x, y \in R^n$, can be defined as the norm of the
difference $x - y = (x_1 - y_1, ..., x_n - y_n)$:

$$d(x, y) = (\sum_{k \in K} (x_k - y_k)^2)^{1/2}.$$

Sometimes this n-dimensional space is referred to as the *variable* space since its
dimensions correspond to the variables.

In Fig. 2.1 (a), the rows from the 6 by 2 data table presented in the first two
columns of Table 2.1 are shown as 6 points of the 2-dimensional Euclidean plane
of the variables x_1, x_2. The other two pictures in Fig. 2.1 show the same entities

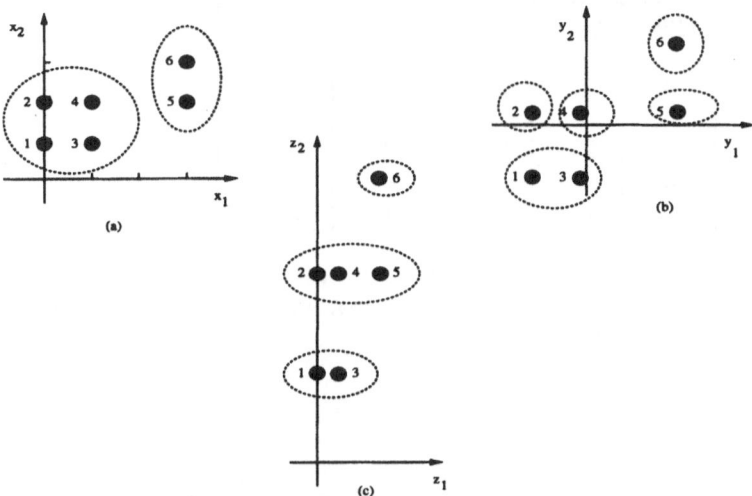

Figure 2.1: The row entities presented at a visual display; tentative clusters are shown with dotted ovals.

as they are represented in the variables y and z which are simple transformations of the variable space: a traditional standardization (along with shifting the space origin to the mean point and scaling the variables according to their standard deviations, see details in Section 2.1.2) is presented in (b), and (c) shows the points in space (z_1, z_2) where $z_1 = x_1$ and $z_2 = 5x_2$ which means that values of x_2 have been increased 5 times, while x_1 has been left unchanged. We can see how the inter-point distances and point clusters vary depending on which space, x, y, or z, is considered.

This instability reflects the principal contradiction between the major feature of the data, column-conditionality, and the nature of the geometrical representation which presumes comparability of all its dimensions. The comparability is involved in the definition of the scalar product and, correspondingly, the distance between two points. The problem of finding a relevant way to aggregate incomparable variables in the scalar product or distance is a major problem of the theory of clustering.

In terms of the space of entities, clusters are "compact", "dense" groups of the row-points which can be determined in terms of the inter-point distances. However, the distances are subject to transformations of the variables.

Point	x_1	x_2	y_1	y_2	z_1	z_2
1	0	1	-1.07	-1.21	0	5
2	0	2	-1.07	0.24	0	10
3	1	1	-0.27	-1.21	1	5
4	1	2	-0.27	0.24	1	10
5	3	2	1.34	0.24	3	10
6	3	3	1.34	1.70	3	15

Table 2.1: **Points:** Six two–dimensional points as given originally (variables x_1 and x_2), traditionally standardized (columns y_1, y_2), and scale-changed (columns z_1, z_2).

The two items — inter-point distances, and transformation of the variables — seem to be the most important mathematical concepts in the space-of-the-entities paradigm.

> Space of the entities, or the variable space, is the most common geometrical representation of data sets: the axes are the variables, the entities are the points (vectors), and the clusters are the "compact" groups of points. The problem is that the inter-point distances heavily depend on the variable measurement scales and the metric chosen.

There is a straightforward extension of the Euclidean space concept in terms of the so-called Minkovski norm family. For every $p > 0$, norm l_p on R^n is defined as follows:
$$l_p(x) = (\sum_{k \in K} |x_k|^p)^{1/p}.$$
Having the Minkovski l_p norm defined, the corresponding distance measure is just $d_p(x, y) = l_p(x - y)$.

The most popular among Minkovski norms are l_2 (Euclidean norm), l_1 (City-block, or Manhattan norm), and l_∞ (Uniform, or Chebyshev norm); corresponding distances have the same names. For any $x \in R^n$, $l_2(x) = ||x||$, $l_1(x) = \sum_{k \in K} |x_k|$, and $l_\infty(x) = \max_{k \in K} |x_k|$ (which is determined as the limit of $l_p(x)$ when $p \to \infty$). The differences of these three norms are illustrated in Fig. 2.2 where, for $x = (4, 3) \in R^2$, $l_2(x) = \sqrt{4^2 + 3^2} = 5$, $l_1(x) = 4 + 3 = 7$, and $l_\infty(x) = \max(3, 4) = 4$.

A distinctive feature of Euclidean space is that, for any $x, y \in R^n$, the following equality holds:
$$||x - y||^2 = ||x||^2 + ||y||^2 - 2(x, y)$$
which can be easily derived just from the definition of scalar product. This equality

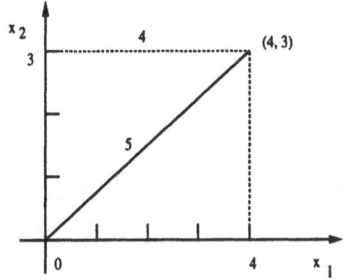

Figure 2.2: Minkovski p-distances 5, 7=3+4 and 4 for $p = 2, 1$ and ∞, respectively.

leads to the following two observations, quite important in data analysis applications.

The first observation concerns the geometrical meaning of the scalar product (x, y) which is just $(x, y) = ||x|| ||y|| \cos \alpha_{xy}$ where α_{xy} is the angle between vectors x and y, and $\cos \alpha$ is the cosine of angle α.

The second one involves the concept of orthogonality. The vectors $x, y \in R^n$ are called *orthogonal* if $(x, y) = 0$. When x and y are orthogonal, the squared norm of their sum $x + y$ is decomposed as $||x + y||^2 = ||x||^2 + ||y||^2$, which can be extended to the sum of any number of mutually (pair-wise) orthogonal vectors.

(2) Space of the Variables.

In this approach, the data matrix is considered to be a set of the variables, $k \in K$, presented by corresponding N-dimensional column-vectors, $x_k = (x_{ik}), i \in I$, $N = |I|$. The vectors are considered along with certain basic statistical coefficients, such as the mean (average), variance, standard deviation, covariance and the correlation. Although the coefficients listed have been introduced in a specific statistical context, they can be defined also in terms of the geometrical space, via the scalar product operation. The mean, or average value of a vector $x = (x_i), i \in I$, is $\bar{x} = \sum_{i \in I} x_i / N = (x, u)/N$ where $u = (1, ..., 1)$ is the vector with all its components equal to 1. The (empirical) variance of x is defined as

$$\sigma^2(x) = ||x - \bar{x}u||^2/N = (x - \bar{x}u, x - \bar{x}u)/N = \sum_{i \in I}(x_i - \bar{x})^2/N$$

where $\bar{x}u$ is the vector with all its components equal to $\bar{x}$. The standard deviation is just $\sigma = \sqrt{\sigma^2} = ||x - \bar{x}||/\sqrt{N}$ which has also a statistical meaning as an average deviation of the variable's values from the mean.

The covariance coefficient between the variables x and y considered as vectors in the space R^N, $x = (x_i)$ and $y = (y_i)$, can be defined as

$cov(x, y) = (1/N)(x - \bar{x}u, y - \bar{y}u).$

Obviously, $cov(x, x) = \sigma^2(x)$. The covariance coefficient changes proportionally when the variable scales are changed. The scale-invariant version of the coefficient is the correlation coefficient (sometimes called Pearson's product-moment coefficient) which is the covariance coefficient normalized by the standard deviations:

$r(x, y) = cov(x, y)/(\sigma(x)\sigma(y)) = (1/N)(x - \bar{x}u, y - \bar{y}u)/(\sigma(x)\sigma(y)).$

A somewhat simpler formula can be obtained when the data are first standardized by subtracting their average and dividing the result by the standard deviation:

$r(x, y) = cov(x, y) = (x', y')/N$

where $x_i' = (x_i - \bar{x})/\sigma(x)$, $y_i' = (y_i - \bar{y})/\sigma(y)$, $i \in I$.

Thus, the correlation coefficient is nothing but the mean of the component-to-component product when both of the vectors are standardized as above, For the variables x_1, x_2 in Table 2.1, their standardized versions x_1', x_2' are the variables y_1, y_2 in the Table.

Relationships between the variables (column-vectors) in Table 2.1 can be illustrated with Fig. 2.3 as featured in the lengths of the vectors and the angle between them since, as was stated above, the cosine of the angle is equal to the scalar product of the corresponding normalized variables; thus, $cos(y_1, y_2) = Nr(x_1, x_2)$.

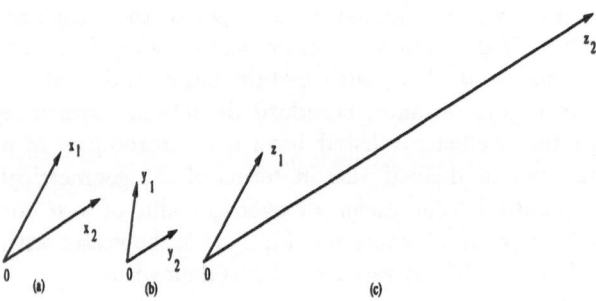

Figure 2.3: Variables x_1 and x_2 as vectors in the space of the variables.

A motivation for the statistical concepts introduced can be suggested in a geometrical setting, also. Let us consider a problem of description of one of the variables, y, as a linear function of the other, x. The question is this: is it possible to represent column y as $y = ax + b$ for some real a and b? Usually not, since the data are empirically observed. This means that the residual vector $e = y - ax - b$ should not be expected to be zero, for any real a and b. Then, the question arises, which values of a and b minimize $||e||$ or equivalently $||e||^2$ (linear regression problem)? Obviously, the minimizing values of a and b make e orthogonal to $ax + b$ (see Fig. 2.4), which implies the Pythagorean decomposition, $||y||^2 = ||ax+b||^2 + ||e||^2$. The optimal values of a and b are easily derived by setting the derivatives of $||e||^2 = ||y - ax - b||^2$ with regard to a and b to zero:

$$a = r(x, y)\sigma(y)/\sigma(x), \ b = \bar{y} - a\bar{x}.$$

With these values substituted, the minimum $||e||^2$ becomes

$$||e||^2 = N(1 - r(x, y)^2)\sigma^2(y).$$

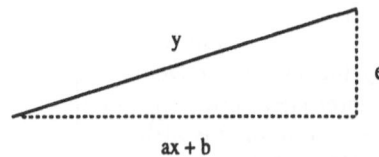

y

e

$ax + b$

Figure 2.4: Geometry of the linear regression.

This means that the so-called residual variance $\delta = ||e||^2/N$ (which is the variance of the residual vector e since its mean $\bar{e} = 0$) may be considered to be the residual part of the variance $\sigma^2(y)$ after y's linear expression via x has been removed. The correlation coefficient squared, $r^2(x, y)$, called the *determination* coefficient, shows the extent of decreasing the variance of y, $\sigma^2(y)$, when y's best linear expression through x is subtracted. The value $r^2(x, y)$ belongs to the interval $[0,1]$. When $r^2(x, y) = 1$ nothing remains, the residual is zero and $y = ax + b$ where sign of a equals the sign of $r(x, y) = \pm 1$; when $r^2(x, y) = 0$, or, equivalently, the scalar product $(x, y) = 0$, there is no decrease of $\sigma^2(y)$ at all (this is the case of the so-called non-correlate variables).

Clusters of the variables can be recognized by the structure of correlation between corresponding column-vectors. As to the row-clusters, there is no way of revealing them straightforwardly. However, at least two not-so-direct approaches can be indicated. One, the traditional approach, is based on modeling the data with a density function, $f(x_1, ..., x_n)$, which is a function of the variables, characterizing, quite loosely speaking, the probability of randomly getting particular row-vectors $x = (x_1, ..., x_n)$. The density function allows defining and determining some particular cluster concepts as related to "high-density" regions of the variable space (see more detail in Section 3.2.7). The other, more recent approach, is based on considering the fact that the cluster structure sought can be thought of as another variable (or a set of variables) which approximates the variables given. If, for example, a row cluster structure which is a two-class partition is sought, the structure can be represented as a Boolean variable, with a Yes value for one class, and a No value for the other. To consider the approximation problem mathematically, one needs a common representation for both the original variables and the sought classification variable(s). This is why the approach has been developed, mostly, for the case when the variables are qualitative, thus allowing a simpler way for comparing them with the cluster structure sought (see, for instance, Breiman et al. 1984). This idea will be elaborated in more detail further (see p. 116 and Sections 3.2.5 and 6.3.2).

> The space of the variables paradigm has been employed mostly for developing the probabilistic clustering approach as based on the multivariate distribution/density function. Though nobody has claimed developing methods for approximating the variables by cluster structures, the conceptual clustering methods can be thought of as filling in this niche.

(3) Matrix Space.

This is the $N \times n$ matrix $X = (x_{ik})$ itself considered as an element of a $N \times n$-dimensional space.

Formal representation of a cluster structure in terms of the matrix space is quite different from both representations above, in terms of the space of the entities and the space of the variables. To be definite, let us consider a two-cluster structure for the subtable (x_1, x_2) in Table 2.1. Each cluster is represented with its list of the entities, S, and a standard element (called also center or centroid or prototype) $c_S \in R^n$. Let the first cluster be $S_1 = \{1, 2, 3, 4\}$, the second, $S_2 = \{5, 6\}$, and their prototypes, just the vectors of the means, $c_1 = (0.5, 1.5)$ and $c_2 = (3, 2.5)$. This structure can be represented by the following $N \times n$ matrix C as an element of the matrix space:

$$
C = \begin{pmatrix}
0.5 & 1.5 \\
0.5 & 1.5 \\
0.5 & 1.5 \\
0.5 & 1.5 \\
3.0 & 2.5 \\
3.0 & 2.5
\end{pmatrix}
$$

Matrix C contains c_1 in all the rows from S_1 and c_2 in all the rows from S_2.

Actually, the following matrix equation can be suggested to maintain the cluster structure:

$$
\begin{pmatrix}
0 & 1 \\
0 & 2 \\
1 & 1 \\
1 & 2 \\
3 & 2 \\
3 & 3
\end{pmatrix}
=
\begin{pmatrix}
0.5 & 1.5 \\
0.5 & 1.5 \\
0.5 & 1.5 \\
0.5 & 1.5 \\
3.0 & 2.5 \\
3.0 & 2.5
\end{pmatrix}
+
\begin{pmatrix}
-0.5 & -0.5 \\
-0.5 & 0.5 \\
0.5 & -0.5 \\
0.5 & 0.5 \\
0 & -0.5 \\
0 & 0.5
\end{pmatrix}
$$

where the last matrix entries are the errors (additive residuals) of representation of the original matrix X through "idealized" cluster structure matrix C. This equation means that $X = C + E$ where E is the residual matrix. Looking at E, we can see that all the residuals are within ± 0.5 range, although only two of them are zero. If we wish to decrease the number of non-zero residuals along with requirement that all of them are non-negative, we may take other centroids, $c_1 = (0, 1)$ and $c_2 = (3, 2)$, which leads to the equation

$$
\begin{pmatrix}
0 & 1 \\
0 & 2 \\
1 & 1 \\
1 & 2 \\
3 & 2 \\
3 & 3
\end{pmatrix}
=
\begin{pmatrix}
0 & 1 \\
0 & 1 \\
0 & 1 \\
0 & 1 \\
3 & 2 \\
3 & 2
\end{pmatrix}
+
\begin{pmatrix}
0 & 0 \\
0 & 1 \\
1 & 0 \\
1 & 1 \\
0 & 0 \\
0 & 1
\end{pmatrix}
$$

where more than half of the residuals are zero, although the non-zero residuals are larger than in the preceding matrix equation.

Such a representation allows formalizing clustering problems as problems of approximation of the data matrix by an "idealized" cluster structure matrix subject to appropriate choices of the residuals. This is the paradigm in which the present author did his research for years (see, for example, Mirkin 1985, 1987a, 1990, 1994, Mirkin and Muchnik 1996 and references therein), and will be described in detail in Chapters 4 through 7.

> Matrix space representation allows treating the clustering problem as just a problem of approximation of a given data matrix with a matrix corresponding to a cluster structure.

The matrix space representation has the advantage of allowing its interpretation to be in either or even both of the preceding representations, the spaces of the entities and variables, which will be employed thoroughly in the corresponding considerations (see Sections 4.4, 6.1.1 and 6.3). This is based on the following observation.

Let us consider a most important geometrical characteristic of the data

$$L_p(X) = \sum_{i \in I} \sum_{k \in K} |x_{ik}|^p$$

which will be referred to as the p-scatter of the data and is nothing but the p-th power of the Minkovski norm of the data matrix, $L_p(X) = l_p(X)^p$.

$L_2(X) = \sum_{i \in I} \sum_{k \in K} x_{ik}^2$, 2-scatter, is usually called the squared data scatter while $L_1(X)$, is the module data scatter. The value L_∞ may be defined, by continuity, as $L_\infty = \max_{i \in I, k \in K} |x_{ik}|$.

The data scatter characterizes the spread of the data entries around zero, which has a statistical meaning when the zero-point represents important characteristics of the data concentration, such as, for instance, the means of the variables; in that latter case, the data scatter is proportional to the total data variance (when $p = 2$) or absolute deviation (when $p = 1$). In matrix equations representing the data, X, through a "theoretical" matrix A and residuals E, $X = A + E$ as above, the scatter, sometimes, can be decomposed in two scatter-wise parts, one depending on A, the other, on E; these parts are interpreted as explained part of the scatter and non-explained, respectively, which allows estimating the contribution of the model A to the data scatter along with further partitioning into smaller parts corresponding to the elements of the model. This plays an important role in interpreting the results.

Obviously, for any finite p, the following decompositions hold:

$$L_p(X) = \sum_{i \in I} d_p^p(0, i)^p = \sum_{k \in K} l_p(x_k)^p \tag{2.1}$$

where $d_p(i, 0)$ is Minkovski norm (distance from zero) of row-vector $i \in I$ and x_k is the column-vector, $k \in K$. These equalities may be employed in reinterpreting the matrix-based contributions in terms of either the variables or entities or both.

There are three types of geometrical representation of the entity-to-variable quantitative data format:
1. Space of the entities: the rows (entities) are considered as points in n-dimensional variable space.
2. Space of the variables: the columns (variables) are considered as vectors in N-dimensional entity space.
3. Matrix space: the data matrix is considered as a point in $N \times n$-dimensional data space.
Clustering problems sound quite different depending on the paradigm accepted.

2.1.2 Standardization of Quantitative Entity-to-Variable Data

Let us consider a quantitative entity-to-variable matrix $X = (x_{ik})$, $i \in I, k \in K$, like that presented in data set Points (Table 2.1), Iris (Table 1.10), or Disorders (Table 1.11).

Before processing such a data set, it is traditionally standardized into a data matrix $Y = (y_{ik})$, $i \in I, k \in K$, where

$$y_{ik} = \frac{x_{ik} - a_k}{b_k}, \ i \in I, k \in K, \tag{2.2}$$

A common example of the Fahrenheit measured temperature x being transformed into Celsius scale y with formula $y = (x - 32)/1.8$ may help in perceiving the direct meaning of the standardization parameters: a_k denotes shift of the origin while b_k, change of the scale factor. In data analysis, the shift of the origin, a_k, usually is recommended to be a central value of vector x_k while the change of the scale factor, b_k, reflects scatter of the components of x_k. Among the central values, the following are known: mean $\bar{x}_k = \sum_{i \in I} x_{ik}/N$, midrange $mr(x_k) = (max_{i \in I}x_{ik} + min_{i \in I} x_{ik})/2$, and median $m(x_k)$ which is the middle term in the ordered series $x^{(1)} \le x^{(2)} \le ... \le x^{(N)}$ of the components of x_k. More precisely, $m(x_k) = x^{(N+1)/2}$ if N is odd or $m(x_k)$ is any real between $x^{(N/2)}$ and $x^{(N/2+1)}$ if N is even; in the latter case, we take $m(x_k) = (x^{(N/2)} + x^{(N/2+1)})/2$, to be definite. Among the scatter measures, there are standard deviation, $\sigma(x_k) = \sqrt{\sum_{i \in I}(x_{ik} - \bar{x}_k)^2/N}$, absolute deviation, $ad(x_k) = \sum_{i \in I} |x_{ik} - \bar{x}_k|/N$, range, $ra(x_k) = max_{i \in I} x_{ik} - min_{i \in I} x_{ik}$, etc.

Although, in experiments, the structures revealed are much more relevant to the substantive meaning of the data when the data have been standardized rather than not, there is still no theory for data standardizing when, as it frequently happens, there is no theory available on how the data have been generated.

In quite vague terms, some may (and do) say that the standardization should be made to make the variables comparable by presenting each of the variables in the most natural scale, with the variable's central point as the origin and its scatter measure as the scale factor. This sounds like an incantation rather than explanation: Why are the central points and the scatters so important? How they can be combined? Is there any connection with the subsequent treatment of the data?

The following account may be suggested to address at least some of these issues.

The variables are considered as measured in the interval scale (Luce, Bush and Galanter 1963, Roberts 1979). Somewhat differently, let us say that a variable $y : I \to R$ is referred to as measured in *interval* scale if any statement involving y does not change its meaning if y is substituted by the transformed variable $x = \phi \circ y$ (which means that $x(i) = \phi(y(i))$, for any $i \in I$) for any function $\phi(y) = by + a$, where a, b are reals and $b > 0$. Such a transformation ϕ is associated with change of both the scale factor (multiplication by b) and the origin (adding of a) of the original variable y. The fact that it is the interval scale which is assigned to a variable may be interpreted as that, actually, no linear mapping $x = by + a$ ($b > 0$) may change meaning of the variable.

We assume that there are some "natural" scale parameters assigned to the variables to make them all compatible with each other. So, the variables observed, x_k, are considered to be linear functions of the "naturally" scaled variables y_k, $x_k = b_k y_k + a_k$ and $y_k = (x_k - a_k)/b_k$ ($k \in K$). To choose values a_k, b_k properly, we suggest that it is the scatter of the data which should be involved, along with its decompositions (2.1). Obviously, $L_p(y_{ik}) = L_p([x_{ik} - a_k]/b_k)$. This makes meaningful the following two principles.

P1. Equal Contribution to the Data Scatter.

The variables y_k ($k \in K$) have equal contributions to the data scatter.

This principle can be considered as a mathematical form of the intuitive idea of the "equal importance" of the variables underlying most of the clustering methodology (see Sokal and Sneath 1973). This form of the "equal importance" postulate, however simple it is, does not appear to be too obvious or unanimously acceptable. Usually, the equal importance principle is claimed to be provided by equal contributions of the variables to the inter-entity similarities/distances calculated from them (see, for instance, Romesburg 1984). Mathematical formulation of this latter requirement must be different; moreover, it is not easy to formalize that in a convenient way: say, if two points are parallel to an axis (related to a variable in the variable space), than, obviously, the distance depends on that variable only, and still it is subject to the scale change.

Based on the right part of decomposition in (2.1), the principle immediately implies that the only b_k which may be used for this purpose is $b_k = C(\sum_{i \in I}(x_{ik} - a_k)^p)^{1/p} = Cl_p(x_k - a_k)$ where C is a positive constant. For $p = 2$ or $p = 1$, this is proportional to the standard or absolute deviation of x_k, respectively.

The second principle concerns the other part of (2.1), decomposition of the scatter by the distances from the origin to the entities. Geometrical considerations (see, for example Fig.2.1) show that the requirement of putting the origin in an equidistant point (to make all $d_p(0, i)$, $i \in I$, equal to each other) may be too challenging a goal. A somewhat weaker, but achievable goal is as follows.

P2. Minimum Data Scatter.

The origin of the variable space should be a minimizer of the data scatter.

The data p-scatter concept is a generalization of the moment of inertia concept in mechanics (which corresponds to the squared data scatter of a system of material points). Rotation of a system requires minimum force when the center of rotation minimizes the moment of inertia since the axes of rotation, in that case, are most natural. Analogously, the principle above can be interpreted as a way of getting most "natural" directions to the axes of the set ("cloud") of the entity points.

The problem of minimizing L_p by a_k (p is supposed to be positive here) may be difficult, in general case. However, it is resolved quite easily when $p = 1$ or $p = 2$. Indeed, the derivative of $L_p([x_{ik} - a_k]/b_k)$ by a_k is equal to $-p/|b_k|^p \sum_{i \in I} f_p(x_{ik} - a_k)$ where $f_2(x) = x$ and $f_1(x) = sgn\ x$ ($sgn\ x = 1$ if $x > 0$, $sgn\ x = -1$ if $x < 0$, and $sgn\ 0 = 0$). This leads to a_k equal to mean or median of x_ks when $p = 2$ or $p = 1$, respectively.

The principles **P1** and **P2** above suggest that both the shift of the origin and the scale factor must be chosen based on p-scatter for the same p. When $p = 2$, this leads to the usual standardization rule: the origin is the mean while the standard deviation is the scale factor, which will be referred to as *square-scatter standardization* (it is called sometimes z-score transformation). When $p = 1$, the origin must be median while the scale factor is the absolute deviation, which will be referred to as *module-scatter standardization*.

A speculation can be provided for the L_∞ case: $L_\infty(Y)$ may be approximately presented as

$$L_{p\infty}(Y) = \sum_{k \in K} \max_{i \in I} |(x_{ik} - a_k)/b_k|^p,$$

for sufficiently large p. Since the items in the sum each depend only on corresponding a_k, they can be minimized (by a_k) independently. This implies that the optimal a_k are midranges, $a_k = (max_{i \in I} x_{ik} + \min_{i \in I} x_{ik})/2$. The principle of equal contribution, applied to the resulting formula, implies that b_k is proportional to the range, $b_k = C(\max_{i \in I} |x_{ik} - \min_{i \in I} |x_{ik})^p)$ where C is a positive constant. The standardization, putting the origin in midrange with the scale factor being proportional to the range, will be called *infinity-scatter standardization*.

Point	x_1	x_2	y_1	y_2	z_1	z_2	u_1	u_2
1	0	1	-1.07	-1.21	-1	-2	-1	-1
2	0	2	-1.07	0.24	-1	0	-1	0
3	1	1	-0.27	-1.21	0	-2	-0.33	-1
4	1	2	-0.27	0.24	0	0	-0.33	0
5	3	2	1.34	0.24	2	0	1	0
6	3	3	1.34	1.70	2	2	1	1

Table 2.2: **Points:** Six two–dimensional points as given originally (variables x_1 and x_2), square-scatter standardized (columns y_1, y_2), module-scatter standardized (columns z_1, z_2), and infinity-scatter standardized (columns u_1, u_2).

To illustrate the three standardizing options suggested, let us refer again to the data set Points (see Table 2.1). The Table 2.2 presents the original data set (x variables) along with its square-, module-, and infinity-scatter standardized versions represented by the variables y, z, and u, respectively. The module-standardization of x_1 and x_2 is done with 1 and 2 as their respective medians and 1 and 0.5 as their absolute deviations. The midranges of the variables are 1.5 and 2, while their ranges are 3 and 2. In the latter case, infinity-standardization, the half-ranges, 1.5 and 1, have been used as the normalizing factors b_k, to have both of the variables' contributions equal to 1.

Corresponding entity-points are presented in Fig. 2.5. It should be pointed out that some experimental observations show that normalizing by range may be superior to normalizing by the standard deviation, in clustering (Milligan and Cooper 1988), at least with some clustering methods. In the present author's opinion, in the clustering procedures that heavily rely upon the scatter of the data (as those discussed in Chapters 4 and 6), the experimental results cannot be considered conclusive since the methods involved in the experiment have not been adjusted to the scatter of the data (see also Sections 6.1.1 and 7.6).

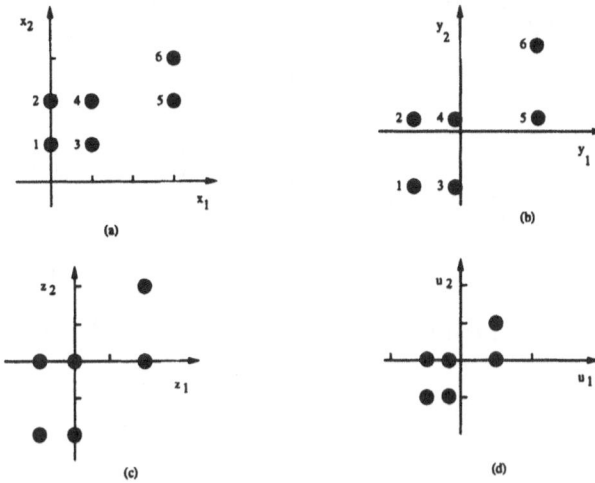

Figure 2.5: Difference in cluster structures depending on the choice of standardization option.

> Standardizing the data is an option which can make revealing the structures in them easier. It can be considered as a way to make all the variable scales compatible with each other. The approach suggested requires the variables to be standardized to make the data scatter minimum while maintaining equal contributions of the variables to the scatter.

2.1.3 Quantitative Representation for Mixed Data

Any nominal variable can be considered as a set of Boolean variables, each corresponding to a category. In particular, variable Presentat in Masterpieces set (Table 1.12) can be presented with three Boolean variables: 1) Is this Direct? 2) Is this Behavior? 3) Is this Thought?

Theoretically, a variable $y : I \rightarrow R$ is referred to as measured in *nominal* scale if any statement involving y does not change its meaning when y is substituted by the transformed variable $x = \phi \circ y$ (where $x(i) = \phi(y(i))$, for any $i \in I$) for any one-to-one function $\phi : R \rightarrow R$. In general, any nominal variable y is associated with a partition of the entity set into non-overlapping classes corresponding to particular categories. In the example considered, the first class ("Direct") consists

Num	LenSent	LenD	NumCh	InMon	Direct	Behav	Thought
1	15.0	16.6	2	0	1	0	0
2	12.0	9.8	1	0	0	1	0
3	11.0	10.4	1	0	0	1	0
4	20.2	202.8	2	1	0	0	1
5	20.9	228.0	4	1	0	0	1
6	29.3	118.6	2	1	0	0	1
7	23.9	30.2	4	1	1	0	0
8	27.2	58.0	5	1	1	0	0

Table 2.3: Quantitative presentation of the Masterpieces data as an 8 by 7 entity-to-variable/category matrix.

of three entities: two novels by L. Tolstoy and "EugOnegin" by A. Pushkin; the second class ("Behav") comprises the other two novels by A. Pushkin; and the third contains the three novels by F. Dostoevski.

An interesting question, not only of theoretical importance, arises from the correspondence between Boolean, two-category nominal and interval scale variables. The point is that any one-to-one mapping of a two-category nominal variable y can be presented in linear form $\phi(y) = by + a$, where a, b are reals (since there are only two values to map). Such a transformation ϕ is associated with change of both the scale factor (multiplication by b) and origin (adding a) of the original variable y. This allows treating such a two-category qualitative variable as measured in the interval scale. On the other hand, any nominal two-category variable corresponds to two mutually complementary Boolean variables, which requires a specific standardization rule and makes the difference between these kinds (Boolean and two-category nominal) of variables.

Nevertheless, based on considerations above, we may think of all the categories as 1/0 quantitative columns. So, any mixed data table involving both quantitative and nominal variables can be presented as a quantitative matrix $X = (x_{iv})$, $i \in I$, $v \in V$, where I is the entity set considered, V is set of the all quantitative/binary variables and of the all categories of the nominal variables; any binary variable is presented in zero-one quantitative format: 1 for Yes, 0 for No (see Table 2.3).

In this work, we limit ourselves to only two kinds of qualitative variables: nominal and Boolean. In particular, we do not deal with two other important kinds of qualitative variables: a) multiple choice, and b) rank order.

The *multiple choice* variable is a nominal variable which may have overlapping categories. For example, in a survey, a question may be asked on the newspaper preferred; some respondents may indicate several of the items, and the sets chosen could have a complicated structure of overlaps. To process this kind of data with the methods presented below, one must consider any of the categories as a particular Boolean variable (though, some results described in Chapter 6 suggest that the multiple choice categories can be treated together, see, for instance, p. 291).

The *rank order* variable is a qualitative variable which has the set of its categories ordered like, for example, college examination grades (from A to F), or answers to an attitude question (from "do not like" or even "hate it" through "indifferent" to "like very much"). Usually, this kind of variable is considered as quantitative (when the variable is similar to testing scores), or as nominal (when the variable is similar to the attitude relations), or, maybe at best, as a set of Boolean variables corresponding to the categories such as this: "Is the entity better than or indifferent to the category a ?" On the other hand, in the context of similarity or interaction matrices, rank-order variables could be treated genuinely, as well as some other, non-standard, qualitative structures of data (which will not be covered in this work, see Mirkin 1985).

After the mixed data have been presented in a quantitative format, they should be standardized according to the two principles in the preceding section.

Let us consider the square-scatter standardization for the nominal variables. The minimum data scatter principle, **P2**, in this case, works the same way as for quantitative variables: the shift of the origin value a_v is determined to be the mean of the corresponding column x_v. Obviously, this is nothing but p_v, the relative frequency of the category, which is just the proportion of ones in the column.

Choosing the scale factor, b_v, is not as unambiguous. Let us calculate the contribution of category v, after $a_v = p_v$ has been subtracted, to the square scatter of the data. Obviously, there are Np_v entries $1 - p_v$ and $N(1 - p_v)$ entries $-p_v$, in the column v. This gives the contribution to the square scatter equal to $Np_v(1 - p_v)^2 + N(1-p_v)p_v^2 = Np_v(1-p_v)$. Thus, if v belongs to a $\#k$-grade nominal variable k, the total contribution of the variable k is $\sum_v Np_v(1 - p_v) = N(1 - \sum_v p_v^2)$.

The value $\delta^2(k) = 1 - \sum_v p_v^2$ is not unknown in data analysis. It is referred to as the *Gini index*, or *qualitative variance* of the variable k and has a nice interpretation as the average error of the so-called proportional prediction rule (see p. 236).

Thus, taking $b_v = \delta(k)$ for any category v of the variable k makes contribution of the variable to the square scatter equal to that of the square-standardized quantitative variable.

However, there are other options leading to the same result. Let us take, for instance, $b_v = \sqrt{(\#k - 1)p_v}$ which differs from $\delta(k)$ in that that it depends on v. The contribution of v to the square scatter of the data, in this case, equals:

$Np_v(1 - p_v)/(p_v(\#k - 1)) = N(1 - p_v)/(\#k - 1)$. The total contribution of the variable k is thus equal to $N(\#k-1)/(\#k-1) = N$ which is exactly the contribution of the other standardized variables.

In Chapter 6, it will be shown that both of the normalizing coefficients b_v suggested for the nominal categories are meaningful in terms of evaluation of association between the clusters sought and the original variables.

> Square-scatter standardization for mixed data:
> (a) the shift parameter a_v is taken as the mean if v is a quantitative variable or as the frequency p_v of v when v is a category or a binary variable;
> (b) the factor scale parameter b_v is taken as the standard deviation $\sigma(x_v)$ if v is a quantitative or binary variable (in the latter case, $\sigma(x_v) = \sqrt{p_v(1 - p_v)}$); and there are two options when v is a category: (1) $b_v = \delta(k) = \sqrt{1 - \sum_v p_v^2}$, the error of proportional prediction, or (2) $b_v = \sqrt{(\#k - 1)p_v}$.

> If the categories of a nominal variable are considered as having an independent meaning, they should each be considered as Boolean variables.

-0.775	-0.816	-0.444	-1.291	0.722	-0.354	-0.433
-1.247	-0.898	-1.154	-1.291	-0.433	1.061	-0.433
-1.404	-0.891	-1.154	-1.291	-0.433	1.061	-0.433
0.041	1.428	-0.444	0.775	-0.433	-0.354	0.722
0.151	1.732	0.976	0.775	-0.433	-0.354	0.722
1.470	0.413	-0.444	0.775	-0.433	-0.354	0.722
0.622	-0.652	0.976	0.775	0.722	-0.354	-0.433
1.141	-0.317	1.686	0.775	0.722	-0.354	-0.433

Table 2.4: Square-scatter standardized 8 by 7 data matrix for Masterpieces mixed data set; the second normalizing option has been applied to the categories of the nominal variable Presentat (the last three columns).

Considering module-scatter standardization for a category v, it can be seen that the median equals 1, 1/2, or 0 when p_v is greater, equal or less than 1/2, respectively. The absolute deviation of the values of Boolean column from this value, b_v, is equal to p_v or $1 - p_v$ depending on whether p_v is less than $1/2$ or not. To make the nominal variable k have its contribution to the scatter L_1 equal to N, this value of b_v must be multiplied by $\#k$. This actually allows us to say, that, for the mixed data case, module-scatter standardization does not differ too much from the quantitative case requiring taking a_v and b_v as, respectively, medians and absolute deviations of columns v corresponding to either quantitative/Boolean variables or categories (based upon what is written above on the category transformation).

> The data-scatter based standardizing rules may be extended to the case when both quantitative and nominal kinds of the variables are presented, just keeping the contributions equal to each other.

Sometimes the data require only a partial standardization (centering or normalizing of some of the variables) or even do not require it at all, which should be considered pragmatically.

2.1.4 Discussion

The problem of preliminary data standardization is a most vague issue in all the disciplines concerned with handling data: data analysis, pattern recognition, image processing, data and knowledge bases, etc. The only relevant discipline where this problem is simple, at least in principle, is mathematical statistics: In a common case, the data are supposed to be a random sample from a theoretical distribution/density function $f(x_1, ..., x_n, \theta_1, ..., \theta_p)$ where $x_1, ...x_n$ are variables while $\theta_1, ..., \theta_p$ are parameters defining the function entirely and estimated from the data. After the parameters are determined, the data can be transformed to have the parameter values standardized (usually, to the most simple format, such as the mean and variance of the normal distribution standardized to be equal to 0 and 1, respectively). Such standardizing helps in comparing different functions and testing statistical hypotheses. However, the assumption underlying this line, that the distribution function exists and is known up to the parameter values, may be inapplicable in many clustering and classification problems.

We suggest using the data scatter concept as the main notion in theoretical thinking on the subject when data are collected/generated based on a vague and imprecise substantive idea rather than on a reproducible rigid mechanism. The data scatter concept is employed here for the following:

1. There are three different approaches to data clustering underscored as related, primarily, to the row-entities, column-variables, and to the matrix-data table itself. Although the approaches feature quite different cluster-analysis modeling concepts, the third one related to the matrix space and matrix norm

(data scatter) seems more general since it usually allows reformulating the models in terms of either of the other two approaches.

2. Two data standardization principles are suggested explicitly in terms of the data scatter: (P1) equal contribution of the variables to the data scatter, and (P2) minimum data scatter. The principles lead to reasonable standardization rules, depending on degree of the scatter.

3. Principle (P1) of equal contribution of the variables may be extended to the case of nominal variables represented by associated sets of binary variables. Standardizing the mixed data is suggested upon the extension, thus providing equal contributions of the quantitatively coded variables to the scatter, without any regard what kind of variable it is originally: quantitative, nominal, or binary.

2.2   Transformation of Comparable Data

2.2.1   Preliminary Transformation of the Similarity Data

It is supposed that a similarity measure between the entities, a_{ij}, $i \in I, j \in J$, monotonically reflects the extent of similarity: the larger the a_{ij}, the more similar i and j. A similarity measure can be obtained as primary data (like Functions in Table 1.13) or as secondary data, usually based on an entity-to-variable data set.

We have not much to say about similarities as the primary data, except for the following three points.

First, the matrix $A = (a_{ij})$ is considered, primarily, as an $N \times N$–dimensional vector if it is asymmetric. When it is symmetric, that is, $a_{ij} = a_{ji}$ for any $i, j \in I$, it should be considered $N \times (N+1)/2$–dimensional, thus involving only the necessary upper (or lower) half-matrix, principal diagonal included. Still, in many occasions, the principal diagonal entries, a_{ii} (similarities of the entities to themselves), are not given or just do not fit (like in data Confusion, Table 1.16, where they are just outliers in comparison to the non-diagonal data entries). Without the principal diagonal, the data table is an $N \times (N-1)$–dimensional vector, if asymmetric, or on $N \times (N-1)/2$–dimensional one, if symmetric.

Second, the user might want to have the matrix symmetrized somehow. There are two options which could be recommended. The first of the options suggests direct symmetrization with the formula $a'_{ij} = (a_{ij} + a_{ji})/2$ (where a'_{ij} is the symmetrized similarity) which can be substantiated in the approximation clustering context considered (see Section 4.5.1). The other option suggests considering the columns of the matrix A as the variables and calculating a secondary similarity index from them as a column-conditional data table.

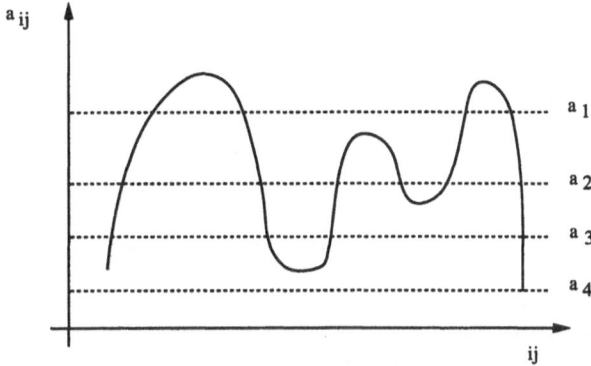

Figure 2.6: Threshold change depending clusters.

Third, since all the entries are comparable across the table, there is no need to change the scale factor. On the other hand, shifting the origin by subtracting a threshold value a, $b_{ij} = a_{ij} - a$ where b_{ij} is the index shifted, may allow a better manifestation of the structure of the data. Fig. 2.6 illustrates the nature of the effect of shifting on the shape of a positive similarity index, a_{ij}, presented along the ordinate while ijs are ordered somehow in the abscissa: shifting by a_4 does not make too much difference since all the similarities remain positive, while shifting by a_2, a_3, or a_1 makes many similarities b_{ij} negative leaving just a couple of the higher similarity "islands" positive. We can see that a better result can be achieved with an intermediate $a = a_2$ manifesting all the three islands on the picture, while increasing it to a_1 or higher values loses some (or all) of the islands in the negative depth.

In the framework discussed, the shift can be determined with the principle of minimizing the scatter of the data, $L_p(A - a) = \sum_{i,j \in I}(a_{ij} - a)^p$. The principle leads to a equal to mean or median of the similarities when $p = 2$ or $p = 1$, respectively.

Let us discuss now the similarity indices derived from the other data formats. There exist many various suggestions (see a review in Gower 1985 or Baulieu 1989 where 42 indices have been indicated for the binary data only!). The only similarity index derived from the entity-to-variable format data which will be recommended here is the scalar product of the row-vectors, defined as

$$a_{ij} = \sum_{v \in V} y_{iv} y_{jv} \qquad (2.3)$$

where $Y = (y_{iv})$ is the quantitative (perhaps standardized) data matrix.

There are two reasons for the recommendation made. First, the measure is compatible with the (multi) linear data analysis and clustering models (that also involve the square scatter of the data) considered in the framework of approximation clustering (see Chapters 4 to 7). Second, the measure is quite general and allows presentation of some popular measures as special cases. For instance, a popular measure, the number of the Boolean attributes (categories) which are common to both of the entities, i and j, can be presented in form (2.3) with the binary 1/0-columns non-standardized.

Obviously, the scalar product similarity coefficient matrix $A = (a_{ij})$ can be expressed algebraically through the entity-to-variable/category matrix $Y = (y_{iv})$: $A = YY^T$.

Let us note that the measure a_{ij} in (2.3) is actually the sum of the column-driven similarity indices $a_{ij}^v = y_{iv} y_{jv}$, $v \in V$, and let us get an intuition on the meaning of the measure with a couple of examples. Let y be a quantitative variable, preliminarily square-scatter standardized. Then $a_{ij} = y_i y_j$ will be positive if both of the values y_i, y_j are larger or lower than the average $\bar{y} = 0$, and it will be negative if one of y_i, y_j is larger and the other is smaller than the average. This makes the average a kind of border between the positivity and negativity areas. Moreover, the value $a_{ij} = y_i y_j$ is not indifferent to the location of the y-values: the closer they to the average, the lesser a_{ij} is. This seems quite compatible with the variables y having a unimodal symmetrical distribution around the mean: the closer to the mean, the more probable, thus having a lesser information content.

Another example concerns a category v (with its frequency p_v), square-scatter standardized due to the second option. This means that, in column v, $C(1 - p_v)/\sqrt{p_v}$ and $-Cp_v/\sqrt{p_v}$ are present, corresponding to Yes or No, respectively (C here is a constant, $C = 1/\sqrt{\#k-1}$). Thus, the category-driven similarity a_{ij}^v will be negative, $C^2(p_v - 1)$, when either of i, j has the category v while the other does not. The similarity will be positive if the entities are compatible with regard to the category; it equals $C^2 p_v$ when none of the entities is in the category, or, $C^2(1 - p_v)^2/p_v$ when the category contains both of the entities. We can see also, that the latter value is larger than the former one when $p_v < 1/2$, which is usually the case with multi-categorical variables.

> A shift of the origin, or subtracting a threshold, is the only prelimi-
> nary transformation admitted here for the quantitative similarity en-
> tries. Such a transformation may provide a better picture of the data
> cluster structure as shown in Fig.2.6. For a secondary similarity data
> table, the origin is determined by the way the original entity-to-variable
> data have been standardized.

However universal and linearly-compatible the formula (2.3) looks, there is no
doubt that it may be quite inapplicable in some situations; moreover, there can be
occasions when a particular other similarity measure may be implied by the nature
of the data.

2.2.2 Dissimilarity and Distance Data

Dissimilarity matrix $D = (d_{ij})$ reflects dissimilarities between the entities $i, j \in I$:
the larger d_{ij}, the less similar i and j are. In contrast to the similarity data
which may be both non-positive and asymmetrical, the dissimilarities are supposed
to be both non-negative, $d_{ij} \geq 0$, and symmetrical, $d_{ij} = d_{ji}$, for all $i, j \in I$.
Moreover, the diagonal dissimilarities, d_{ii}, are zeros, though $d_{ij} = 0$ does not
necessarily means that $i = j$. If it does, the dissimilarity D is referred to as
definite. A dissimilarity is called *even* if it satisfies the following, weaker, condition:
$d_{ij} = 0 \to d_{ik} = d_{jk}$, for all $k \in I$. A dissimilarity is called a *semi-metric* if
it satisfies the so-called *triangle inequality*, $d_{ij} \leq d_{ik} + d_{jk}$, for any $i, j, k \in I$.
Obviously, any semi-metric is even. If a semi-metric is definite, it is called a
metric or *distance* (see more detail in Van Cutsem 1994). Some particular kinds of
dissimilarities associated with classification structures will be discussed in Chapter
7 (ultrametrics, pyramidal indices, etc.).

Dissimilarities can be observed empirically (especially in psychological experiments). A dissimilarity $D = (d_{ij})$ can be obtained by a transformation of a
similarity $A = (a_{ij})$, with a formula such as $d_{ij} = C - a_{ij}$ or $d_{ij} = 1/(C + a_{ij})$ or
$d_{ij} = \exp(-a_{ij})$.

The entity-to-entity distances are calculated from entity-to-variable tables; the
most popular distances in the variable space are l_p metrics, of which the most
popular is Euclidean distance, $d_{ij} = \sqrt{\sum_{k \in K}(y_{ik} - y_{jk})^2}$.

There exists an evident connection between the Euclidean distance and the
scalar product similarity measure derived from the same entity-to-variable table:

$$d_{ij}^2 = (y_i, y_i) + (y_j, y_j) - 2(y_i, y_j) \tag{2.4}$$

which allows converting the scalar product similarity matrix $A = YY^T$ into the
distance matrix $D = (d_{ij})$ rather easily: $d_{ij}^2 = a_{ii} + a_{jj} - 2a_{ij}$. The reverse

transformation, converting the distance matrix into the scalar product matrix, can be defined when the data matrix Y is arranged in such a way that all its columns are centered, which means that the sum of all the row-vectors is equal to the zero vector, $\sum_{i \in I} y_i = 0$. In this case, the following equality holds:

$$(y_i, y_j) = -\frac{1}{2}(d_{ij}^2 - d_{i.}^2 - d_{.j}^2 + d_{..}^2) \tag{2.5}$$

where $d_{i.}^2$, $d_{.j}^2$, and $d_{..}^2$ denote the row-mean, column-mean, and the grand mean, respectively, in array (d_{ij}^2).

Based on these equations, the Euclidean distance and scalar product data could be converted into each other, allowing for the user the most preferable format.

A formula resembling (2.4) can be suggested for l_1-metric, $d_1(x, y) = \sum_{i \in I} |x_i - y_i|$, based on the equality

$$|a - b| = |a| + |b| - |sgn\ a + sgn\ b|min(|a|, |b|),$$

which holds for any real a, b.(Function $sgn\ a$ equals 1 or -1 depending on the sign of a, positive or negative, respectively; $sgn\ 0 = 0$.)

Let us define l_1-scalar product as

$$[x, y] = \sum_{i \in I} |sgn\ x_i + sgn\ y_i|min(|x_i|, |y_i|)/2.$$

Let us point out that the corresponding components, x_i and y_i, having their signs different, $x_i y_i < 0$, give no contribution to the l_1-scalar product at all; only the minimums of same-sign components matter in $[x, y]$. The definition leads to the equality

$$d_1(x, y) = [x, x] + [y, y] - 2[x, y]$$

which is a complete analogue of (2.4) implying that the concept of l_1-scalar product defined should be put under a thorough theoretical and experimental analysis. Obviously, it does not satisfy the basic linear space axiom of additivity since, in general, $[x, y + z] \neq [x, y] + [x, z]$; however, $||x||_1 = [x, x] = \sum_i |x_i|$ still is a norm (see any textbook on linear algebra; K. Janich 1994 is a most recent reference).

> Secondary scalar product similarity and Euclidean distance data are mutually convertible, which can be employed when the data are raw data, also.

2.2.3   Geometry of Aggregable Data

Although all the content of this section applies to any aggregable data, we consider here a contingency table, like the Worries data set, Table 1.6, for convenience (since most of the terms have been introduced so far for the contingency data specifically).

Let us consider a contingency data table $P = (p_{ij})$ $(i \in I, j \in J)$ where $\sum_{i \in I} \sum_{j \in J} p_{ij} = 1$, which means that all the entries have been divided by the total $p_{++} = \sum p_{ij}$ which is permissible by the meaning of aggregability. Since the matrix is non-negative, this allows us to treat the shares p_{ij} as frequencies or probabilities of simultaneously occurring row $i \in I$ and column $j \in J$ (though, no probabilistic estimation problems are considered here). Please note that although the same capital letters, I and J, are involved here to denote sets of the rows and columns as were used for preceding kinds of data sets, their meaning might be completely different: the rows and columns of such a table are usually some categories while the entities observed have been counted in the entries (like the respondents in Worries or the families in Mobility data sets).

The only transformation we suggest for the aggregable data is

$$q_{ij} = \frac{p_{ij} - p_{i+}p_{+j}}{p_{i+}p_{+j}} \tag{2.6}$$

where $p_{i+} = \sum_{j \in J} p_{ij}$ and $p_{+j} = \sum_{i \in I} p_{ij}$ are so-called marginal probabilities (frequencies) expressing the total weights of the corresponding rows and columns.

As we have seen in Section 1.3.4, q_{ij} means the relative change of probability of i when column j becomes known (due to the probabilities in P), RCP(i/j)=(p(i/j)-p(i))/p(j). Symmetrically, it can be interpreted also as RCP(j/i).

In the general setting, when matrix $P = (p_{ij})$ has nothing to do with the frequencies or probabilities, p_{ij} may be interpreted as amount of flow, or transaction from $i \in I$ to $j \in J$. In this case, $p_{++} = \sum_{i,j} p_{ij}$ is the total flow, $p(j/i)$ is defined as $p(j/i) = p_{ij}/p_{i+}$, the share of j in the total transactions of i, and $p(j) = p_{+j}/p_{++}$ is the share of j in the overall transactions. This means that the ratio $p(j/i)/p(j) = p_{ij}p_{++}/(p_{i+}p_{+j})$ compares the share of j in i's transactions with the share of j in the overall transactions. Then,

$$q_{ij} = p(j/i)/p(j) - 1$$

shows the difference of transaction p_{ij} with regard to "general" behavior: $q_{ij} = 0$ means that there is no difference in $p(j/i)$ and $p(j)$; $q_{ij} > 0$ means that i favors j in its transactions while $q_{ij} < 0$ shows that the level of transactions from i to j is less than it is "in general"; value q_{ij} expresses the extent of the difference and can be called *flow index*. The same flow index value can be interpreted in the backward direction, as the difference between the share of i in j's inflow and its "general" share, $p(i) = p_{i+}/p_{++}$, as a flow issuer.

The table $Q = (q_{ij})$ for the Worries set is in Table 2.5.

The minimum value of q_{ij} is -1 corresponding to $p_{ij} = 0$. Equation $q_{ij} = 0$ is equivalent to $p_{ij} = p_{i+}p_{+j}$ which means that row i and column j are *statistically independent*. In classical statistics, statistical independence is the basic notion;

	EUAM	IFEA	ASAF	IFAA	IFI
POL	222	280	-445	-260	36
MIL	168	-338	-129	-234	72
ECO	145	-81	-302	239	487
ENR	28	-40	11	-58	-294
SAB	19	-77	22	-66	-129
MTO	186	-252	-53	-327	-1000
PER	-503	-269	804	726	331
OTH	-118	582	-65	149	181

Table 2.5: Values of the relative changes of probability (RCP), multiplied by 1000, for the Worries data.

much of the theoretical development in contingency data analysis has been made around the concept. In the context presented, the information content is important: $q_{ij} = 0$ means that knowledge of j adds nothing to our ability in predicting i, or, in the flow terms, that there is no difference between the pattern of transactions of i to j and the general pattern of transactions to j. Such a negligible, at the first glance, reinterpretation leads to many important advances in understanding the nature of the subsequent models.

The maximum value of q_{ij} goes to infinity. We can see from (2.6) that the smaller p_{i+} and/or p_{+j}, the larger q_{ij} grows. For instance, when p_{i+} and p_{+j} are some 10^{-6}, q_{ij} may jump to million while the other entries will be just around unity. This shows that the transformation, along with the analyses based on that, should not be applied when the marginal probabilities are too different. To cope with such a nonuniform situation is not difficult: we can aggregate the rare rows/columns or exclude them from the data.

Aggregability of the data matrix P manifests in that an aggregate value may be calculated for any pair of subsets, $A \subset I$ and $B \subset J$, to show the relative change of probability of B, $p_B = \sum_{j \in B} p_{+j}$, when A becomes known, $q_{AB} = (p_{AB} - p_A p_B)/(p_A p_B)$, where $p_A = \sum_{i \in A} p_{i+}$ and $p_{AB} = \sum_{i \in A} \sum_{j \in B} p_{ij}$. For example, taking $A = \{POL, ECO\}$ and $B = \{ASAF, IFAA\}$ for the data set Worries, we have $p_{AB} = 0.029, p_A = 0.137$, and $p_B = 0.345$ which gives $q_{AB} = -0.405$, 40.5% less than the average behavior.

Taking into account aggregability of the data (to unity), the distance between the row (or column) entities should be defined by weighting the columns (or rows) with their "masses" p_{+j} (or, respectively, p_{i+}), as follows,

$$\chi^2(i, i') = \sum_{j \in J} p_{+j}(q_{ij} - q_{i'j})^2. \qquad (2.7)$$

The name of this distance, the chi-squared distance, is due to the fact that it is equal to the distance considered in the correspondence analysis theory (see, for example, Benzécri 1973, Greenacre 1993, and Section 6.4), and is defined, in that theory, with the so-called *profiles* of the conditional probability vectors $y_i = (p_{ij}/p_{i+})$ and $y_{i'} = (p_{i'j}/p_{i'+})$:

$$\chi^2(i, i') = \sum_{j \in J} (p_{ij}/p_i - p_{i'j}/p_{i'})^2/p_{+j}.$$

The proof of equivalence of these two distance formulas is quite straightforward and thus omitted.

The concept of the data scatter for the aggregable data is introduced also as a weighted one (for $p = 2$ only):

$$X^2(Q) = \sum_{i \in I} \sum_{j \in J} p_{i+} p_{+j} q_{ij}^2 \qquad (2.8)$$

The notation reflects the fact that this value is closely connected with the so-called Pearson chi-squared index, which is defined as a measure of deviation of the data in matrix P from the statistical independence:

$$X^2 = \sum_{i \in I} \sum_{j \in J} \frac{(p_{ij} - p_{i+} p_{+j})^2}{p_{i+} p_{+j}} \qquad (2.9)$$

Elementary arithmetic shows that $X^2(Q) = X^2$.

Under the statistical hypothesis that the data p_{ij} are based on a random independent sampling of the observations from a bivariate statistically independent distribution (with the marginal probabilities fixed), distribution of the value NX^2 has been proven asymptotically convergent to the distribution χ^2 with $(|I|-1)(|J|-1)$ degrees of freedom. Although this fact is tremendously important in statistical hypotheses testing, in this book, we maintain only the geometrical data scatter meaning of X^2 as based on (2.8).

It can be easily proven that a decomposition, analogous to (2.1), holds:

$$X^2(q_{ij}) = \sum_{i \in I} p_{i+} \chi^2(i, 0) = \sum_{j \in J} p_{+j} \chi^2(j, 0).$$

> Aggregable data is a newly emerging concept; its important notions
> (developed in the contingency data context) such as "statistical inde-
> pendence" or "contingency coefficient" become "zero flow index" and
> "square data scatter", in the general framework. Flow index transfor-
> mation of the data, (2.6), will be heavily employed in the consequent
> aggregable data analysis models.

2.2.4   Boolean Data and Graphs

Boolean data, as in Tables Digits 1.9 and Tasks 1.8, are supposed to give, basically, set-theoretic information. Due to such a table $X = (x_{ik})$, any row $i \in I$ is
associated with the set W_i of columns j for which $x_{ij} = 1$ while any column $j \in J$
is associated with the row set V_j consisting of those i for which $x_{ij} = 1$. There is
no other information in the table beyond that.

However, consideration of the Boolean 1/0 data as quantitative has its advantages, allowing linear algebraic computations instead of comparisons or counts. Let
us denote the 10 by 7 matrix in Table 1.9 as $X = (x_{ik})$, $i = 1, ..., 10$; $k = 1, ..., 7$.
Then, the (k, l)-th entry $\sum_i x_{ik} x_{il}$ of $X^T X$ is equal to the number of entities
(digits) having both of the attributes $k, l = 1, ..., 7$. For instance, entry (2,3) in
$X^T X$ is 4 since the attributes 2 and 3 are present simultaneously in four of the
numerals (4, 8, 9, and 0), and entry (3,3) is equal to 8 which is the number of the
digits which have attribute 3 presented. Analogously, $X X^T$ has its (i,j)-th entry,
$\sum_k x_{ik} x_{jk}$, equal to the number of the attributes which are present in both of the
digits, $i, j = 1, ..., 9, 0$. In particular, if an entry of $X X^T$ or $X^T X$ is zero, that
means that the corresponding rows (or columns) are mutually orthogonal, that is,
the corresponding numerals have no common attributes present, or, respectively,
the attributes are present on the non-overlapping subsets of the numerals. This
shows that the quantitative operations, applied to the set-theoretic meaning of the
data, can catch (and count) the overlapping/non-overlapping properties of subsets.

Still there are other set-theoretic relations and operations, such as union and
intersection, which must be reflected in algebraic transformations. This can be
done through introducing Boolean operations between the vectors and matrices.
For instance, conjunction $x \wedge y = (\min(x_i, y_i))$ corresponds to intersection of the
subsets corresponding to Boolean x and y. Boolean product $X \circ X^T$, with its
(i, j)-th element defined as $\max_k \min(a_{ik}, b_{kj})$, shows whether a pair of rows in X
is orthogonal (non-overlapping) or not.

Boolean square matrices $s = (s_{ij})$ can be represented in graph theory terms: a
graph corresponding to s has I as its vertex set while its edges correspond to pairs
ij such that $s_{ij} = 1$ (when s is a symmetrical matrix; the pairs ij are considered
ordered and represented by arcs if s is asymmetric). Let us introduce some useful

concepts of graph theory.

Let $G = (V(G), E(G))$ be a directed graph. A sequence $v_1, ..., v_k$ of its distinct vertices is called a path connecting vertices v_1 and v_2 if every two neighbor vertices v_l, v_{l+1} are joined in G. In an undirected graph, the analogous notion is referred to as a chain. Two vertices, v and u, are *connected* if there is a chain connecting them in G. In a directed G, vertices u and v are called *biconnected* if there are paths both from u to v and from v to u. These two paths form what is called *cycle* through u and v. Obviously, every two vertices in a cycle are biconnected. Both of the defined relations, connectedness and biconnectedness, are equivalence relations on the set of the vertices, thus, defining partitions of $V(G)$ into nonempty classes called *components* and *bicomponents*, respectively. No connection exists between components, and there can be only one-way connection between bicomponents (see Fig. 2.7). A graph is called connected (biconnected) if it consists of only one component (bicomponent).

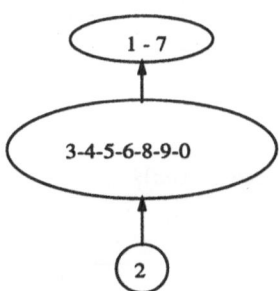

Figure 2.7: Graph of bicomponents of the threshold Confusion data graph in Fig.1.11.

There are two extreme kinds of connected graphs: complete graphs and trees. G is called a *complete* graph if every pair of distinct vertices is joined by an edge in G. G is called a *tree* if it is connected and has no cycles (see Fig. 7.1). The graph of bicomponents of a directed graph G is a graph whose vertices are the bicomponents of G, and an arc joins two of the bicomponents if there is at least one arc in G joining some vertices in these bicomponents. The graph of bicomponents of a

connected graph is a tree. A graph H is a *subgraph* of G if $V(H) \subseteq V(G)$ and
$E(H) \subseteq E(G)$. A *spanning subgraph* of G is a subgraph H with $V(H) = V(G)$.
Spanning tree is a spanning subgraph of G, being a tree; spanning trees exist if and
only if G is connected.

The concept of spanning tree is especially interesting when G is a *weighted*
graph, that is, a weight function $w : E(G) \to R^1$ is defined for its edges (arcs).
The weights, actually, correspond to similarities/dissimilarities between the vertices. The length of a spanning tree H is defined as the sum of the weights of its
edges: $l(H) = \sum_{uv \in E(H)} w(uv)$. A spanning tree is called *minimum (maximum)*
spanning tree (MST) if its length is the minimum (maximum) over the set of all
spanning trees in G. The MST concept was originally invented, in the beginning
of this century (Wrozlaw, Poland, see K. Florek et al. 1951), as a tool in biological
taxonomy and it has become an important concept in combinatorial optimization.
It turns out rather simple; finding a MST (either minimum or maximum) requires
rather simple, so-called "greedy" calculations. Any greedy algorithm, in this context, starting from a vertex, consists of sequential steps; at each step, a best (at the
given step) vertex is picked from those who have been unselected so far, which explains the term "greedy" applied to this kind of computation. A particular greedy
algorithm for the minimum spanning tree problem starts from arbitrary vertex to
be included in the tree constructed. At each other step, there are three disjoint
categories of the vertices:

(a) Tree vertices (in the tree constructed so far);

(b) Fringe vertices (not in the tree, but adjacent to a tree vertex);

(c) Other vertices.

Building Minimum Spanning Tree
Every step consists of selecting an edge of minimal weight between a tree
vertex and a fringe vertex and adding the edge and the fringe vertex to
the tree, repeating the step until no fringe vertices remain (see, for
example, Baase 1991).

The same greedy construction can be applied to the problems of finding the
components or bicomponents of a graph.

There are some other important concepts in graph theory involved in mathematical classification; two of them will be considered here.

A subset $H \subset V(G)$ is called a *clique* if the corresponding subgraph is complete,
that is, every two vertices from H are joined in G. A subset $H \subset V(G)$ is called
independent if no two vertices from H are joined in G. This means, that the cliques
are independent subsets (and vice versa) in the complementary graph obtained
from G by changing zeros to ones and ones to zeros in the adjacency matrix. A

partition of $V(G)$ is called a coloring if its classes are independent subsets. In a coloring, no vertices of the same color (belonging to the same class) can be joined, which formalizes a known problem of coloring countries in a map in such a way that all adjacent countries are distinguishable (getting different colors). In the complementary graph, coloring becomes a partition with all its classes being cliques.

The maximum clique is that one having maximum number of vertices; the minimum coloring is that one having minimum number of classes (colors). Neither of the problems finding a maximum clique or minimum coloring has any simple solution.

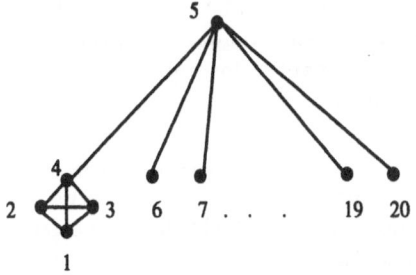

Figure 2.8: A graph with a four-element clique and a vertex joined with 19 neighbors.

The 20-vertex graph shown in Fig.2.8 has an obvious maximum clique comprising vertices 1 through 4. However, it is not clear how to find it algorithmically. A reasonable idea that the vertices having maximum neighbors are most probable to belong to a maximum clique proves false since the vertex 5 in the picture has obviously maximum number of the adjacent vertices but does not belong to the maximum clique. Some most recent developments in the problem can be found in McGuinness 1994, Xue 1994. There is a strong belief that this is a "Non-Polynomial", or "NP-complete" problem, which, basically, means that resolving the problem requires a more or less complete enumeration of all the subsets of the set of vertices $V(G)$; that is, the number of operations (each involving a consideration of a subset) required is of the order of the number of subsets, $2^{|V(G)|}$, which is greater than any polynomial function of $|V(G)|$ (see, for instance, Papadimitriou and Steiglitz 1982). The minimum coloring problem also belongs to the set of NP-complete problems.

2.2.5   Discussion

1. There are two kinds of similarity data: primary and secondary; the latter form is calculated from the data in entity-to-variable format. The only form of similarity data considered here is a quantitative entity-to-entity coefficient format.

2. The secondary similarity data are suggested to be compatible with the data-scatter framework developed in the previous Section. This leads to the row-to-row scalar product as the only similarity measure (compatible with the square scatter). When applied to the data previously standardized, the scalar product measure gives a larger weight to rarer values; for quantitative variables, this usually corresponds to the extremes.

3. The sum-of-the-absolute-values scatter L_1 implies an entity-to-entity similarity measure having the feature that the opposite-sign coordinates do not contribute to it at all! There is nothing known about the similarity measures corresponding to other forms of the data scatter.

4. The Euclidean distance and scalar product matrices are convertible into each other.

5. The only standardizing option to be applied to the similarity data under consideration is shift of the origin, that is, subtracting of a threshold from all the similarity entries. Depending on the threshold value, it allows for more or less revealing the similarity structure assumed (on the level of positive or negative similarities), as shown in Fig. 2.6. Such a revealing effect, known to the author for more than twenty years, still remains almost missed in cluster analysis considerations.

6. The aggregable data have been successfully treated for quite a long time in the framework of so-called Correspondence Analysis based on weighting the columns and rows by their totals (marginals) (see Section 2.3.3). The basic features of the techniques, RCP transformation and chi-squared data scatter, are extracted and extended to the other data analysis strategies based on contingency data as well as on any other aggregable data tables. The feature of the RCP transformation, an overestimation of the rare events, may be overcome with aggregating/excluding the rare items.

7. In mathematical analysis of Boolean data, there are two lines: (1) set- or graph-theoretic approach based on meticulous analysis of connections between the entities, and (2) quantitative approach based on algebraic operations. These approaches have an overlap, which allows us to use algebraic operations for counting or just observing set-theoretic overlaps.

2.3 Low-Rank Approximation of Data

2.3.1 SVD and Principal Component Analysis

In quantitative data analysis, a quite elegant and powerful tool, low-rank approximation of the data, has been developed. In this section, a review of the techniques will be done and an extension will be developed for employment in clustering and related applications.

The primary techniques is called principal component analysis, which was invented by K. Pearson (1902) and put in a modern matrix format by G. Hotelling (1937). Amazingly, the method is usually introduced as a heuristic technique to construct the linear combinations of the variables which contribute the most (see, for example, Everitt and Dunn 1992 and Krzanowski and Marriott 1994 as most recent references). Here, a more model-based approach is described, emphasizing, in particular, that the fact that the principal components are linear combinations of the variables follows from the model, instead of presupposing it.

Let us consider a $N \times n$ matrix $X = (x_{ik})$ having a very peculiar format: there exist two vectors, $c = (c_1, ... c_n) \in R^n$ and $z = (z_1, ..., z_N) \in R^N$, such that $x_{ik} = c_k z_i$, for every $i \in I$ and $k \in K$ (this can be written also as $X = zc^T$ since every Euclidean space element is considered as a column-vector). As an operator applied to the other vectors, this matrix acts quite uniformly: for any $a \in R^n$, $Xa = \mu z$ where μ is a constant defined as $\mu = (c, a)$; for any $y \in R^N$, $y^T X = \nu c$ where $\nu = (z, y)$. Matrix X, as an operator, maps all the n or N dimension vectors into the unidimensional space defined by z or c, respectively; this is why it is assigned with rank 1.

The problem is this: for an arbitrary $X = (x_{ik})$, find its best one-rank least-squares approximation matrix zc^T, that is, find $z \in R^N$ and $c \in R^n$ minimizing $L_2 = \sum_{i \in I} \sum_{k \in K} (x_{ik} - z_i c_k)^2$. Amazingly, the solution to the problem is a linear transformation of the variables. However, there is an ambiguity in the problem since the entries are approximated by the products $c_k z_i$ that do not vary when c and z are substituted by αc and z/α, respectively, for any real α, and, thus, the norm(s) of c or/and z must be previously specified somehow. Usually, the vectors sought, c and z, are required to be normed (that is, $||z|| = ||c|| = 1$) while the optimal norm scale factor (for both of them) is put in the problem explicitly.

Another option which can be applied: require $||z|| = 1$ while no conditions on c are imposed.

Let us set a more general low-rank approximation problem: for a given number m, find unknown normed vectors $z_1, ..., z_m \in R^N$ and $c_1, ..., c_m \in R^n$, and reals $\mu_1, ..., \mu_m$ satisfying equations

$$x_{ik} = \sum_{t=1}^{m} \mu_t c_{tk} z_{it} + e_{ik} \tag{2.10}$$

and minimizing the Euclidean norm of the residual matrix $E = (e_{ik})$, $L_2(E) = \sum_{i \in I, k \in K} e_{ik}^2$.

The equations connecting the data x_{ik} with the sought vectors c_t, z_t sometimes are referred to as a *bilinear model*, since there are products, $c_{tk} z_{it}$, of two sought values involved, within otherwise quite a linear setting.

The problem can be resolved in the framework of the so-called singular-value decomposition (SVD) theory in linear algebra (see, for example, Golub and Van Loan 1989). Given an $N \times n$ real-valued matrix $X = (x_{ik})$ of the rank $p \leq \min(N, n)$, its singular triple (μ, z, c) is defined as a positive $\mu > 0$, called the singular value, an N-dimensional normed vector $z = (z_i)$, and an n-dimensional normed vector $c = (c_k)$ (vector x is called normed if $||x|| = 1$) satisfying the following equations:

$$Xc = \mu z, \quad X^T z = \mu c.$$

It is well known that number of singular values is equal to the rank p. In data analysis, all the singular values are traditionally considered different since an observed data matrix X can hardly have a specific structure. Then, there exist exactly p singular triples (μ_t, z_t, c_t); the vectors z_t are mutually orthogonal, in R^N, as well as the vectors c_t, in R^n $(t = 1, ..., p)$. Moreover, the singular value decomposition (SVD)

$$X = ZMC^T \tag{2.11}$$

holds, where Z is $N \times p$ matrix having z_t as its columns, C^T is $p \times n$ matrix having c_t as its rows, and M is $p \times p$ diagonal matrix having μ_t as its diagonal entries $(t = 1, ..., p)$ while the other entries are zeros. In terms of the matrix entries, equation (2.11) can be written as follows:

$$x_{ik} = \sum_{t=1}^{p} \mu_t z_{it} c_{tk}.$$

Let the singular values be indexed so that $\mu_1 > \mu_2 > ... > \mu_p$. Then, as it is well known, a solution to the low-rank approximation problem is provided with the first m singular triples; the one-rank approximation problem is resolved with just the first singular triple, corresponding to the maximum singular value.

By definition, each of the vectors z_t is a linear combination of the columns of matrix X while the other singular vector, c_t, is employed in a dual capacity of collection of the linear combination coefficients. Denoting by Z_m, C_m and M_m parts of the matrices Z, C, and M, respectively corresponding to the first m triples, we have $Z_m = X M_m^{-1} C_m^T$, by the definition of the singular triples. The property that vectors $c_1, ..., c_m$ are mutually orthogonal means that vectors z_t can be obtained by rotating and distorting the original base of the variable space (see Fig. 2.9 where the axes corresponding to z_t are shown by dotted lines; the picture presents a typical ellipsoid-shaped cloud of the entity points).

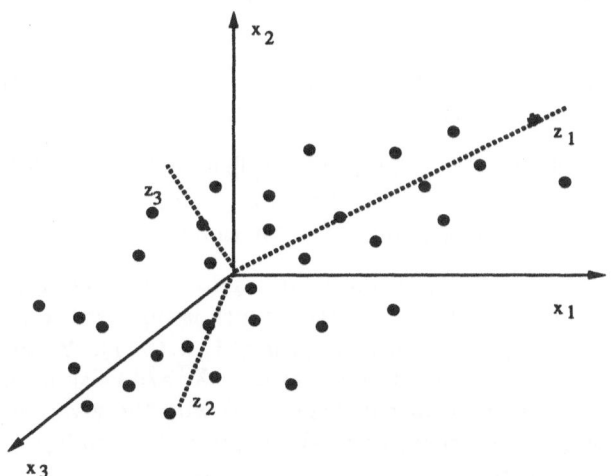

Figure 2.9: A typical pattern of the principal components.

In data analysis, vectors $z_t = (1/\mu_t) X c_t$ are referred to as the principal components having coordinates z_{it} as their (normalized) "factor scores" for the entities $i = 1, ..., N$ while the coordinates c_{tk} are called (standardized) "factor loads" as coefficients of the linear representation of the components z_t through the variables $k = 1, ..., n$ (Jolliffe 1986). The singular value squared μ_t^2 represents contribution of the t-th principal axis (component) to the total data scatter $L_2(X) = Tr(X^T X) = \sum_{i,k} x_{ik}^2$ ($t = 1, ..., p$). Here, $Tr(A) = \sum_{i \in I} a_{ii}$ is the so-called *trace* of a square matrix A.

Traditionally, the principal component analysis is associated with sequentially finding the elements of the spectral decomposition of the (covariance/correlation) matrix $Y^T Y$, $Y^T Y = A^T M^2 A$, since vectors a_t are its eigenvectors corresponding to the eigenvalues μ_t^2, $t = 1, ..., p$.

However, a similar step-by-step sequential fitting procedure based on the matrix Y itself can be employed for finding the singular-value decomposition (2.11). At each step, the procedure finds one singular triple (μ_t, f_t, a_t) (in the decreasing order of μ_t), processing a residual form of the data matrix X. At the first step, $t = 1$, matrix X itself is processed to find the largest singular value μ_1. Any step t consists of the following computations.

Find any pair of vectors $c = (c_k)$ and $z = (z_i)$ minimizing criterion

$$D_2(c, z) = \sum_{i=1}^{N} \sum_{k=1}^{n} (x_{ik} - z_i c_k)^2 \tag{2.12}$$

Compute norms $\|z\|$ and $\|c\|$ and let $\mu_t = \|z\|\|c\|$, $z_t = z/\|z\|$, and $c_t = c/\|c\|$. Calculate the residual entries $x'_{ik} = x_{ik} - \mu_t c_{tk} z_{it}$ and put them as x_{ik} for the next step which is performed with t increased by 1 unless t becomes larger than m.

To find a pair of vectors minimizing (2.12), the following iterative algorithm can be applied starting with an arbitrary vector $c = (c_k)$ (which is supposed to be linearly independent of the previously found vectors c_u, $u = 1, ..., t - 1$). Vector $z = X^T c/\|X^T c\|$ is used for iteratively recalculating c as $c \leftarrow X^T z/\|X^T z\|$ until the vectors c and z coincide (up to a small threshold specified) with those vectors at the previous iteration. The process converges to the triple corresponding to maximum (at the given step) singular value μ_t (note that it is supposed to be strictly greater than the subsequent singular values). Thus, the vectors c and z found can be taken as c_t and z_t, respectively, while $\|X^T z\|$ approximates μ_t.

Applied to the Masterpieces standardized 8×7 data matrix, the method has produced four singular triples, each contributing more than 1 per cent to the scatter of the data. The normed vectors z_t are columns in the following matrix:

$$
Z_4 = \begin{pmatrix}
0.30 & 0.24 & -0.15 & 0.82 \\
0.51 & -0.08 & 0.05 & -0.32 \\
0.52 & -0.09 & -0.02 & -0.32 \\
-0.20 & -0.50 & -0.05 & 0.21 \\
-0.35 & -0.35 & -0.59 & -0.12 \\
-0.26 & -0.22 & 0.78 & 0.07 \\
-0.19 & 0.47 & 0.05 & -0.09 \\
-0.33 & 0.54 & -0.08 & -0.25
\end{pmatrix}
$$

while vectors $\mu_t c_t$ (c_t are normed, too) are the columns in

$$C_4 M_4 = \begin{pmatrix} -2.54 & 0.55 & 1.08 & -0.03 \\ -1.95 & -1.92 & -0.69 & 0.15 \\ -2.20 & 1.44 & -0.96 & -0.37 \\ -2.75 & -0.13 & 0.24 & -0.37 \\ -0.25 & 1.44 & -0.20 & 0.56 \\ 1.46 & -0.25 & 0.04 & -0.90 \\ -0.94 & -1.23 & 0.17 & 0.18 \end{pmatrix}$$

The values μ_t squared give their relative contributions to the data scatter: 64.37, 24.34, 6.71, and 3.63 per cent, respectively (thus covering 99.05 percent of the data scatter altogether). To approximately reconstruct an entry x_{ik} in the original matrix, matrix $Y_4 = Z_4 M_4 C_4^T$ is calculated along with subsequent transformation of its every entry into $b_k y_{ik} + a_k$ where a_k, b_k are the parameters of the origin shifting and scale factoring employed for standardization of the data. We don't supply the reconstructed matrix since it is almost coincides with the original one.

The standardized components of c_t (factor loads) are employed for interpreting purposes. For instance, the first principal component distinguishes Pushkin's novels since the only variable/category which got a positive factor load (see 1.46 in the first column of $M_4 C_4$), is Behav (6-th column in the data table) featuring this writer.

Let us point out some features of the principal component analysis method:

1. Although the method finds the principal components sequentially, the vectors found give a solution to the original "parallel" low-rank approximation problem.

2. The sequential extraction strategy decomposes computations into two levels: the first level deals with organization of the whole process as the sequence of optimizing (2.12) steps while the second level resolves the one-rank approximation problem.

3. Each principal component z_t is accompanied by the corresponding value μ_t^2 (2.12) of its contribution to the square data scatter, which can be employed for sequential decision making on the number m of the principal components to be found. Such a decision rule may be based on either or all of the following conditions:

a) t becomes higher than a prefixed value $\tilde{m}$;

b) the relative contribution, $\mu_t^2/Tr(X^T X)$, of the component t to the data scatter becomes equal to or less than a prefixed small $e_1 > 0$;

c) contribution of the "unexplained" part, $L_2(E)$, to $L_2(X)$ becomes equal to or less than a prefixed value e_2.

4. Principal components heavily depend on the preliminary data transformation. With the standard z-scoring (square-scatter standardization), minimizing the scatter $L_2(X)$, the ellipsoid-shape of the entity point cloud will not be as elongated as it may be under a different standardization (obviously, the more "elongated" is the cloud, the greater is the contribution of the first component).

5. Actually, the solution to the low-rank approximation problem is the space generated by the singular vectors $z_1, ..., z_m$ as its base, thus any other base of the space, $Z_m A$ where A is an $m \times m$ nonsingular matrix, provides the same values for the residuals e_{ij} with the vectors c_t changed as $C_m A^{-1}$. This property allows for developing specific strategies of "rotation" (or even distortion) of the base Z_m to find the corresponding loading matrix $C_m A^{-1}$ of a "simple" structure. In this book, the original structure of the singular vector matrix C_m is considered only.

6. Though the bilinear approximation model involves no requirement that the approximating components are connected with the data matrix analytically, it turns out that the optimal components are linear combinations of the data matrix columns, with the coefficients being the components of the dual vector.

2.3.2   Ordination of the Similarity Data

The problem of embedding similarity/dissimilarity data into a variable space is known as the problem of multidimensional scaling (see, for example, Kruskal and Wish 1978, and Arabie, Carroll, and De Sarbo 1987). There exists a technique, called *ordination* (Gower 1966, Krzanowski and Marriott 1994), which does such a job in the manner of principal component analysis. The underlying bilinear model is this.

Let $A = (a_{ij})$, $i, j \in I$, be symmetric similarity data. A one-rank similarity data matrix has its entries equal to $a_{ij} = x_i x_j$ for some $x = (x_i) \in R^N$; multiplying such a matrix by any $y \in R^N$ gives $Ay = \lambda x$ where $\lambda = (y, x)$. This means that such a one-rank matrix maps all the N-dimensional vectors into the unidimensional space of vectors λx ($\lambda \in R$), which is its only "principal" axis.

The problem of low-rank approximation for an arbitrary A can be formulated as follows: for an m pre-specified, find normed vectors $z_1, z_2, ..., z_m \in R^N$ and corresponding real $\lambda_1, \lambda_2, ..., \lambda_m$ minimizing the residuals squared, $L_2(E) = \sum_{i,j \in I} e_{ij}^2$, in the following equations:

$$a_{ij} = \sum_{t=1}^{m} \lambda_t z_{it} z_{jt} + e_{ij} \qquad (2.13)$$

Solution to this problem is based on the theory of eigenvalues and eigenvectors for symmetrical square matrices (Golub and Van Loan 1989, Janich 1994, Jolliffe 1986). A normed vector z is referred to as an eigenvector of A if it satisfies equation $Az = \lambda z$, for some λ, real or complex, which is called the eigenvalue corresponding to z. It is well-known that any symmetric A has p (p is rank of A) non-zero eigenvalues, λ_t, $t = 1, ..., p$, all of which are real. Provided that all the eigenvalues are different, the corresponding eigenvectors, z_t, are pair-wise orthogonal and satisfy the following equality: $a_{ij} = \sum_{t=1}^{p} \lambda_t z_{it} z_{jt}$ (for any $i, j \in I$) or, in matrix form, $A = Z\Lambda Z^T$ where Z is the $N \times p$ matrix of the eigenvectors and Λ is the $p \times p$ diagonal matrix containing λ_t in its principal diagonal. The square-scatter $L_2(A)$ can be decomposed by the eigenvalues as follows: $L_2(A) = \sum_{t=1}^{p} \lambda_t^2$.

When A is $A = XX^T$ for a matrix X, that is, A is a matrix whose entries are scalar products, all the non-zero eigenvalues are positive (moreover, $\lambda_t = \mu_t^2$ ($t = 1, ..., p$) where μ_t is a singular value of matrix X and p is its rank); actually, in this case, the eigenvectors are just the principal components of X. However, it is not a big deal to get all the eigenvalues positive, for any symmetric A, since adding a large constant $a > 0$ to its diagonal entries shifts all the eigenvalues from λ_t to $\lambda_t + a$ without changing the corresponding eigenvectors.

The theory outlined leads to the following solution to the low-rank approximation problem.

Order eigenvalues $\lambda_1^2 > \lambda_2^2 > ... > \lambda_p^2 > 0$. Pick up the first m eigenvalues and corresponding eigenvectors (provided that $m < p$); they give the solution along with the minimum value of the least squares criterion equal to $L_2(E) = \sum_{t=m+1}^{p} \lambda^2$.

Finding the first (maximum) eigenvalue and corresponding eigenvector may be done by sequentially multiplying any vector z by A since matrix A^q has λ_t^q as its eigenvalues corresponding to the same eigenvectors z_t: when q is large enough, λ_1^q dominates all the other eigenvalues λ_t^q. An iteration of the sequential process can be formulated as follows: with a normed z given, find $z' = Az/||Az||$ and check whether the difference $||z' - z||$ is small enough. If not, take z' as z and reiterate. If yes, either of z, z' is the eigenvector while $||Az||$ is the corresponding maximum eigenvalue.

To find the next eigenvalue and eigenvector, calculate the residual matrix $A' = A - \lambda zz^T$ (with its entries $a'_{ij} = a_{ij} - \lambda z_i z_j$) and apply the procedure above to this residual matrix as A. Continuing residuation of the data, we can find as many of the eigenvalues as would be sufficient (either to cover a desired part of the data scatter, which is estimated by the cumulate sum of λ_t^2, or to reach a prior value of m, or to get the latest λ_t^2 as small as a pre-fixed "noise" level).

Letting the entities $i \in I$ be the points $y_i = (\sqrt{\lambda_1} z_{i1}, ..., \sqrt{\lambda_m} z_{im})$ of the space generated by the first m eigenvectors (usually, $m=2$ for visualization purposes) is referred to as ordination of the similarity data. Due to the model (2.13), a_{ij} is approximately (y_i, y_j), for any $i, j \in I$.

> Although the ordination model works for any symmetric matrix, it works
> even better when the similarities are scalar products of the columns of a
> data table: the solution found is equally good for both of the matrices,
> due to the models in (2.10) and (2.13).

2.3.3   Correspondence Analysis Factors

Correspondence Analysis (CA) is a method for presenting an integral picture of
interactions between I and J (basing on contingency table $P = (p_{ij})$, $i \in I$ and
$j \in J$) in a geometric factor space (see, for example, Benzécri 1973, Greenacre 1993,
Lebart, Morineau and Warwick 1984, Nishisato 1994). Two sets of "underlying"
factors, $\{F_t\}$ and $\{G_t\}$, $t = 1, ..., p$, with I and J as their respective domains, are
calculated in such a way that the following two conditions hold:

A. *Reconstruction formula:*

$$p_{ij} = p_i p_j (1 + \sum_{t=1}^{p} \mu_t F_t(i) G_t(j)) \qquad (2.14)$$

where $\mu_1 > \mu_2 > ... > \mu_p > 0$, and p is the rank of matrix $F = (f_{ij})$ with
$f_{ij} = (p_{ij} - p_i p_j)/(p_i p_j)^{\frac{1}{2}}$ as its entries;

B. *Weighted orthonormality:*

$$\sum_{i \in I} p_i F_s(i) F_t(i) = \sum_{j \in J} p_j G_s(j) G_t(j) = \delta_{st} \qquad (2.15)$$

where $\delta_{st} = 1$ if $s = t$, otherwise $= 0$.

The factors, in fact, are determined by the singular-value decomposition of
the matrix $F = (f_{ij})$ defined above. More explicitly, the values μ_t in (2.14) and
vectors $f_t = \{F_t(i)p_i^{\frac{1}{2}}\}$ and $g_t = \{G_t(j)p_j^{\frac{1}{2}}\}$ are the corresponding singular values
and vectors defined by the equations: $F g_t = \mu_t f_t$, $f_t F = \mu_t g_t$. (It is assumed that
all the values μ_t are different, which is almost always satisfied when the contingency
data are of empirical nature.)

In the p-dimensional Euclidean CA factor space each item $i \in I$ (or $j \in J$)
is represented by its vector $F(i) = \{F_t(i)\}$ (or, respectively, by $G(j) = \{G_t(j)\}$).
Frequently, the factor axes t are scaled by the factors μ_t; then the vectors are $F(i) =
\{\mu_t F_t(i)\}$ and $G(j) = \{\mu_t G_t(j)\}$. Within the sets I and J these representations
reflect the similarities between corresponding conditional probability profiles: it has
been proven that the squared Euclidean distance between CA scaled factor space
points $F(i)$ and $F(i')$ equals the chi-squared distance χ^2 between corresponding
rows, and that the symmetric equality holds for the distances between arbitrary
column items $j, j' \in J$.

Use of the distances between different sorts of the items, is and js, in the CA factor space is considered as justified with the so-called *transition formulas*:

$$F_t(i) = \frac{1}{\mu_t} \sum_{j \in J} p(j/i) G_t(j), \ \ G_t(j) = \frac{1}{\mu_t} \sum_{i \in I} p(i/j) F_t(i) \tag{2.16}$$

showing that, up to a scaling factor of $1/\mu_t$, any i-point could be considered as the weighted average of the j-points and vice versa.

The formula in (2.14) can be rewritten in the following equivalent form:

$$q_{ij} = \sum_{t=1}^{p} \mu_t F_t(i) G_t(j)$$

which shows that CA can be thought of as a bilinear model of the RCP (flow index) values q_{ij} rather than of the primary contingency data.

This allows us to bring to light a data approximation model which can be thought of as underlying the CA method.

Let us minimize the following weighted least-squares criterion

$$L^2 = \sum_{i,j} p_i p_j e_{ij}^2 \tag{2.17}$$

with regard to the following constraints expressing given q_{ij} through the sought μ_t, $F_t(i)$, $G_t(j)$, and e_{ij} (based on the reformulated form of the equality in (2.14)):

$$q_{ij} = \sum_{t=1}^{m} \mu_t F_t(i) G_t(j) + e_{ij} \tag{2.18}$$

using the same sequential fitting procedure as in the preceding two sections, though with modifications due to differences in the criteria. A convenient way is just finding the singular triples of matrix F step-by-step consequently transforming them in vectors F_t, G_t sought, as explained above.

The visualization strategy of CA is based on the following considerations. Obviously,

$$Tr(F^T F) = \sum_{i,j} (p_{ij} - p_i p_j)^2 / (p_i p_j) = \sum_{i,j} q_{ij}^2 p_i p_j \tag{2.19}$$

which is the Pearson chi-squared contingency coefficient Φ^2. Thus, the eigenvalues μ_t^2 of $F^T F$ show which part of the contingency coefficient value can be considered as "explained" by the factors t ($t = 1, ..., m$). In a common situation, the first two eigenvalues account for a major part of Φ^2; this is considered to justify use of the plane of the first two factors to display the interrelations between I and

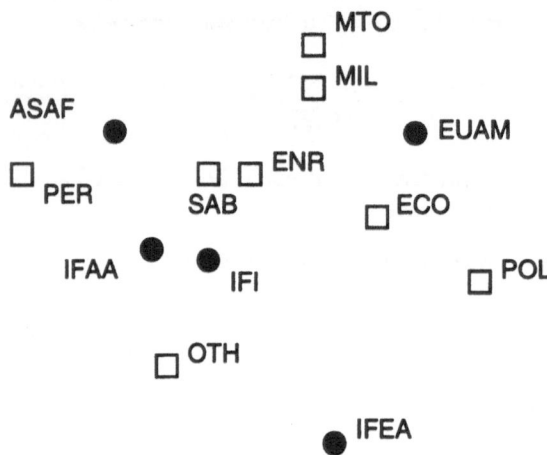

Figure 2.10: Visual display of the correspondence analysis for the Worries data.

J, with the items i and j presented by two-dimensional points $(F_1(i), F_2(i))$ and $(G_1(j), G_2(j))$ of the plane. Such a joint display for the Worries data of Tables 1.6 and 1.19 is shown on Fig. 2.10.

Based on equivalence between the chi-squared distances in the contingency table and the physical distances in the CA factor space, one can see, in Fig.2.10, that the conditional probability profiles of MTO and MIL are close to each other, as are the profiles of IFAA and IFI, or one can see that IFEA and EUAM profiles are very distant, and so on. From the same picture, one can also conclude that the living place EUAM relates to the MTO, MIL, ENR and ECO worries located around it, and ASAF is close to the PER and SAB worries.

But such an interpretation needs to be examined because it is based on transition formulas in (2.16), where the column points are expressed through the row points (and vice versa) up to the scaling factors $1/\mu_t$ which are different from 1 in a typical situation.

The following advantages of the CA method should be pointed out. The method:

1. shows the structure of interrelations between row and column items based on well interpreted RCP values;

2. is based on the double normalization of the contingency data by both row and column totals;

3. decomposes the chi-squared contingency coefficient value by the factors obtained;

4. visualizes the data with a joint display based on chi-squared distances between the conditional probability profiles.

> Correspondence factor analysis provides a joint display for the rows and columns, which is a most helpful feature of the method, as well as a most controversial one: in spite of the fact that the row-to-row and column-to-column distances in the plot are well defined within the model, the "mixed" row-to-column distances cannot be defined this way.

2.3.4 Greedy Approximation of the Data: SEFIT

The sequential approximation approach described in the three sections above can be extended to any additive model in a Euclidean space. A general formulation is this (Mirkin 1990). Let $x \in R^l$ be a given vector ($l = N \times n$ with $x = X$ in Section 2.3.1, $l = N \times N$ with $x = A$ in Section 2.3.2, and $l = |I| \times |J|$ with $x = Q$, in Section 2.3.3).

The problem is to represent x as

$$x = \sum_{t=1}^{m} \mu_t z_t + e \tag{2.20}$$

where z_t are chosen from given subsets D_t of R^l, μ_t are chosen from the set of all reals R, and residual e is "small". In the examples above, D_t was a set of one-rank matrices of a corresponding format.

The sequential fitting strategy (referred to as SEFIT) extended to this general setting consists of the iterations (t=1, 2,...) involving two major steps:

(1) for given x_t, minimize the least-squares criterion $\|x_t - \mu z\|^2$ with regard to arbitrary $\mu \in R$ and $z \in D_t$ and let $\mu_t = \mu^*$ and $z_t = z^*$ where μ^*, z^* are minimizers found;

(2) calculate the residual vector $x_{t+1} = x_t - \mu_t z_t$.

In the beginning, $t = 1$, we set $x_1 = x$. After an iteration is completed, we check whether the process has to be stopped or not. The stop condition involves the following three parts:

(i) the number of members, $z_1, ..., z_t$, found, t, exceeds a number pre-set, m;

(ii) the relative contribution of the t-th solution, $\mu_t^2(z_t, z_t)$, to the data scatter, (x, x), becomes too small;

(iii) the relative contribution accumulated, $\sum_{s=1}^{t} \mu_s^2(z_s, z_s)$ to (x, x) becomes large enough.

When any of them is satisfied, the process ends, and equation (2.20) holds with $e = x_{t+1}$.

As it was described in the preceding three sections, such a greedy process works well due to the particular nature of the singular- and eigen-vectors. However, even when applied with D_t being not as regular (when the minimization problem at step (1) cannot be resolved globally), the procedure still bears some good properties.

First, for every z in the minimization problem, the optimal μ can be determined easily as $\mu(z) = (x_t, z)/(z, z)$. Then the following statement holds:

Statement 2.1. *If $\mu_t = \mu(z_t)$ (which is the optimal solution for z_t fixed), then independently of the selection of the set $z_1, ..., z_m$, the standard decomposition of the data scatter holds:*

$$(x, x) = \sum_{t=1}^{m} \mu_t^2(z_t, z_t) + (e, e) \tag{2.21}$$

Proof: Indeed, vector $x_{t+1} = x_t - \mu_t z_t$ is orthogonal to z_t, and, by Pythagoras' theorem, $(x_t, x_t) = (\mu_t z_t, \mu_t z_t) + (x_{t+1}, x_{t+1})$. Summing over $t = 1, ..., m$ and noting that $x_1 = x$ and $x_{m+1} = e$, we obtain (2.21). □

The equality (2.21) decomposes scatter $||x||^2$ into an "explained" part, $\sum_{t=1}^{m} \mu_t^2(z_t, z_t)$, and an "unexplained" part $||e||^2$. It allows calculation of contributions of the elements of solution to the data scatter and justifies the stop-condition above. Since $\mu_t = (x_t, z_t)/(z_t, z_t)$, the contribution of t-th element of the model (2.20) is equal to $g_t = (x_t, z_t)^2/(z_t, z_t)$ which is, actually, the function of z_t maximized in step (1) of t-th iteration.

A question arises about the correctness of the algorithm SEFIT described. Does the residual become or tend to zero sometime, or might it grow in the process? The answer depends on the sets D_t and the quality of the minimizing algorithm (at step 1) involved. The following condition assumes a kind of extensiveness of D_t and, simultaneously, a quality of the algorithm.

Condition E. Every l-dimensional base vector $u_k = (0, 0, ..., 0, 1, 0, ..., 0)$, having the only 1 in the k-th position, belongs to D_t, and the vector z_t chosen in step (1) of SEFIT procedure is no worse than u_k, for any $k = 1, ..., l$; that is, $g_t \geq (x_t, u_k)/(u_k, u_k) = x_{tk}^2$ where x_{tk} is k-th entry in the vector x_t.

Statement 2.2. *If Condition E holds, then x_t converges to 0 as t increases.*

Proof: Let $|x_{kt}| = \max_{k'=1,...,l} |x_{k't}|$. Then, $(x_t, x_t)/l \leq x_{kt}^2$. Thus, by Condition E, $g_t \geq (x_t, x_t)/l$, which implies:

$$(x_{t+1}, x_{t+1}) = (x_t, x_t) - g_t \leq (x_t, x_t)d,$$

where $d = 1 - 1/l < 1$. Thus, $(x_t, x_t) \leq (x, x)d^{t-1}$, where d^{t-1} converges to 0 when t increases, and, therefore, $(x_t, x_t) \to 0$. $\qquad\square$

The procedure described, SEFIT, will be a basis of many other approximation algorithms in the remainder of this book (for more detail on SEFIT see in Mirkin 1990).

> In general, SEFIT leads to a nonoptimal solution in the "parallel" approximation problem. However, it exhausts the data and additively decomposes the data scatter with the solutions found. To expect the algorithm recover the real structure underlying the data y, the contributions of its constituents must be quite different.

2.3.5 Filling in Missing Data

Sometimes, and in some application areas quite frequently, the data table comes with some entries missing: in medicine, a patient is gone although all his symptoms are known except for two of them added quite recently; in geology, a datum was not collected because of impossibility of reaching a site; in a survey, a respondent filled in a questionnaire leaving some questions unresponded, etc.

There are three major options for dealing with the tables containing missing data: 1) excluding corresponding entities/variables, 2) dealing with the data as they are, adopting a particular strategy for handling the missing values within each particular algorithm involved in the data processing, 3) filling in the missing data before the data processing. The first option has nothing to do with any special considerations; it is just quite a generous cleaning of the data set. The second option relies heavily on the nature of the data processing problem and the procedure involved; we will not consider that in this book. The third option has received an attention in the data analysis literature along with suggestions on filling in the missing data in a regression-wise or principal-component-wise style.

The following is a universal method extending the Principal component analysis strategy to the case when some data entries are missing, due to SEFIT strategy above. For convenience, let us restrict ourselves to data in the entity-to-variable format.

Let $X = (x_{ik})$, $i \in I, k \in K$, be a data matrix with some of its entries missing. Let matrix $M = (m_{ik})$ indicate the missing data so that $m_{ik} = 0$ if x_{ik} is missing, and $m_{ik} = 1$ if x_{ik} is present.

Let us consider the bilinear model equations

$$x_{ik} = \sum_{t=1}^{m} c_{tk} z_{it} + e_{ik} \qquad (2.22)$$

only for those of the entries $(i, k) \in I \times K$ where $m_{ik} = 1$, thus minimizing the least-squares criterion defined only for them, which can be written as $L_2(E, M) = \sum_{i \in I, k \in K} e_{ik}^2 m_{ik}$ to show that the items for the missing entries are zeros without any regard to what kind of values can be put there.

The problem is to find approximating values c_{tk}, z_{it} in the restricted model formulated. After the model values are fit, we can "extrapolate" the missing values x_{ik} (for those (i, k) where $m_{ik} = 0$) with the bilinear model equations, this time without any residuals since they are unknown for the missing entries:

$$x_{ik} = \sum_{t=1}^{m} c_{tk} z_{it}.$$

Let us apply the SEFIT strategy to the problem. That means that, initially, we find only one axis, $t = 1$, values z_{i1} and c_{1k}, minimizing the following criterion (of the step (1) in SEFIT procedure):

$$L_2(c, z, M) = \sum_{i \in I, k \in K} (x_{ik} - c_k z_i)^2 m_{ik} \qquad (2.23)$$

with regard to arbitrary c_k and z_i satisfying a supplementary norming requirement $\|z\|^2 = \sum_{i \in I} z_i^2 = 1$ (to have a unique solution as discussed in Section 2.3.1).

The first-order optimality condition (applied to Lagrange function $L = L_2(c, z, M) - \lambda(1 - \|z\|^2)$) leads us to the following equations:

$$\sum_{k \in K} x_{ik} m_{ik} c_k = z_i [\sum_{k \in K} c_k^2 m_{ik} + \lambda],$$

$$\sum_{i \in I} x_{ik} m_{ik} z_i = c_k \sum_{i \in I} z_i^2 m_{ik}.$$

$$\lambda = \sum_{i \in I} \sum_{k \in K} [x_{ik} m_{ik} z_i c_k - z_i^2 c_k^2 m_{ik}]$$

These lead to the rules for iterative recalculation of the values c_k and z_i:

$$z_i \leftarrow \sum_{k \in K} x_{ik} m_{ik} c_k / (\|c\|_i^2 + \lambda),$$

$$c_k \leftarrow \sum_{i \in I} x_{ik} m_{ik} z_i / ||z||_k^2],$$

where $||c||_i^2 = \sum_{k \in K} c_k^2 m_{ik}$ and $||z||_k^2 = \sum_{i \in I} z_i^2 m_{ik}$ are varying analogues of the norm squared. Supplemented with norming z after each iteration, this gives us an iterative method for resolving the problem.

Although the present author has no proof that this method converges to a minimizer of $L_2(c, z, M)$ or, equivalently, to a maximizer of

$$\sum_{i \in I, k \in K} c_k z_i m_{ik} x_{ik} + \lambda,$$

which is the solution's contribution to the data scatter, the procedure converges quite fast, in experimental computations (with a relatively small number of zeroes in M).

After the first factor is estimated, we proceed to the residual data $x_{ik} \leftarrow x_{ik} - c_k z_1$ for $m_{ik} = 1$, to reiterate the process. Correctness of the SEFIT method follows from the fact that condition E in Statement 2.2. applies here. The method was suggested by the author in 1989 for the program he managed (see Mirkin and Yeremin 1991); recently, the author learned that its step (1) (finding a pair (c, z)) is known, in data analysis, as the Ruhe-Wiburg method, suggested back in mid-seventies (see Shum, Ikeuchi, and Reddy 1995).

Let us remove the following four of the entries from Masterpieces data: LenSent and Presentat for Crime&Punishment, LenDial for Captain's Daughter and NChar for Eug.Onegin, which leads to six missing entries, in the quantitative version of the data, since none of the three Presentat categories is known for Crime&Punishment (there are four, not two, missing entries in row 4 of Table 2.6).

After square-scatter standardization of the data (the means and scale factors are calculated within the columns by the entries available), the procedure above leads to the solution which seems quite similar to that found in Section 2.3.1 in the framework of principal component analysis (which is, basically, the same method applied when no $m_{ik} = 0$). The four vector pairs found account for 63.98, 24.16, 7.44, and 3.17 per cent of the data scatter each (totaling to 98.75% of the data scatter).

The matrix reconstructed (see Table 2.7) shows that the missing values have been estimated rather satisfactorily, except for the entry $x_{32} = 10.2$ estimated as 19.00. One should remember that there is a requirement in reconstructing the nominal categories: only one of them must have unity in the corresponding row while the others have zeroes. In the case considered, this is easy to do since the entries (4,5) and (4,6) reconstructed are quite close to zero while the entry (4,7) is correctly 1.

2.3.6   Discussion

1. The low-rank approximation methods considered in Sections 2.3.1 through

Num	LenSent	LenD	NumCh	InMon	Direct	Behav	Thought
1	15.0	16.6	-	0	1	0	0
2	12.0	9.8	1	0	0	1	0
3	11.0	-	1	0	0	1	0
4	-	202.8	2	1	-	-	-
5	20.9	228.0	4	1	0	0	1
6	29.3	118.6	2	1	0	0	1
7	23.9	30.2	4	1	1	0	0
8	27.2	58.0	5	1	1	0	0

Table 2.6: Quantitative presentation of the Masterpieces data as having six entries missed.

Num	LenSent	LenD	NumCh	InMon	Direct	Behav	Thought
1	14.76	19.67	2.14/2	-0.01	1.04	0.00	-0.04
2	11.51	11.36	0.93	0.02	0.05	0.99	-0.04
3	10.39	18.99/10.4	0.92	-0.03	0.02	1.01	-0.04
4	23.23/20.2	201.02	2.03	0.99	-0.32/0	0.32/0	1.00/1
5	21.12	241.30	3.77	1.02	0.15	0.03	0.82
6	29.15	124.77	1.74	1.04	0.09	0.02	0.89
7	25.18	42.01	4.06	0.90	1.00	0.08	-0.08
8	26.52	63.86	4.84	1.08	1.12	-0.03	-0.09

Table 2.7: Reconstructed Masterpieces data: in the missing entries, the estimate is accompanied with the original datum.

2.3.2 are an important part of the data analysis techniques presented here in a unified way. The basic kinds of problems treated with these techniques can be listed as follows:

(a) *Compression of the data.* Applied to an $N \times n$ rectangular table, SVD decomposition allows substituting it with $m(N + n)$ numbers where m is the number of components (singular triples) extracted. The compression index, $nN/[m(n + N)]$, may be quite large depending on the sizes involved. If, for instance, $N = 100000$, $n = 100$, and $m = 10$, the ratio is about 10, which indicates the coefficient of decreasing the storage space. Decompression is made based on the bilinear model for the original data.

(b) *Transformation of the data.* There are two features of the principal component (or ordination, or correspondence analysis) factors, that are important for the subsequent analysis: (a) a smaller number of the components (m rather than n); (b) mutual orthogonality of the components, to be employed further, in the regression analysis especially, for obtaining non-biased estimates.

(c) *Visualization*. When presented as points on the screen where the space of the most salient two or three components is depicted, the entities may show a particular kind of structure in the data cloud, and/or a pattern of spatial location of a particular subset of the entities, etc. Joint row/column display is quite helpful, especially in Correspondence analysis (though, there is still a controversy about the extent of theoretical support for such a representation).

(d) *Factor analysis*. The principal component, ordination, and correspondence analysis factors are frequently used as "latent" variables (such as talent or agressivity of individuals), which are computed from the manifested "indirect" data matrix.

2. Mathematical theories of the singular- and eigen-vectors, underlying the methods, belong to the most profound and beautiful part of mathematics. The factor variables can be found quite efficiently, while the data scatter can be decomposed into the sum of their contributions plus the unexplained part which can be decreased as much as necessary. On the other hand, the solutions heavily depend on the preliminary data transformation, both the origin shifting and scale factoring, which makes it an issue to develop an understanding of what kind of data standardization should be employed. Luckily, the aggregable (contingency) data low-rank approximation may be considered as that one where standardizing is completely determined by the choice of the data scatter measure (Pearson's chi-squared contingency coefficient).

3. Mathematical structure of the low-rank approximation problems allows for sequential solution of them (though they are "parallel" optimization problems). Sequential strategy of the singular-value and eigen-value decomposition methods is picked out as a particular SEFIT procedure which is potentially applicable to many other approximation problems. The procedure is applied to an important problem in data analysis: filling in the missing entries, as an extension of the principal component analysis.

Chapter 3

Clustering Algorithms: a Review

FEATURES

• A review of clustering concepts and algorithms is provided emphasizing: (a) output cluster structure, (b) input data kind, and (c) criterion.

• A dozen cluster structures is considered including those used in either supervised or unsupervised learning or both.

• The techniques discussed cover such algorithms as nearest neighbor, K-Means (moving centers), agglomerative clustering, conceptual clustering, EM-algorithm, high-density clustering, and back-propagation.

• Interpretation is considered as achieving clustering goals (partly, via presentation of the same data with both extensional and intensional forms of cluster structures).

3.1   A Typology of Clustering Algorithms

3.1.1   Basic Characteristics

To discuss the variety of clustering concepts and algorithms, a classification of them is necessary. Typically, such a classification involves the following three binary oppositions presented by Sneath and Sokal 1973, among others which turned out not to be important in the subsequent development of the discipline.

1. *Hierarchic versus Nonhierarchic Methods.* This is a major distinction involving both the methods and the classification structures designed with them. The hierarchic methods generate clusters as nested structures, in a hierarchical fashion; the clusters of higher levels are aggregations of the clusters of lower levels. Nonhierarchic methods result in a set of unnested clusters. Sometimes, the user, even when he utilizes a hierarchical clustering algorithm, is interested rather in partitioning the set of the entities considered.

2. *Agglomerative versus Divisive Methods.* This refers to the methods of hierarchical clustering according to direction of generating the hierarchy, merging smaller clusters into the larger ones bottom-up (agglomerative) or splitting the larger ones into smaller clusters top-down (divisive). Agglomerative methods have been developed for processing mostly similarity/dissimilarity data while the divisive methods mostly work with attribute-based information, producing attribute-driven subdivisions (conceptual clustering).

3. *Nonoverlapping versus Overlapping Methods.* This item divides the set of nonhierarchic clustering methods into two parts according to the resulting cluster structure: one, partitioning, is well defined while the other, overlapping clusters, still has been not systematized.

These three characteristics usually are presented as the nested classification in Fig.3.1.

To present an updated classification of the methods, we accept the following three general bases:

1. *Kind of Input Data.* This subject has been discussed in Chapters 1 and 2 where we limited ourselves to two-way table data divided into three major categories:

   (a) Column-conditional data;

   (b) Comparable data;

   (c) Aggregable (mostly Contingency) data.

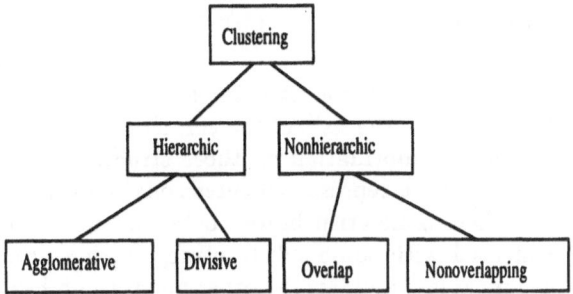

Figure 3.1: Classical taxonomy of clustering methods.

2. *Kind of Output Cluster Structure.* The following major categories of cluster structures employed can be outlined:

   (a) Subset.

   (b) Partition.

   (c) Hierarchy.

   (d) Association Structure.

   (e) Biclustering Structure.

   (f) Nonstandard Clusters.

   (g) Concept.

   (h) Separating Surface.

   (i) Neural Network.

   (j) Probabilistic Distribution.

3. *Kind of Criterion.* There can be external or internal criteria; the internal criteria are of the four kinds:

   (a) Within-Algorithm Criteria (Direct Clustering).

   (b) Optimization.

   (c) Definition.

   (d) Consensus.

In the following two subsections, two latter base sets will be discussed in more detail.

3.1.2   Output Cluster Structure

There is an important distinction between the cluster structures. Some of them [(a) to (f)] are of mostly empirical nature since the lists of the constituent entities of the clusters present the most important information on those structures (extensional approach). The others [(f) to (i)] are represented intensionally by some formulas in mathematical language. There is no crisp border between them: the empirical structures can be supplemented with some substructures (such as the cluster centroid) having a theoretical meaning, and, on the other hand, the "theoretical formulas" may heavily depend on the data set from which they have been derived.

1. **Subset.**

   A subset $S \subseteq I$ is a simple classification structure which can be presented in any of three major forms:

   a) Enumeration, when the subset is given by the list of its elements, $S = \{i_1, i_2, ..., i_m\}$, or, as we usually shall put it, $S = i_1 - i_2 - ... - i_m$;

   b) Boolean vector form: $s = (s_i)$, $i \in I$, where $s_i = 1$ if $i \in I$ and $s_i = 0$ otherwise; it is usually referred to as the *indicator* of S;

   c) Intensional predicate $P(i)$ defined for $i \in I$, which is true if and only if $i \in S$.

   As it was seen in Chapter 1, subsets are a significant part of the classification universe, though, in cluster analysis, this hardly was explicitly acknowledged.

2. **Partition.**

   A set of nonempty subsets $S=\{S_1, ..., S_m\}$ is called a partition if and only if every element $i \in I$ belongs to one and only one of these subsets called classes; that is, S is a partition when $\cup_{t=1}^m S_t = I$, and $S_t \cap S_u = \emptyset$ for $t \neq u$. The partition concept is a basic model for classifications in the sciences and logic related, mainly, to typology, taxonomy and stratification. Also, a partition can be considered as another form of representation of the nominal scale. Every nominal scale variable is defined up to any possible one-to-one recoding of its values, which means that the variable, really, is defined only up to the classes corresponding to its different values.

3. **Hierarchy.**

   A hierarchy is a set $S_H = \{S_h : h \in H\}$ of subsets $S_h \subseteq I$, $h \in H$, called *clusters* and satisfying the following conditions: 1) $I \in S_H$; 2) for any $S_1, S_2 \in S_H$, either they are nonoverlapping ($S_1 \cap S_2 = \emptyset$) or one of them includes the other ($S_1 \subseteq S_2$ or $S_2 \subseteq S_1$), all of which can be expressed as $S_1 \cap S_2 \in \{\emptyset, S_1, S_2\}$. Throughout this book, yet one more condition will be assumed:

(3) for each $i \in I$, the corresponding singleton is a cluster, $\{i\} \in S_H$. This latter condition guarantees that any non-terminal cluster is the union of the clusters it contains. Such a hierarchy can be represented graphically by a rooted tree: its nodes correspond to the clusters (the root, to I itself), and its leaves (terminal or pendant nodes), to the minimal clusters of the hierarchy, which is reflected in the corresponding labeling of the leaves. Since this picture very much resembles that of a genealogy tree, the immediate subordinates of a cluster are called frequently its children while the cluster itself is referred to as their parent.

This concept corresponds to the genuine Aristotelean notion of classification. Moreover, the hierarchy corresponds to the basic way in which the human mind handles all kinds of complex natural or societal or technical phenomena.

4. **Association Structure.**

Association is a generic name which is suggested to refer to all kinds of sets of subsets $S = \{S_t\}$, $t \in T$, for which a supplementary relation (graph) $\kappa \subset T \times T$ is given to represent "close" association between corresponding subsets S_t, S_u when $(t, u) \in \kappa$. The following particular associations are considered in this book:

1) threshold graph Γ_π defined as a structure on the set of all singleton clusters $\{i\}$, $i \in I$, based on a similarity matrix $A = (a_{ij})$ and a threshold value π, $\Gamma_\pi = \{(\{i\}, \{j\}) : a_{ij} > \pi\}$; this can represent a rude picture of interrelation among the entities;

2) structured partition (block model) (S, κ) where S is a partition of the set of the entities and κ is a structure of associations between its classes. There are two major interpretations of such a structure: (a) a set of qualitative categories with a complicated structure of association among them (for instance, there is a partial, not linear, order between possible answers to a preference question: "like", "indifferent", "do not like", "difficult to say"); (b) a picture of interrelation between subsystems of an industrial or biological or social system such as interaction between small groups of humans/animals or between subunits of an enterprise.

3) ordered partition which is a particular kind of a structured partition having an order structure. The concept relates to ordered or stratified classifications. It is associated also with the so-called order (or rank) variables, like "degree of preference", whose values are considered as defined up to any monotone transformation.

5. **Biclustering Structure.**

This concept is defined for two-mode data only, meaning that there are interconnected cluster structures on both rows and columns as represented by their index sets I and J, respectively. The following biclustering structures involve single subsets, partitions, and hierarchies:

1) box (V, W) is a pair of associated subsets $V \subseteq I$ and $W \subseteq J$ whose elements are "highly connected";

2) bipartition is a pair of partitions having the same number of classes, (S, T), with S being defined on the row set I while T on the column set J, along with a one-to-one correspondence κ between the highly connected classes $(t, u(t)) \in \kappa$;

3) bihierarchy is a pair of hierarchies, (S_F, T_H), S_F being defined on the row set I while T_H on the column set J, along with a one-to-one correspondence κ between some of their classes $(S_f, T_{h(f)}) \in \kappa$.

6. Nonstandard Structures.

There have been some empirical structures considered in the literature, that will be touched, though not covered extensively, in this monograph. Let us mention some of them.

1) Fuzzy clusters and partitions.

Any fuzzy cluster is represented by its membership function $z = (z_i)$, $i \in I$, where z_i $(0 \leq z_i \leq 1)$ is the degree of membership of the entity i in the cluster. In contrast to the usual, hard (crisp) clusters, where membership degree z_i may be only 1 or 0, fuzzy membership can be any quantity between 0 and 1.

Fuzzy partition is represented by a matrix $Z = (z_{it})$ having the membership functions of clusters t as its columns. The total degree of membership for any entity is 1, $\sum_t z_{it} = 1$, to allow transferability of membership among the clusters.

This concept, especially when supplemented with the standard points (centroids), seems close to the concept of typology where it is not uncommon to have the entities fitting into different clusters with various membership degrees.

2) Extended hierarchies.

There are few examples.

An *additive tree* is a hierarchy along with a set of weights assigned to its edges (joining parents with children) which is employed in molecular evolutionary trees to reflect the differences in the number of inconsistencies between the parents and their different "children".

A *pyramid* is a hierarchy where the clusters are permitted to be overlapping, although in a restricted manner, along an order of the entities, being thus intervals of the ordered entity sequence. Such an ordering may reflect various additional parameters as, for instance, chronology of the burial sites in archaeology.

A *stratified clustering* is a nested set of threshold graphs, with the actual clusters represented as cliques of these graphs; such a structure is based on the concept of an equivalence relation in the hierarchy being substituted by the more general concept of the threshold graph.

A *weak hierarchy* is a set of overlapping clusters S_H in which the nestedness condition is substituted by the following more general condition: for every three distinct clusters, $S_1, S_2, S_3 \in S_H$, their intersection, if not empty, coincides with intersection of some two of them.

3) Standard point typology.

Such a typology is represented just with a set of standard points in the variable space; the points are to be used as the patterns to compare with. Given such a set, any other vector can be compared with some (or all) of them to be related to that one which represents the region containing the vector. Probably, Kohonen 1989 was the first to consider such a form of classification as a classification structure (calling the points *reference* or *codebook* vectors to be learned).

4) Overlapping clusters.

In many applications, especially in those connected with semantic relations, the structure of overlapping seems completely unpredictable, which makes considering arbitrarily overlapping cluster sets meaningful.

On the other hand, the concept of faceted classification (as well as some of the typology structures discussed) requires getting such a set of clusters, which contains all the intersections occurring as proper clusters.

7. **Concept.**

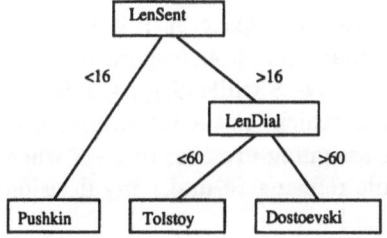

Figure 3.2: A concept tree for Masterpieces.

The concept (or conceptual cluster, or classification tree, or decision tree), as a notion in clustering, is a hierarchical tree as above, with the particular feature that any of its clusters is presented as described intensionally. Since the children of a cluster form its partition, such an intensional description is usually made with some nominal or categorized variables; for any cluster, its children correspond to the categories of such a variable as shown in Fig.3.2 for the Masterpiece data. We can see, with this example, that any conceptual cluster relates to an aspect of the world as intensionally structured (in a hierarchical fashion). In contrast to the extensional set hierarchy, the conceptual tree gives a directly interpretable pattern and may not depend too much on empirical data. On the other hand, the extensional hierarchy can be associated with a wider set of data; it can be designed based on a similarity matrix which may involve too many variables to be reflected in a conceptual cluster, or not involve these variables at all, being primary data (as mobility or industrial flows).

It is convenient to supplement the notion of hierarchical concept with the other kinds of cluster structures considered above — first of all, the subsets and partitions.

An intensionally described subset, in the simplest case, corresponds to an interval of a quantitative variable such as the "senior citizen" category, comprising all the people older than 60 years. More complex kinds of conceptual subsets correspond to logical operations over unidimensional intervals. For instance, a conjunctive concept (such as "young female college graduate") relates to the conjunction of several unidimensional ones ("young" [age], "female" [sex] and "college graduate"[education]); the other logical operations (such as disjunction or implication) could be involved also. The conjunctive concepts are most popular; they have a simple geometrical interpretation as the Cartesian product of corresponding unidimensional intervals represented by a (hyper)rectangle in the variable space.

As to the intensional partition, it still has not received any special attention, perhaps, because it can be represented with a conceptual cluster hierarchy in an adequate manner.

8. **Discriminant Function (Separating Surface).**

This is an intensional construction in the variable space R^n: a function $G(x)$, $x \in R^n$, is referred to as a discriminant function (separating surface) for a subset $S \subset I$ if $G(y_i) \geq \pi > 0$ for all $i \in S$ while $G(y_i) \leq \pi$ for all $i \in I - S$. Sometimes the surface is considered "thick" in the following sense: function $G(x)$ gives a threefold decision rule assigning i to S or to $I - S$ when $G(y_i) \geq \pi$ or $G(y_i) \leq -\rho$, respectively, while refusing to make any decision when $-\rho \leq G(y_i) \leq \pi$ (for some positive ρ and π).

Usually, the discriminant function is a hyperplane $G(x) = \sum_k c_k x_k$. Linear functions can separate only convex sets, which relates the theory of discriminant hyperplanes to the theory of convex sets and functions. On the other hand, it is connected with some important clustering methods such as K-Means.

The theory of discriminant functions, developed by R. Fisher, is part of the mathematical multivariate statistics. The theory of use of the discriminant surfaces in clustering was initially developed, mainly, in the framework of pattern recognition; currently, it is being shifted smoothly into the area of neural networks.

9. **Neural Network.** A formal neuron is a model for the neuron, a nerve cell working like an information-transforming unit. The neuron provides a transformation of an input vector x into the output signal $y = \theta(\sum_k c_k x_k - \pi)$ where $\theta(v) = 1$ if $v > 0$ and $\theta(v) = 0$ if $v \leq 0$. Actually, the neuron discriminates between two half-spaces separated by the hyperplane $\sum_k c_k x_k = 0$. Frequently, the threshold output function θ is substituted by the so-called *sigmoid* function $\theta(v) = 1/(1 + e^{-v})$ which is analogous to the threshold function but is smooth and more suitable for mathematical derivations. The interpretation of the formal neuron is straightforward: the components k represent synapses excited on the level x_k; the weight c_k shows relative importance of the synapse to the neuron; the neuron fires output if the total charge $\sum_k c_k x_k$ is higher than the neuron threshold π.

Sometimes the neuron is considered to have the identity output function $\theta(v) = v$ thus performing just linear transformation $\sum_k c_k x_k$; this is called *linear neuron.*

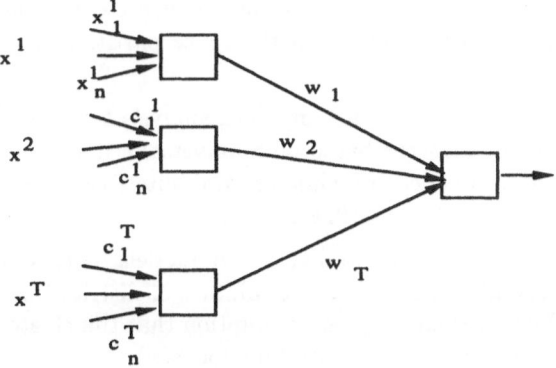

Figure 3.3: A neural network with a hidden layer.

A single hidden layer neural net (see Fig.3.3) defines a transformation $\theta_1[\sum_{t=1}^{T} w_t \theta(\sum_k c_k^t x_k^t - \pi_t)]$ of the input vectors x^t, $t = 1, ...T$, into an output through T "hidden" neurons. Such a net has two important properties: 1) it can be used to approximate any continuous function (having $\theta_1(v) = v$) (Cybenko 1989); 2) it can be used to separate any subset S of a given set of normed vectors y_i, $i \in I$. To resolve the latter problem, let us take $T = |I|$ to consider any $t = 1, ..., T$ as a corresponding element of I; then, for any neuron $t \in I$, let its weight vector $c^t = y_t$. Obviously, the maximum of the scalar products (c_t, y_i) with regard to $i \in I$, in this case, is reached for the only $i = t$. Thus, fixing π_t between this maximum value and the second maximum, we have $\phi(t, y_i) = \theta(\sum_k c_k^t y_k^i - \pi_t) = 1$ if and only if $i = t$; $\phi(t, y_i) = 0$ when $i \neq t$. Then, taking $w_t = 1$ for $t \in S$ and $w_t = -1$ for $t \in I - S$, we get the desired output.

A similar construction can be made to learn an arbitrary partition; this time, a third layer is necessary to distinguish among the different classes.

10. **Probabilistic Distribution.**

In the probabilistic paradigm, we can distinguish between two approaches to clustering which could be referred to as: 1) Probabilistic environment and 2) Probabilistic clusters. Let us consider them in turn.

1. *Probabilistic environment* is a term to be used when a reformulation of a data analysis technique in terms of a probabilistic space has been done for investigating the properties of the technique as based on a random sample, in its relation to the "theoretic" solution. For example, the minimum square-error clustering problem was considered by Shlezinger 1965 in the following setting. The entities are a random sample from an unknown distribution $p(x)$ in the Euclidean space. The problem is to find a hyperplane $G(x) = 0$ minimizing the weighted variance of $p(x)$ with regard to the two subspaces separated by the hyperplane. The clustering structure here is a partition of the space (not just of a sample). The problem is to suggest such a clustering method for the samples, which is consistent with the space partition, in a strictly defined sense.

This kind of modeling can be accomplished for any data analysis technique. Recently, it received an appealing impetus based on a universal criterion for deriving summaries of data, the so-called principle of Minimum Description Complexity (Rissanen 1989, see also Briant 1994).

2. *Probabilistic Clusters* are a set of probability based models of cluster structures. Two major models developed so far are mixture of distributions and high density clusters. Both of them rely on assumption that the cluster structure can be represented in terms of a density function $p(x)$.

The mixture of distributions model assumes that the density function has the form $p(x) = \sum_{t=1}^{m} \pi_t f(x, a_t)$ where $f(x, a)$ is a cluster model, a family of density functions over x defined up to the parameter vector a. Usually, the density $f(x, a)$ is considered unimodal (the mode corresponding to a cluster standard point), such as the normal density function defined by its mean vector μ and covariance matrix Σ; shape of the clusters depends upon the properties of Σ (see Section 3.2.7). Acceptance of such a theoretical model must be based on preliminary knowledge about the universe classified. The mixture of distributions concept can be considered for qualitative variables as well (Celeux and Govaert 1991).

The other model, high density clusters, requires not so much prior knowledge. Let us consider a level set $P_c = \{x : p(x) \geq c\}$, for some $c > 0$. Any maximal connected subset $S_c \subseteq P_c$ is called high density cluster (c-cluster). Obviously, high density clusters form a hierarchy, being nested for different c.

To define a partition-wise concept, let us consider the following notion: a high-density cluster S_c is referred to as a *unimodal* cluster if, for every $c' > c$, there exists no more than one c'-cluster included in S_c (Kovalenko 1993). For sufficiently large $c_1, .., c_m$, let $S_{c_1}, ..., S_{c_m}$ are mutually nonoverlapping high density clusters such that $p(x) \leq \min_t c_t$ for all x outside these clusters. Then, along with the rest of the space, they form a partition consisting of m high density clusters and a "swamp", or "ground" cluster.

Various constraints on the final result (such as the site-to-site adjacency structure on a geographic map) should supplement the list of admitted cluster structures. However, this topic will not be discussed in this work.

3.1.3 Criteria

Clustering criteria can be 1) internal, based only on the data involved in clustering, or 2) external, defined in terms of the variables which are not involved in clustering directly. If a set of respondents is classified by the demographic variables (sex, marital status, etc.) to have the clusters homogeneous with regard to, for instance, the respondents' traveling habits, this would be clustering with the external goal of traveling-habits homogeneity. Yet, such a homogeneity could be observed in the clusters created internally, based upon the demography only. In this latter case, the homogeneity (obtained as a by-product, not as a goal to achieve) will be a regularity revealed via clustering. Although the external-goal-based classification methods seem a subject of a great interest, there is little to say on that, currently. Weighting of the variables involved in clustering with regard to the external goal seems to be the only idea elaborated in literature (something on this is presented in Chapter 6).

Internal clustering criteria could be put in the following four major categories:

1. Within-Algorithm Criterion (Direct Clustering).

2. Optimization.

3. Definition.

4. Consensus Approach.

Let us discuss them in turn.

Within-Algorithm Criterion (Direct Clustering)

Clustering usually involves iterative recalculations of the clusters to have them rearranged in more-and-more "cluster-like" fashion. The recalculations can be based upon a set of natural optimality criteria applied to between-entity distances, such as the rule "put the entity in the nearest class". The ad hoc structure of such an algorithm reflects the clustering goal without any explicit criterion, while the formal criteria are used within its particular iterative steps.

Although, in a theoretical discipline, no direct heuristics would be appropriate, still so many structures to be revealed are uncovered by the clustering theories available (as, for instance, geometrical bodies of an elongated shape) that the availability of the direct clustering algorithms provides a possibility to reveal and deal with strange structures in exploratory data analysis.

Optimization

There are several sources for emerging optimization criteria in clustering:

1) heuristics:

The most explicit clustering criteria involve some or all of the following requirements: 1) cohesion ("the entity-to-entity distances within clusters shall be small"), 2) isolation ("the clusters shall be spaced apart"), 3) uniformity ("other things being equal, the cardinalities of the clusters must not differ from each other significantly"). Each of these requirements can be formalized in many ways. For instance, to obtain a cohesive cluster, we can minimize: the sum, the maximum, the average, the median of the within-cluster distances. Then, combining the heuristic criteria defined for the particular goals, an heuristic clustering criterion can be produced.

2) probabilistic modeling:

Having assumed a particular probabilistic model for clustering, the maximum likelihood approach may lead to a set of working criteria emerging under different hypotheses about the parameters of the model. If, for example, the population of interest consists of m different subpopulations represented by their multivariate normal density functions $f(x, \mu_w, \Sigma_w)$ (μ_w and Σ_w are the mean vector and covariance matrix, respectively) and $x_1, ..., x_N$ is a random sample with unknown assignment of the observations x_i to the clusters (subpopulations) $S = \{S_w\}$, then the likelihood function has the form

$$L(\mu, \Sigma, S) = C \prod_{w=1}^{m} \prod_{i \in S_w} |\Sigma_w|^{-1/2} exp\{-(x_i - \mu_w)^T \Sigma_w^{-1}(x_i - \mu_w)/2\}$$

which becomes

$$l(\mu, \Sigma, S) = D - \frac{1}{2} \sum_{w=1}^{m} \{tr(R_w \Sigma_w^{-1}) + n_w log|\Sigma_w|\}$$

after the maximum likelihood estimator $\bar{x}_w = \sum_{i \in S_w} x_i/n_w$ of μ_w is put in the formula. R_w here is the cross-product matrix defined as $R_w = \sum_{i \in S_w}(x_i - \bar{x}_w)(x_i - \bar{x}_w)^T$.

Let us consider different hypotheses on cluster shapes reflected in the properties of Σ_w: (1) clusters are spheres of the same sizes, $\Sigma_w = \sigma^2 I$ ($w = 1, ..., m$); (2) clusters are ellipsoids of the same size and of the same orientation, $\Sigma_w = \Sigma$ ($w = 1, ..., m$); (3) clusters are spheres of different sizes, $\Sigma_w = \sigma_w^2 I$; (4) clusters are ellipsoids of possibly different sizes and orientations, no constraints on Σ_w ($w = 1, ..., m$). Respective criteria that are equivalent to the maximum likelihood function are these:

1. $\sum_{w=1}^{m} Tr(R_w)$;

2. $\sum_{w=1}^{m} |R_w|$;

3. $\sum_{w=1}^{m} n_w \log Tr(R_w/n_w)$;

4. $\sum_{w=1}^{m} n_w \log |R_w/n_w|$,

to be minimized by the sought partition S (see Aivazian et al. 1989, Banfield and Raftery 1993 for these and other criteria and for references).

Let us discuss the first of the criteria in more detail. Obviously, $Tr(R_w) = \sum_{i \in S_w} \sum_{k=1}^{n}(x_{ik} - \bar{x}_{wk})^2$. This is why the criterion is usually referred to as the within group sum of squared errors (WGSS) or simply the *square-error clustering* criterion. On the other hand, it is equal to $n_w s_w^2$ where $s_w^2 = \sum_{k=1}^{n} \sum_{i \in S_w}(x_{ik} - \bar{x}_{wk})^2/n_w$ is the total variance of the variables k ($k = 1, ..., n$) in cluster S_w. Thus, the WGSS criterion can be expressed either as the sum of the errors squared or as the sum of the within cluster variances weighted,

$$D(S) = \sum_{w=1}^{m} \sum_{i \in S_w} \sum_{k=1}^{n}(x_{ik} - \bar{x}_{wk})^2 = \sum_{w=1}^{m} n_w s_w^2 \qquad (3.1)$$

The criterion has been used in many important developments in clustering (see Sections 4.4, 6.1, 6.2 and 6.3). It much resembles the variance criterion employed in statistical discipline of analysis of the variance (ANOVA) (see Ward 1963, Edwards and Cavalli-Sforza 1965, Shlesinger 1965, McQueen 1967). However, in cluster analysis, the classes are to be found, while they are pre-given in ANOVA. Criterion (3.) (Banfield and Raftery 1993) is much like WGSS (though logarithms of the variances are taken in (3.) rather than the variances themselves), but, to the author's knowledge, it has been never tried in real-world clustering problems.

3) data approximation:

This is also a statistical approach, related to representation of the data in the matrix space. The sought cluster structure is presented in the format of a data matrix. The clustering criterion is to minimize the difference between these two data tables, measured, usually, by the sum of the entry-to-entry differences squared (least-squares approach). This is the approach which will be discussed in all the subsequent Chapters.

We will not consider a more general approximation approach also based on presenting both the data and the cluster structure in the same format, which may be different from the original data format (Gifi 1990, Aven, Muchnik and Oslon 1988).

4) substantive classification problems:

Every particular engineering problem involving classification leads to a particular criterion for classification, which can be utilized also in general clustering context.

Let us consider the problem of stratified sample design as an example of such an engineering problem. As it is well known, measuring any parameter of a population such as the "average income" or "public opinion on an issue", can be made using a relatively small, the so-called stratified, sample based on a preliminary partitioning of the population. A known Dalenius' model for stratified sampling design can be presented as the following equation: $x = af + e$ where x is a variable, the mean of which is to be estimated (a usual goal of sampling), f is a variable (factor) with known distribution to be used for stratifying, and e is a random error with mean zero; the conditional variance of e with respect to f is assumed constant. In practice, factor f is presented empirically with a set of the variables; stratifying such a factor is equivalent to partitioning the population to minimize the variance of the estimate. Varying the sampling assumptions, different clustering criteria are obtained as being equivalent to the engineering criterion of minimizing the variance. For instance, when the sample is of a fixed size and sampling is proportional, the criterion is exactly minimization of WGSS, (3.1), which emerged in the context of probabilistic modeling. If, in contrast, the sampling scheme takes only one respondent from every strata designed, then the clustering criterion is minimizing $\sum_t (p_t^2(s_t^2 + 1))$ (for $a = 1$) (Braverman et al. 1975).

Definition

In contrast to the approaches above based on the assumption that the sought classification structure cannot exactly fit into the "imperfect" original data, this approach involves concepts defined to fit perfectly into any feasible data. For instance, for a given dissimilarity matrix $d = (d_{ij})$, $i, j \in I$, a subset $S \subset I$ is called strong cluster if $d_{ij} < d_{ik}$ for every $i, j \in S$ and $k \notin S$ (Apresian 1966, Diatta and Fichet 1994). Having such a definition, traditional problems arise: characterizing the strong clusters and finding them. Fortunately, this case is rather simple. The set of all strong clusters is a hierarchy since if strong clusters S and T are overlapping, one of them includes the other. Indeed, if $S \cap T \neq \emptyset$ and $S \not\subset T$ and $T \not\subset S$, then there exist $i \in S \cap T$ and $j \in S - T$ and $k \in T - S$ such that $d_{ij} < d_{ik}$ (since $i, j \in S$ and $k \notin S$) and $d_{ik} < d_{ij}$ (since $i, k \in T$ and $j \notin T$), which is impossible. Finding strong clusters is also simple. The problem with this definition is that, typically, every strong cluster is rather small comprising very few entities and, thus, provides no aggregation to the data.

In the sequel, some more cluster definitions will be considered, leading to more intriguing cluster structures.

Consensus

In this approach, a classification method is considered as a mapping $F : D \to C$ where D is set of all feasible data and C set of all classification structures of a given kind. For instance, in the problem of partitioning of set I based on information provided by a set of nominal variables on I, C can be considered as set of all partitions on I, and D as the set of all n-tuples $\{R^1, R^2, ..., R^n\}$ (n may be unfixed) of the partitions R^l on I. Or, D may be a set of square entity-to-entity dissimilarity matrices while C is the set of all set-hierarchies on the set of the entities. Mapping F transforms the initial data into a final classification structure. Mapping F is referred to as a "consensus" function if it satisfies some natural properties (usually called "axioms"). For instance, such a method should map any n-tuple having all its partitions coinciding into this coinciding partition. Thus, in the consensus approach, one considers a set of some "natural" requirements to F and tries to find out an explicit form of the corresponding consensus function(s).

Such an approach takes its origin in the theory of social choice initiated by work of K. Arrow 1951 and continued, in the clustering framework, by Jardine, Jardine and Sibson 1971 (see also Mirkin 1975, Margush and McMorris 1981, Barthélemy, Leclerc, and Monjardet 1986, and Section 7.2).

There is also a particular approach, index-driven consensus, which should be shared with the optimization approach. Let, for example, $\mu(S, R)$ be an index of similarity between partitions of I. Then, S will be an index-driven consensus partition if it maximizes $\sum_{l=1}^{n} \mu(S, R_l)$ (see Section 5.3.4).

3.1.4 Algorithmic Aspects of Optimization

Let us discuss the algorithmic aspect of optimization approach for its importance in mathematical clustering. The optimization problems arising may have a quite general character involving either continuous or discrete or even mixed variables. Although some criteria may lead to quite nice and simple solutions, the mainstream problems are related to the least-squares criteria and their extensions that lead, in general, to quite difficult problems. Combinatorial clustering problems belong to the core problems of combinatorial optimization (see, for instance, Section 4.2) and thus can be treated with all the tools of combinatorial optimization available. Some general working optimization techniques such as genetic algorithms and simulated annealing (Laarhoven and Aarts 1987; see also a somewhat simpler approach by Charon and Hudry 1993), are quite applicable to the clustering problems and are used quite extensively. We do not discuss general global or approximation algorithms, but refer mostly to algorithms based on local iterative optimizing steps (see a review in de Leeuw 1994). In particular, two major optimization techniques will be employed through:

(1) local search algorithms, and

(2) alternating optimization.

To discuss these techniques, let the problem be to maximize $f(x)$ over $x \in D$ where D is a set of admissible solutions. To define a local search algorithm, a neighborhood system in D must be supposed. The neighborhood system is a mapping $N(x)$ assigning a subset $N(x) \subset D$ to any $x \in D$, such that $x \in N(x)$. The subset $N(x)$ consists of the "neighbors" of x, and $f(x)$ is supposed to be easily maximized in every $N(x)$.

Local search algorithm
It starts from an $x_0 \in D$ referred to, usually, as the *initial setting* (it can be, for example, an initial partition of the entity set). Then, let x_1 be a maximizer of $f(x)$ in $N(x_0)$. If $x_1 \neq x_0$, find x_2 which is a maximizer of $f(x)$ in $N(x_1)$. Carrying on the computation, a sequence $x_0, x_1, ..., x_t, ...$ is obtained where x_t is a maximizer of $f(x)$ in $N(x_{t-1})$. The process stops when x_t is equal or close enough to x_{t-1}.

Obviously, $f(x_0) \leq f(x_1) \leq ... \leq f(x_t)....$ When the algorithm of maximizing $f(x)$ in $N(x)$ is arranged in such a way that it takes x as a maximizer in $N(x)$ if $f(x) \geq f(y)$ for all $y \in N(x)$, then, obviously, the algorithm satisfies the stopping condition, $x_t = x_{t-1}$, in that point x_t where $f(x_t) = f(x_{t-1})$. This implies that in the inequalities above only strict "less" $<$ holds, which proves that the algorithm converges (to a local extremum).

As an important example of a local search algorithm, let us consider the agglomerative optimization technique suggested by Ward 1963. Set D of admissible solutions here is the set of all partitions on I. For every partition, $S = \{S_1, ..., S_m\}$, its neighborhood, $N(S)$, consists of partitions $S(t, u)$ obtained from S by merging clusters S_t and S_u into $S_{tu} = S_t \cup S_u$, $t, u = 1, ...m$.

Agglomerative optimization

The computation starts with an N-class partition consisting of singletons. At each local search step, merging is made to optimize the increase of the clustering criterion, $\delta(u, t) = f(S(u, t)) - f(S)$. The computation ends when the number of clusters becomes equal to a prior fixed number, m_0 (usually, $m_0 = 1$).

In the case when square-error criterion (3.1) is minimized, it is quite easy to prove that every merging can only increase its value. Thus, the problem, at every agglomeration step, is to minimize increase $\delta(u, t) = D(S(u, t)) - D(S)$. Let us denote c_t the center of gravity of cluster S_t, $c_t = \sum_{i \in S_t} y_i / |S_t|$. It appears,

$$\delta(u, t) = D(S(u, t)) - D(S) = \frac{|S_t||S_u|}{|S_t| + |S_u|} d^2(c_t, c_u) \tag{3.2}$$

where d^2 is Euclidean distance squared, which is proved with elementary arithmetic just from the definition of criterion D (3.1).

With this formula, computations become quite fast; the procedure is known as Ward's method. Gower 1967 provided an example demonstrating a peculiarity of the criterion reflecting the fact that factor $|S_t||S_u|/(|S_t| + |S_u|)$ in (3.2) favors equal distribution of the entities among the clusters and, thus, the criterion may fail to separate immediately some outliers. Though for a long time treated as a shortcoming (see, for instance, Sokal and Sneath 1973), the peculiarity does not appear to actually be so: in many clustering studies, tendency of the cluster cardinalities to the same number has been claimed a criterion of clustering (see, for example, Braverman and Muchnik 1983, Mirkin 1985).

An alternating optimization algorithm can be considered as a specific case of the local search techniques developed for the case when there are two (or more) groups of the variables, that is, $x = (y, z)$, and $f(y, z)$ is easy to minimize by any of y or z when the other part of x is fixed.

Alternating minimization

Computations at iteration t: for every z_t fixed, y_{t+1} is found as a minimizer of $f(y, z_t)$. Then z_{t+1} is determined as a minimizer of $f(y_{t+1}, z)$, after which new iteration, $t + 1$, is carried out.

At each iteration, the value of $f(x)$ may decrease only, which guarantees that the algorithm converges when f is curved enough (as the least-squares criteria are).

Let us consider the partitioning problem with the square-error clustering criterion, (3.1), taken in the form

$$D(c, S) = \sum_{t=1}^{n} \sum_{i \in S_t} \sum_{k=1}^{n} (x_{ik} - c_{tk})^2 \qquad (3.3)$$

so that there are two groups of variables: standard points, c_t, and cluster membership lists, S_t. In this case, the alternating minimization algorithm can be reformulated as a clustering method, as follows.

Alternating Square-Error Clustering
Starting with a list of tentative centers c_t, the following two steps are iterated until the partition is stabilized:
Step 1. Cluster membership. Having c_t fixed, find clusters S_t minimizing $\sum_{t=1}^{n} \sum_{i \in S_t} d^2(x_i, c_t)$ where $d^2(x_i, c_t)$ is Euclidean distance squared, as in (3.3). To do this S_t must consist of all those entities i that are closer to c_t than to any other c_u, $u \neq t$.
Step 2. Standard points. Having clusters S_t fixed, find standard points c_t minimizing (3.3), that is, equal to the gravity centers of S_t, $t = 1, ..., m$.

These kinds of algorithms are quite reasonable in clustering because:

(1) typically, the local algorithm may be considered as a model of the construction of a classification by a human (for instance, agglomerative optimization may be thought of as a formalization of a systematization process, while alternating square-error clustering could be a model for typology making);

(2) in some cases, it can be proved that the local solution found still satisfies some requirements for "cohesive" clustering;

(3) the computations are easy, fast and memory-efficient.

There is another local optimization technique, employed quite frequently, though not in this monograph: hill-climbing. The hill-climbing algorithm is defined with a so-called gradient $g(x)$ of $f(x)$. The gradient shows the direction where $f(x)$ grows maximally, which allows us to define the following procedure.

Hill-climbing
It starts from an arbitrary $x_0 \in D$. Then, the following recurrent equation, $x_{t+1} = \alpha g(x_t) + (1 - \alpha)x_t$, is employed for computing a sequence of solutions approaching a maximizer of $f(x)$.

A small positive value $\alpha > 0$ is proportional to the length of a step from the previous location, x_t, in the gradient direction, to get a new position, x_{t+1}. This is why the procedure is called hill-climbing. When $f(x)$ is curved enough (as, say, the least squares criterion), value α may be invariant along all the procedure; in the other cases, it is required to be decreasing, though slowly, as $1/t$ (see, for example, Polak 1983). The hill-climbing technique is quite important in many subdisciplines dealing with continuous optimization, such as neural networks and discriminant analysis.

3.1.5 Input/Output Classes

The variety of clustering methods developed, with regard to the data and cluster structure types, is presented in Table 3.1.

Cluster Structure	Kind of Table		
	Col.-Cond.	Comparable	Aggregable
Subset	+	+	+
Partition	++	++	+
Hierarchy	+	++	+
Bicluster	+	+	+
Association	+	+	
Concept	+	na	
Discriminant	+	na	
Neural Net	+	na	
Distribution	++		

Table 3.1: Basic output/input table for clustering algorithms; the symbols mean: there are some algorithms (+) or many of them (++), the author does not know about anything (empty place) or doubts that the algorithms are ever possible (na).

The follow-up review is based on this Table.

3.1.6 Discussion

1. It should be underscored quite clearly that clustering does not exhaust the entire subject of mathematical classification. Clustering is devoted to revealing classification structures in the data. Still there are some other problems concerning classifications. How does a classification structure emerge and evolve in a real-world situation? What are the classification functions and

how they work? These kind of questions seem quite important, though they are never asked in such a general setting; they are left untouched here, too.

2. Our classification of clustering approaches involves most general attributes related, actually, to any algorithmical development: (1) the input data, 2) output data, and (3) the criteria for data processing.

3. Among the wealth of data that can be acquired somehow – pictures, texts, sounds, etc. – we restrict ourselves to the table data only, thus excluding all the interesting but more specific subjects concerning image processing, medicine diagnostics, chemical analyzing, biomolecular evolution, etc. Moreover, among tables, we consider mainly two-way (matrix) tables putting them in the three major categories: column-conditional, comparable and aggregable data. Entity-to-variable, distance and contingency data are the respective prototypical examples of these three categories.

4. Usually, just a few kinds of classification structures are included in clustering: partitions, hierarchies, individual clusters and the like, but neural networks and potential functions have been considered so far much beyond the subject. A dozen classification structures enlisted is far from being exhaustive, though. It seems quite evident that the list will be extended with more detailed structures as well as with mixed ones; completely new kinds of classification structures must come eventually, too.

5. Mathematical thinking in clustering must provide a set of theoretical connections between the four kinds of criteria indicated.

3.2   A Survey of Clustering Techniques

The survey will closely follow the clustering structures enlisted (output) while loosening its relation to the input data format and criteria involved.

3.2.1   Single Cluster Separation

To discuss some of the existing concepts and methods in greater detail, let us take data sets Primates and Mobility (Tables 1.5 and 1.7 in Chapter 1) as presented with the following matrices, D and P, respectively:

$$D = \begin{pmatrix} 2 & 1.45 & & & \\ 3 & 1.51 & 1.57 & & \\ 4 & 2.98 & 2.94 & 3.04 & \\ 5 & 7.51 & 7.55 & 7.39 & 7.10 \\ \hline & 1 & 2 & 3 & 4 \end{pmatrix}$$

$$P = \begin{pmatrix} \begin{array}{c|ccccc} 1 & 1,414 & 521 & 302 & 643 & 40 \\ 2 & 724 & 524 & 254 & 703 & 48 \\ 3 & 798 & 648 & 856 & 1,676 & 108 \\ 4 & 756 & 914 & 771 & 3,325 & 237 \\ 5 & 409 & 357 & 441 & 1,611 & 1,832 \\ \hline & 1 & 2 & 3 & 4 & 5 \end{array} \end{pmatrix}$$

The simplest single cluster concepts have been formulated in terms of the threshold graph $G_\pi = \{(i,j) : d_{ij} \leq \pi\}$, the vertices of which correspond to the entities, and edges to those of the distances that are smaller than the threshold π chosen somehow.

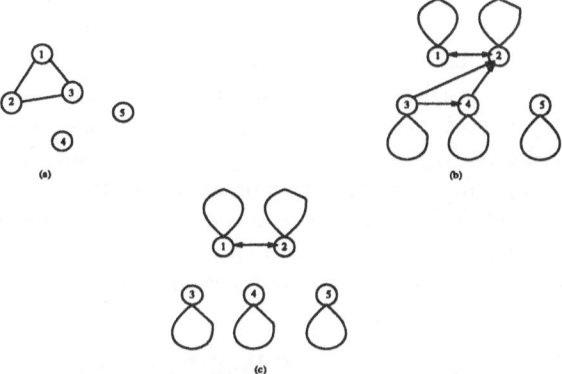

Figure 3.4: Primates and Mobility threshold graphs

Such a graph G_π for D with $\pi = 2.0$ is presented in Fig.3.4 (a). It can be seen that it consists of three connected components; and the non-singleton one is more than just a component: it is a clique, all its vertices are mutually connected, supporting, on the molecular level, Darwin's hypothesis that Humans originated in Africa.

To analyze the contingency data P, let us standardize it with the formula recommended, $q_{ij} = n_{ij}n/n_{i+}n_{+j} - 1$, where n_{ij} are the entries in P, along with the row and column totals n_{i+}, n_{+j} and the grand total $n = 19,912$. The matrix $Q = (q_{ij})$ is as follows:

$$Q = \begin{pmatrix} 1.35 & 0.20 & -0.22 & -0.45 & -0.88 \\ 0.56 & 0.56 & -0.14 & -0.22 & -0.81 \\ -0.05 & 0.07 & 0.59 & 0.03 & -0.77 \\ -0.39 & 0.02 & -0.03 & 0.39 & -0.65 \\ -0.57 & -0.48 & -0.28 & -0.13 & 2.46 \end{pmatrix}$$

For the threshold level $\pi = 0$, the threshold graph (Fig.3.4 (b)) looks rather complicated (each vertex having a loop), although for $\pi = 0.1$ (Fig.3.4 (c)) it is simple, with the only non-singleton cluster $1-2$ to support the hypothesis that the nonmanual professions relate to the same class while the other three occupation classes are isolated.

To discuss a couple of other single cluster concepts, let us make yet another transformation of the mobility data P to present it as a similarity matrix: for this purpose, let us eliminate the diagonal values while taking half-sums $a_{ij} = (n_{ij} + n_{ji})/2$ as (i, j)-th entries:

$$A = \begin{array}{c|ccccc} & 2 & 3 & 4 & 5 \\ 1 & - & 618 & 550 & 700 & 225 \\ 2 & & - & 451 & 809 & 202 \\ 3 & & & - & 1224 & 275 \\ 4 & & & & - & 924 \end{array}$$

Some of the simplest concepts of clusters, although formulated in terms of distance/similarity, still can be reformulated in terms of threshold graphs. A ball of radius r with center $i \in I$ is the subset $B(i, r)$ consisting of all the entities (i included) whose distance from (similarity to) i is not greater (respectively, smaller) than r. For matrix A, $B(i, r) = \{j : a_{ij} \geq r\} \cup \{i\}$. For example, the set of occupations $\{1, 2, 4\}$ is the ball $B(1, 600)$ since both a_{12} and a_{14} are larger than 600 while each of the other two, a_{13} and a_{15}, are smaller than 600. Obviously, the ball $B(i, r)$ coincides with the set of the vertices which are incident to i (including i itself) in the threshold graph with $\pi = r$.

In distance terms, a *clump* cluster is such a subset S that, for every $i, j \in S$ and $k, l \in I - S$, $d_{ij} < d_{kl}$. Obviously, any clump is a clique which is simultaneously a connected component in a threshold graph G_π where π is taken between $\max_{i,j \in S} d_{ij}$ and $\min_{l,k \notin S} d_{lk}$, as it is in Fig.3.4 (a) for $S = 1 - 2 - 3$.

There are some more elaborate concepts. Subset $S \subset I$ is a strong cluster if for any $i, j \in S$ and $k \notin S$, $d_{ij} < \min(d_{ik}, d_{jk})$. This condition involves local interrelations between the entities and cannot be reformulated in terms of the threshold graph. As noted in Diatta and Fichet 1994, S is a strong cluster if and only if, for every $i, j \in S$, $B(i, d_{ij}) \subseteq S$. For the similarity matrix A, there are three non-trivial strong clusters: $3-4$, $1-2-3-4$ and $5-6$; they form a hierarchy, which is a characteristic feature of the strong clusters, as was shown in p. 123.

A more general concept of a weak cluster $S \subset I$ is defined by the inequality $d_{ij} < \max(d_{ik}, d_{jk})$ for all $i, j \in S$ and $k \notin S$ (Bandelt and Dress 1989). It appears,

S is a weak cluster if and only if, for every $i, j \in S$, S includes intersection of the balls $B(i, d_{ij})$ and $B(j, d_{ij})$ (Diatta and Fichet 1994). This concept will be considered also in Section 7.4.

Subset $3-4-5$ is a weak cluster (by the similarity table A) since $a_{35} = 275$ is greater than $a_{51} = 225$ and $a_{52} = 202$, in spite of the fact that a_{35} is less than each of a_{31} and a_{32}. Analogously, subsets $2-4$ and $1-2-4$ are also weak clusters to be added to the former list of the strong clusters which are, obviously, weak clusters also.

Somewhat more practical concepts of clusters appeared in analysis of the variables as, for instance, the concept of B-cluster (Holzinger and Harman 1941), defined in terms of a similarity (actually, correlation) matrix $A = (a_{ij})$. To specify the concept, let $a(S) = \sum_{i,j \in S} a_{ij}/|S|(|S| - 1)$ and $a(k, S) = \sum_{j \in S} a_{kj}/|S|$ be the average similarity within S and between $k \notin S$ and S, respectively; the ratio $B(S) = \max_{k \notin S} a(k, S)/a(S)$ is referred to as B-coefficient to decide whether any new entity, k, is to be added to S or not, depending on how large $B(S)$ is.

To show how this concept works, let us try it with matrix A above. We start with $S = \{3, 4\}$ since $a_{34} = 1224$ is the maximum similarity. Then we compare $a(1, S) = 625$, $a(2, S) = 630$, and $a(5, S) = 600$, which leads to $B(S) = a(2, S)/a(S) = 630/1224 = 0.501$. Let us check what happens if we add 2 to S. Now, $S = \{2, 3, 4\}$ and $a(S) = 828$; $a(1, S) = 623$ and $a(5, S) = 467$. This leads to $B(S) = 623/828 = 0.75$ and adding 1 to S. With $S = \{1, 2, 3, 4\}$ and $a(S) = 725$, $a(5, S) = 406$ and $B(S) = 0.56$. Thus, the universal cluster consisting of all the entities is a B-cluster when the threshold equals 0.5. When the threshold is fixed at a higher level, 0.6, the process stops in the very beginning: $S = \{3, 4\}$.

With the mean, $a = 598$, subtracted, matrix A has the following form:

	2	3	4	5
1 —	20	−48	102	−373
2	—	−147	211	396
3		—	626	−323
4			—	326

$$A =$$

With threshold 1/2, the process stops immediately at $S = \{3, 4\}$ since $B(S) = 0.051$, in this case.

The latter method, actually, belongs to the set of methods based on entity-to-set linkage functions discussed in more detail in Chapter 4.

We can see that there are many possible definitions of the cluster concept, each yielding to a particular method. One should wish to have such a set of cluster concepts which is ordered somehow in itself and, as well, is compatible with other clustering approaches.

3.2.2   Partitioning

It is claimed that there are several dozen distinguishable algorithms for partitioning (Mandel 1988). We'll take into account the following four methods that seem rather distinct while covering a significant number of more specific algorithms:

1. Moving Centers (K-Means).

2. Exchange.

3. Seriation.

4. Graph Partitioning.

Let us consider them in turn.

Moving centers (K-Means) This is a major clustering technique developed in different countries in many particular versions and programs. Its history can be traced back into the 60's when different versions of the method were developed and partly published as (sequential) K-Means by MacQueen 1967 and ISODATA by Ball and Hall 1967. The principal parameters of the moving-centers method are: (1) a method for computing a centroid point $c(S)$ for every particular subset of the entities $S \subseteq I$, and (2) a metric $d(i, c)$ between entities $i \in I$ and centroid points c. In general, the centroid and entity points might belong to different spaces (Diday et al. 1979, Aivazian et al. 1989). Here are some choices for centroids from the literature:

1. When the entities $i \in I$ are presented as the rows $y_i \in R^n$ of a quantitative entity-to-variable matrix:

    a. the gravity center $y(S) = \sum_{i \in S} y_i / |S|$,
    b. the coefficients of the first principal component of the n variables in S;
    c. the coefficients of the linear regression equation of one of the variables with respect to the others (within S);

2. When the entities $i \in I$ are presented as the rows y_i of a nominal entity-to-variable matrix:

    d. the vector of the modal (most frequent) categories of the variables in S;
    e. the vector of the (relative) frequencies of all the categories;

3. When the data are represented by a dissimilarity matrix:

    f. an entity $i \in I$ minimizing the total dissimilarity $d(i, S) = \sum_{j \in S} d_{ij} / |S|$ or $d(i, S) = \max_{j \in S} d_{ij}$ or $d(i, S) = \min_{j \in S} d_{ij}$ (the average or farthest or nearest neighbor, respectively);

g. an entity $i \in S$ maximizing the total dissimilarity $d(i, I - S)$ with the complement of S, among all the entities from S.

Some of the centroid definitions require that a particular distance $d(i, c)$ be defined as in the items (b) and (c) where the centroid actually represents a hyperplane: the distance of a row-vector from that hyperplane is taken as $d(i, c)$; the other definitions may be employed with the ordinary distances/dissimilarities.

The algorithm starts with choosing a set of m (in many publications, k, which explains the method's title) tentative centroids $c_1, ..., c_m$. Usually, these are taken randomly, though a set of suggestions for more reasonable choice can be made as follows:

(a) *Threshold:* Starting from a (most distant from the others) entity-point as c_1, the entities are observed one-by-one unless a y_i occurs having its distance $d(y_i, c_1) > R_b$ where R_b is a pre-fixed threshold; this y_i is taken as c_2. The observation process then continues, adding a new center each time an entity occurs having its distance from each of the centers already chosen larger than R_b.

(b) *Bi-Threshold with Smoothing:* There are two thresholds pre-fixed: R_b, the least admissible distance between cluster centers, and R_w, the maximum admissible radius of a cluster. Then, the entities are observed sequentially. The first is taken as c_1 with its weight 1. The general step: there are some centers $c_1, ..., c_l$ found along with their weights $w_1, ..., w_l$. Then, for any next entity, y_i, the following options are applied depending on the distance $d(y_i, c(i))$ where $c(i)$ is the closest to y_i among the centers $c_1, ..., c_l$. If $d(y_i, c(i)) < R_w$ then $c(i)$ is recalculated as the center of gravity of $c(i)$ with its weight $w(i)$ and y_i with its weight 1 (smoothing). If $d(y_i, c(i)) > R_b$, a new center is defined $c_{l+1} = y_i$ as having weight $w_{l+1} = 1$. Else, a next entity is considered. This process gets a somewhat more meaningful set of centers than the previous one.

(c) *Anti-Cluster Seriation:* Using a linkage function $f(i, S)$ ($S \subset I$, $i \in I - S$) (see Section 4.2) where S denotes a set of the tentative centers selected at a general step, and starting with two entities farthest from each other as centers, apply the seriation procedure to obtain a pre-fixed number m of centers.

(d) *Expert:* Based on prior knowledge of the problem, take some variable space points expressing the expert opinion on the typical combinations of the variable values as the tentative centers. Since the expert-given centers might have no observed entity-points around, this option can lead to a nice combination of the theory with reality, suggesting possible correction of the data set and/or the theoretical understanding.

Then, in both of the major options used, parallel and sequential, two updating steps are reiterated until a stopping condition is satisfied. The steps are: (1) updating of the clusters, (2) updating of the centers (centroids).

K-Means: An Iteration
(1) Updating of the clusters
Parallel K-Means: All the entities are considered available on any step; updating of the clusters is made based on the so-called *minimal distance rule*, which, for any $t = 1, ..., m$, collects in S_t all the entities which are the nearest to c_t (if an entity is at the same distance from several centers, an arbitrary decision is made, say, putting it in the cluster with the minimal index t among the competitors).
Sequential K-Means: Only one entity is available each time to be assigned to the nearest center's cluster.
(2) Updating of the centroids
In both of the options, this is done by taking the centroids of the clusters S_t found. In the sequential option, only two clusters are changed: the one the entity was added to, and the one the entity was taken from; only those clusters' centers are changed which can be done incrementally, without total recomputation.

The process stops, usually, when updating step does not change the clusters found at the previous iteration. When the sequential option is employed, the entities are observed in a pre-fixed order; the total number of repeated runs overall through set I may be used for stopping the process.

It is quite obvious that the alternating square-error minimization procedure is equivalent to the K-Means algorithm when it involves Euclidean distance and centers of gravity of the clusters as standard points.

A somewhat heuristically enriched method is suggested by the name of ISO-DATA (Iterative Self-Organizing Data Analysis Techniques) (Ball and Hall 1967): after the moving centers are stabilized, either of two options is performed: the splitting of too large clusters or the merging of too close clusters. The method, currently, is not in great use since it involves too many heuristic parameters.

Bezdec 1974 suggested a fuzzy version of K-Means. It involves a fuzzy m-class partition which is represented with an $N \times m$ membership matrix (z_{it}) $(i \in I, t = 1, ...m)$ where z_{it} is degree of membership of entity i in cluster t satisfying conditions: $0 \le z_{it} \le 1$ and $\sum_{t=1}^{m} z_{it} = 1$ for every $i \in I$. The calculations are based on an analogue of the square-error criterion (3.3),

$$B(z, c) = \sum_{t=1}^{m} \sum_{i \in I} z_{it}^{\alpha} d^2(y_i, c_t) \tag{3.4}$$

where $\alpha \ge 1$ is a parameter and d is Euclidean distance.

By analogy with the parallel K-means method, which is an alternating optimization technique, a fuzzy K-Means can be defined as the alternating minimization technique for function (3.4). The centroids, actually, are weighted averages of the entity points, while the memberships are related to the distances between the entities and the centroids.

Jawahar, Biswas and Ray 1995 extended criterion (3.4) to the case when every cluster may be represented with several kinds of centroids simultaneously (the gravity center as the zero-degree centroid, the normal vector to a central hyperplane, as the first-degree centroid, etc.) along with particular weights assigned to the centroid of each kind in each of the clusters. Minimizing the criterion with regard to the three groups of the variables (membership functions, centroid characteristics, centroid weights) has been demonstrated to work well for revealing clusters of distinct geometry.

Let us analyze matrix A (based on the similarities rather than on dissimilarities) with the parallel K-Means method, using the centroid definition (g) from the list above (the average neighbor), and taking 1 and 3 as the tentative centroids of the two clusters to obtain.

To assign the other entities to the centroids, we can see that 2 goes to 1 while 4 and 5 to 3: for example, $a_{53} = 275 > a_{51} = 225$ implies that 5 is assigned to centroid 3, not 1. To update the centers, we see that, in cluster $1 - 2$, the average similarity of 2 with the others (the entities 3, 4, 5) is a bit less than that of 1, which makes 2 the centroid. Analogously, 5 becomes the unanimous centroid being much less similar to 1 and 2, than the other elements of the second cluster.

Next updating iteration: 1 and 3 go to 2 while 4 is more close to 5; 1 becomes the centroid since its similarity to 4 and 5 is less than that of 2 and 3; in cluster $4 - 5$, 5 remains the centroid. Next updating iteration: 2 and 3 go to 1 while 4 remains in the second cluster; the clusters coincide with those found at the previous iteration. This stops the process: the clusters found are $S_1 = 1 - 2 - 3$ and $S_2 = 4 - 5$.

The sequential and parallel options of the moving centers method may lead to different results, since the centers are changing permanently depending on the ordering of the entities observed, in the sequential version. Also, the sequential version appears to have a somewhat larger set of solutions estimated since it checks the situation after every particular entity observed, which is supported with some experimental evidence (see, for example, Zhang and Boyle 1991). On the other hand, the parallel version works faster.

An important feature of the moving centers method is that some of its versions can be interpreted as local search methods for optimizing some clustering criteria. The most popular is the version employing the gravity center (a) concept as the centroid and Euclidean (squared) distance $d(y_i, y(S))$ as $d(i, c(S))$. As was noted above, the moving centers method so specified is nothing but an alternating minimization technique applied to the square-error (WGSS) criterion (this subject will be further elaborated in Section 6.2).

Exchange Algorithm.

Let $f(S)$ be a criterion defined for partitions $S = \{S_1, ..., S_m\}$ of I; a maximizer of $f(S)$ is the clustering sought. The family of exchange algorithms can be defined as local search algorithms based on the neighborhoods generated by moves of one or two entities. More explicitly, let $S(i, t)$ be a partition obtained from S after entity i has been moved into class S_t; the criterion change equals $\Delta(i, t) = f(S(i, t)) - f(S)$: the larger the $\Delta(i, t)$, the better the move. The local search algorithm based on the neighborhood $N(s) = \{S(i, t) : i \in S, t = 1, .., m\}$ is called the exchange method.

Let us consider, in the example of similarity matrix A, criterion $g(S) = \sum_t A(S_t)/n_t$ where $A(S_t)$ is the sum of all the similarities within S_t and $n_t = |S_t|$. When entity i is moved from S_1 into S_2, the criterion change is equal to

$$\Delta_i = 2(a(i, S_2) - a(i, S_1 - i)) + b_i$$

where $a(i, S)$ is the average similarity between i and the elements of S, and $b_i = a(S_1) - a(S_2) + (2/(n_2 + 1))(a(S_2) - a(i, S_2))$ where $a(S)$ is the average similarity within S. Let the initial partition be taken from the preceding computation, $S = \{1 - 2 - 3, 4 - 5\}$. The averages are: $a(S_1) = 540, a(S_2) = 924, a(4, S_1) = 911, a(5, S) = 234, a(1, S_2) = 463, a(2, S_2) = 506, a(3, S_2) = 750$. Obviously, to check the partition S, we must consider the largest similarity $a_{34} = 1224$ to be put within a cluster by either moving 3 into S_2 or 4 into S_1. For the first move, $\Delta_i = 2(750 - 500) + 540 - 924 + (2/3)(924 - 750) = 232$, and $\Delta_i = 2(911 - 924) + 924 - 540 + (2/4)(540 - 911) = 173$, for the second move. Although both of the moves make the criterion value increasing, the first increase is larger. Partition $S = \{1 - 2, 3 - 4 - 5\}$ is the result of the first iteration of the local search. All the averages recalculated (the general formulas can be easily derived for the criterion presented) are: $a(S_1) = 618, a(S_2) = 808, a(1, S_2) = 492, a(2, S_2) = 487, a(3, S_1) = 500, a(4, S_1) = 755, a(5, S_1) = 213$. Let us consider the criterion change when 4 is moved in S_1: $\Delta_4 = 2(755 - 1074) + 808 - 618 + (2/3)(618 - 755) = -539$; the other moves also lead to negative changes, which makes the computation stop.

When applied to the square-error criterion, the exchange method becomes almost equivalent to the sequential K-Means method (with gravity center as the centroid and Euclidean distance as the distance) since the latter, in fact, also involves move of i to a closest cluster. The difference is that, in K-Means method, the "cost of a move" is evaluated with the centers unchanged; they are updated after the move has been completed while in the exchange method the centers are considered as corresponding to the clusters updated. Actually, the criterion $g(S)$ considered in the example above, is equivalent to the square-error clustering criterion when Euclidean distance squared is taken as the dissimilarity measure (see p. 293); the criterion, in this case, is minimized, but the increment formulas remain true. The change of the K-Means WGSS criterion when i is moved from S_t into S_u is expressed with a much simpler formula:

$$\Delta(i, t, u) = n_u/(n_u + 1)d^2(y_i, c_u) - n_t/(n_t - 1)d^2(y_i, c_t)$$

since the centers c_u, c_t are updated at the next step, not simultaneously with the move.

The exchange method is especially suitable when the cardinalities of the clusters are fixed and may not be changed in the classification process (which is the case in some engineering problems related to module structuring). In this case, switching pairs of entities is the elementary move in the exchange algorithm. The local search method with the neighborhood defined accordingly is known as the Kernighan-Lin 1972 heuristic, though, in clustering, the exchange method (including the latter version) had been known a couple of years before (see, for example, review by Dorofeyuk 1971).

Seriation

This group of methods is based on a preliminary ordering of the entities with subsequent cutting of the ordering to produce the clusters.

Seriation

The procedure employs a linkage function $l(i, S)$ expressing degree of similarity between subset $S \subseteq I$ and the elements $i \in I - S$. This function determines the ordering as obtained by adding entities one-by-one: that i is added to the initial piece S of the ordering, which maximizes the linkage $l(i, S)$, $i \notin S$.

The most natural linkage functions are the nearest, furthest and the average neighbor as being the maximum, minimum, or the average similarity between i and S, respectively. The last neighbor linkage takes into account only the last element in the ordering of S. The bond energy algorithm employs the similarities with two last elements of the order (Arabie, Hubert, Schleutermann 1990).

Let us find such an order by matrix A starting with 1 and using the average linkage function: the subsequent pieces of the order designed are: 14 (since 4 is the nearest neighbor to 1 with the similarity 700), 143 (since 3 has the maximum average similarity (887) to 1 and 4), 1432 (average similarity 626), and 14325 (average similarity 406). The ordering does not look natural since there is an increase in the average similarity (when 3 was added to 1 and 4) indicating that the initial piece should be changed. Let us start with 3 and 4 having the maximum similarity; any order of them can be considered. One-by-one addition of the entities leads to: 342 (average similarity 630), 3421 (average similarity 623) and 34215 (average similarity 406). This time, the average similarities have been decreasing, although not always smoothly. The sharp drops (when 2 and 5 were added) indicate the points to cut, which leads to the following three-cluster solution: $S = \{3 - 4, 2 - 1, 5\}$.

Obviously, another starting point may give a different result. For example, starting with 5, we obtain the following sequence of the intervals: 54 (924), 543 (808), 5431 (492), 54312 (520) (the average similarities are in parentheses). This leads to only the sharp drop when 1 is added to produce the two-cluster partition $S = \{3 - 4 - 5, 1 - 2\}$, although someone may consider this sequence unnatural too since the last addition caused an increase the average similarity. Others may consider the increase unimportant since it is not high. This is the problem with heuristic algorithms: the heuristics can be expanded very easily.

Yet another, not as local approach to ordering is based on the so-called Robinson form of the similarity matrix (see Section 7.5).

In general, the seriation approach seems better fitting in the single cluster clustering framework than in partitioning since it allows cutting a cluster out immediately after a corresponding interval has been found.

Graph Partitioning.

Graph theory suggests several natural partitions of the set of vertices of a graph with their classes being connected components (maximal subsets of chain-connected vertices), bicomponents (maximal subsets of mutually attainable vertices), cliques (maximal subsets of mutually adjacent vertices), and coloring classes (maximal subsets of mutually nonadjacent vertices). This allows use of these concepts for producing partitions by the threshold graphs associated with similarity/dissimilarity matrices.

The concept of threshold graph can be extended to permit having different thresholds π_i for each entity $i \in I$, which allows to widen the set of graphs associated with a given similarity matrix.

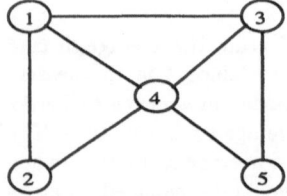

Figure 3.5: A non-standard threshold graph.

For example, let us take into account two maximal similarities in each row i of matrix A, which produces the graph presented in Fig.3.5. This graph contains three three-vertex cliques (subsets $1-2-4$, $1-3-4$, and $3-4-5$) of which two are overlapping in one point only ($1-2-4$ and $3-4-5$) representing 1-clusters by Jardine and Sibson 1968. If we want to have a partition, either of the possibilities, $S = \{1-2-4, 3-5\}$ or $S = \{1-2, 3-4-5\}$, can be taken as the solution (though, from the previous computations, we know that the latter partition is somewhat better at reflecting the similarities).

We can see why using "hard" thresholds to cut out every "small" similarity makes too rough impact to the data, leading, in many cases, to quite non-interesting solutions. It would be nice to have a "soft" threshold concept to allow sometimes even a small individual similarity to be included within a cluster if it is "surrounded" by the larger similarities. Such a "soft threshold" concept appears in the approximation model context considered in Chapters 4 and 5.

Each of the methods considered can be utilized differently by being embedded in a more general strategy; we can distinguish four such general strategies: parallel, divisive, separative, and replicative, as follows.

1. A *parallel* clustering strategy is, basically, one which has been presented so far in this section: all the clusters are constructed simultaneously.

2. A *divisive* clustering strategy works step by step, each step (except for the first one) processing a cluster, not all the entity set: the cluster is partitioned by a selected method into smaller sub-clusters, until the final partition is obtained. This can be done by sequential partitioning of the clusters found on preceding steps using either similarity/dissimilarity data or just dividing them by the categories of "essential" variables (conceptual clustering).

3. A *separative* clustering consists of separating clusters one-by-one from the main body of the data. In contrast to divisive clustering, the parts split here are not symmetrical: one of them is a cluster in its final form while the other represents the "main body" to be "cut from" again and again. This strategy somehow still remains almost missed by the researchers, although it may be considered a model for a typology making process. In Sections 4.3.2, 4.4 and 4.6 a few methods in the main line of the theories developed in this volume will be presented as belonging to this paradigm, the principal cluster analysis method among them.

4. *Replicative* clustering is a data analysis strategy involving independent clustering of two arbitrarily selected data parts along with subsequent comparison of the results; this is an important but understudied area of research (see Breckenridge 1989, Ivakhnenko et al. 1985).

3.2.3 Hierarchical Clustering

Logically, several approaches are possible to find a hierarchy associated with the data. Surprisingly, the only methods widely utilized belong to the family of sequential fission (agglomeration) or fusion (division) methods that construct the

hierarchy level-by-level, from bottom to top (agglomerative clustering) or from top
to bottom (divisive clustering).

Agglomerative clustering can be presented in the following unified way (Lance
and Williams 1967, Jambu 1978). Let (d_{ij}) be a dissimilarity entity-to-entity
matrix. Initially, each of the cases is considered as a single cluster (singleton). The
main steps of the algorithm are as follows.

Agglomerative Clustering

Step 1. Find the minimal value $d_{i^*j^*}$ in the dissimilarity matrix, and
merge clusters i^* and j^* .

Step 2. Transform the distance matrix, substituting one new row (and
column) $i^* \cup j^*$ instead of the rows and columns i^*, j^*, with its dissimilarities defined as

$$d_{i,i^* \cup j^*} = F(d_{ii^*}, d_{ij^*}, d_{i^*j^*}, h(i), h(i^*), h(j^*)) \qquad (3.5)$$

where F is a fixed (usually linear) function and $h(i)$ is an *index* function
defined for every cluster recursively: $h(i^* \cup j^*) = d(i^*, j^*)$, $h(\{i\}) = 0$
for all $i \in I$. If the number of clusters obtained is larger than 2, go to
Step 1, else End.

The result of the agglomerative procedure can be represented as a tree (Fig.3.6):
the singletons are in the lowest level, and every merging is shown by a node of a
higher level connected with the two cluster nodes merged. Height of the merged
cluster $i^* \cup j^*$ node is proportional to the index function $h(i^* \cup j^*)$.

There are several popular specifications of the method:

1. Nearest Neighbor (Single Link): the between-cluster distance $d_{i^*j^*}$ is de-
    fined as the minimum of the distances d_{ij} by all $i \in i^*, j \in j^*$; $d_{i,i^* \cup j^*} =$
    $\min(d_{ii^*}, d_{ij^*})$, in formula (3.5).

2. Farthest Neighbor (Complete Link): the between-cluster distance $d_{i^*j^*}$ is
    defined as the maximum of the distances d_{ij} by all $i \in i^*, j \in j^*$; $d_{i,i^* \cup j^*} =$
    $\max(d_{ii^*}, d_{ij^*})$, in formula (3.5).

3. Average Neighbor (Average Link, UPGMA): the between-cluster distance
    $d_{i^*j^*}$ is defined as the average of the distances d_{ij} by all $i \in i^*, j \in j^*$;
    $d_{i,i^* \cup j^*} = (n_{i^*} \cdot d_{ii^*}/(n_{i^*} + n_i) + n_j \cdot d_{ij^*}/(n_{j^*} + n_i))$, in formula (3.5).

4. Centroid Method: the between-cluster distance $d_{i^*j^*}$ is defined as the distance
    between centroids of the clusters i^* and j^*, centers of gravity, usually; in this,
    latter case, $d_{i,i^* \cup j^*}$ can be expressed with the linear form of function F in
    formula (3.5) when the distance used is Euclidean.

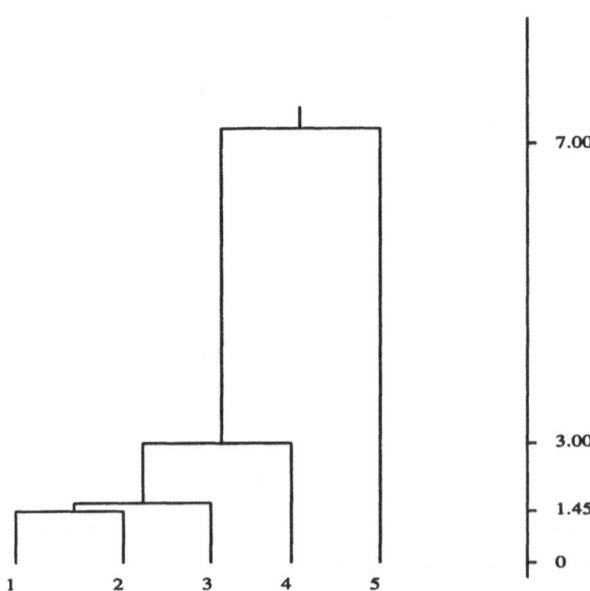

Figure 3.6: Various link clustering results for Primates (well structured data).

5. **Ward's (Incremental Sum of Squares) Method:** the between-cluster distance $d_{i^* j^*}$ is defined as the increment of the within cluster square-error criterion $\Delta(i^*, j^*) = E_{i^* \cup j^*} - E_{i^*} - E_{j^*}$ where $E_S = \sum_{i \in S} d^2(y_i, c(S))$ is the sum of the squared Euclidean distances between all the entities $i \in S$ and the cluster center of gravity $c(S) = \sum_{i \in S} y_i / |S|$; it appears, $\Delta(i^*, j^*) = n_{i^*} n_{j^*} d^2(c(i^*), c(j^*))/(n_{i^*} + n_{j^*})$, weighted distance squared between the gravity centers.

The first three of the methods can be applied to any dissimilarity/similarity data while the centroid and Ward's methods are developed for the entity-to-variable data (using between-centroid distances). When applied to well-structured data, all the methods give the same or almost the same result as presented in Fig.3.6 for matrix D. When the structure is somewhat hidden or complicated, the methods may give quite different results. In Fig.3.7, the hierarchies found for matrix A with complete link (a), single link (b), and average link (c) methods are presented along with corresponding similarity matrices obtained after first merging the closest entities, 3 and 4 (note that the data is a similarity, not dissimilarity, matrix, which makes corresponding changes in the nearest and farthest neighbor algorithms). Each of the methods leads to a quite different structure, which makes the user quite unconfident in what he can get with them.

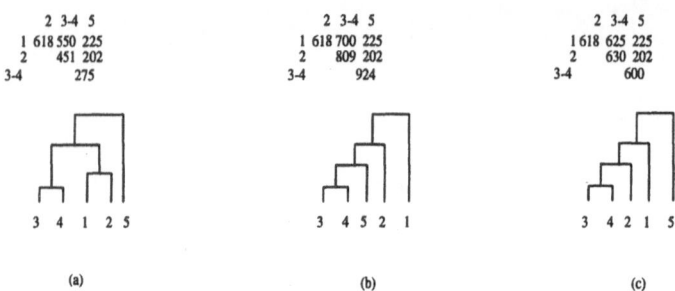

Figure 3.7: Agglomerative clustering for controversial data: complete (a), single (b) and average (c) link trees.

Lance and Williams (1967) suggested an almost linear form for function F in (3.5) to include the main agglomerative clustering methods:

$$d_{i,i\bullet\cup j\bullet} = \alpha_{i\bullet}d_{ii\bullet} + \alpha_{j\bullet}d_{ij\bullet} + \beta d_{i\bullet j\bullet} + \gamma|d_{ii\bullet} - d_{ij\bullet}| \qquad (3.6)$$

The coefficients for the methods listed above are presented in Table 3.2. For more, see Jambu 1978, Gordon 1996.

Among these methods, single and complete linkage methods are especially simple; in terms of the threshold graph, the former finds connected components while the latter, some cliques. This is why these two methods may be considered to represent two extremes of the generally accepted requirement that the "natural" clusters must be internally cohesive and, simultaneously, isolated from the other clusters: single linkage clusters are isolated but can have very complex chained and noncohesive shape; in contrast, complete linkage clusters are very cohesive, but may be not isolated at all. The other three methods represent a "middle way" and are rather close to each other: it is obvious for Ward's and centroid methods since Ward's criterion is just a weighted version of the latter one; the average link method, actually, can be considered a version of the centroid method when the distances, as it frequently happens, are just linear functions of the scalar products between the entity-vectors (the average between-cluster scalar product equals the scalar product of the cluster gravity centers).

To overcome the disadvantages of single and complete linkage methods, Lance and Williams 1967 suggested the so-called β-flexible clustering strategy which requires β be between -1 and 1 while $\alpha_{i\bullet} = \alpha_{j\bullet} = (1 - \beta)/2$ and $\gamma = 0$, in their recurrence formula (3.6). However, value of β can be adjusted in such a way that the flexible strategy can model both single and complete linkage methods, though with a non-linear form of the function F in (3.6), that is, it requires changing β at each amalgamation step. Let the following inequalities hold: $d_{i\bullet j\bullet} < d_{ii\bullet} < d_{ij\bullet}$.

Method	α_{i^*}	α_{j^*}	β	γ
Single linkage	1/2	1/2	0	-1/2
Complete linkage	1/2	1/2	0	1/2
Average linkage	$\dfrac{n_{i^*}}{n_{i^*}+n_{j^*}}$	$\dfrac{n_{j^*}}{n_{i^*}+n_{j^*}}$	0	0
Centroid	$\dfrac{n_{i^*}}{n_{i^*}+n_{j^*}}$	$\dfrac{n_{j^*}}{n_{i^*}+n_{j^*}}$	$\dfrac{-n_{i^*}n_{j^*}}{(n_{i^*}+n_{j^*})^2}$	0
Ward method	$\dfrac{n_{i^*}+n_i}{n_{i^*}+n_{j^*}+n_i}$	$\dfrac{n_{j^*}+n_i}{n_{i^*}+n_{j^*}+n_i}$	$\dfrac{-n_i}{n_{i^*}+n_{j^*}+n_i}$	0

Table 3.2: Lance-Williams coefficients for most known agglomerative clustering methods.

Denoting $\delta = d_{ij^*} - d_{ii^*}$ and $\varepsilon = d_{ii^*} - d_{i^*j^*}$ and taking $\beta = \pm\delta/(\delta + 2\varepsilon)$, one obtains single linkage or complete linkage depending on whether $+$ or $-$ is taken, respectively (Oshumi and Nakamura 1989). Different values β may lead to some intermediate clustering strategies.

Yet another parametrization of the problem is considered in (Aivazian et al., 1989) as based on the so-called (Kolmogoroff) K-distance between clusters, which is actually the average between-cluster link defined, using the p-power of the between-entity distances:

$$d^p(i^*, j^*) = \sum_{i \in i^*} \sum_{j \in j^*} d_{ij}^p / n_{i^*} \cdot n_{j^*}.$$

Although the formula seemingly involves the average linkage only, it behaves differently for different p: it follows the maximum between-cluster distance when p is large positive, the minimum between-cluster distance when p is large negative, and it gives the average link distance when $p = 1$. These three cases correspond, respectively, to the single, complete and average linkage methods. In the author's experiments with low-dimensional geometric point data using Euclidean distance, the single and complete linkage solutions were found with p equal to -10 or $+10$, respectively.

With regard to these possibilities, we cannot help but agreeing with the following remark: "There is, however, some danger in adjusting the parameters until one obtains what one likes, rather than choosing some prior criterion and sticking to the results." (Sneath and Sokal 1973, p. 227-228). Diday and Moreaux 1984 suggest an interesting application of the flexibility of the family of Lance-Williams-Jambu agglomerative algorithms (3.5), (3.6): learning the parameters' values from a model-based hierarchy presented along with all its between-cluster distances. Having such a data set, the parameters' values are adjusted with ordinary linear regression analysis technique; the examples reported by Diday and Moreaux involve successful learning of quite complicated cluster shapes.

3.2.4 Biclustering

The term biclustering refers to simultaneous clustering of both row and column sets in a data matrix. Biclustering addresses the problems of aggregate representation of the basic features of interrelation between rows and columns as expressed in the data. We distinguish between the following bicluster counterparts to the methods discussed above: single cluster biclustering, partition biclustering, and hierarchical biclustering, which will be considered in turn.

Single Cluster Biclustering

It seems, Hartigan 1972, 1975, 1976 was the first who considered the problem explicitly. Given an entity-to-variable matrix with its row set I and column set J, a *block* is defined as a submatrix with the row set $V \subseteq I$ and column set $W \subseteq J$ having a standard vector $c_V = (c_v)$, $v \in V$, of the column values within the block ("block code"). The standard vector may consist of the means (when the variables are quantitative) or the modal values (when the variables are qualitative). Hartigan's block clustering algorithms involve: 1) "elimination" of the grand means or modes, initially; 2) adding/excluding the rows and columns one by one for better adjusting the block and its block code.

Similar concept of box as a submatrix characterized by a typical similarity value λ_{VW} (not vector) applied to similarity data was considered by Rostovtsev and Mirkin 1978, Eckes and Orlik 1993, and Mirkin, Arabie and Hubert 1995. They considered square-error-like criteria and applied similar one-by-one adding/excluding algorithms.

A typical result was reported in Mirkin 1985 based on analysis of the variable-to-variable correlation Activities data in Table 1.17, p. 47. It appeared that three small boxes (A: Sex and Job (rows) to the time spent for Eating, Cooking and Cleaning (columns); B: Age and Children to the time spent for Nursing children; C: Number of working days a week, Farmyard, Sex and College education level to the time spent working at Work or Farmyard) accounted for 64% of the variance of the data. This was used in the subsequent analysis of the activities as restricted to the within-box associations (Mirkin 1985).

In Boolean data analysis, a bicluster notion, Galois connections or (box) concepts, have attracted considerable effort. Given a Boolean matrix, $r = (r_{ij})$, $i \in I$, $j \in J$, a submatrix $r_{VW} = (r_{ij})$, $i \in V$, $j \in W$, is referred to as a *concept* (Wille 1989) or *Galois connection* (Flament 1976) if it is a maximal (by set-theoretic inclusion) submatrix having all its entries equal to unity. A polynomial time graph-matching algorithm for finding the box concepts (introduced under a different name) having maximal perimeter (the total number of rows and columns) is described in Levit 1988.

Partition Biclustering

Partition biclustering, when considered for similarity data, usually is regarded as a tool for rearranging the data matrix by its row-column permutation in such a way that the larger similarities in the matrix are concentrated among the nearest neighbors in the sequences, an idea proposed by Robinson 1951 in a particular problem and then extended to more general framework (see Hubert and Arabie 1994 and the references therein, and also Section 7.5).

For example, the bipartitioning algorithm proposed in Arabie, Schleutermann, Daws, and Hubert 1988 as an extension of the so-called "bond energy" algorithm, primarily finds row and column seriations to maximize the function

$$\sum_{i \in I} \sum_{j \in J} a_{ij}[a_{i,j-1} + a_{i,j+1} + a_{i-1,j} + a_{i+1,j}] \tag{3.7}$$

which heavily depends on the concentration of larger similarity values among neighbors in sought sequences of the elements of I and J.

Marcotorchino 1987 is motivated by the same idea, though he considers a genuine bipartitioning problem of simultaneously finding ordered partitions $S = \{S_1, ..., S_T\}$ of I and $R = \{R_1, ..., R_T\}$ of J having the same number of clusters to contain the largest similarities within the corresponding clusters $S_t \times R_t$ connections as expressed with criterion

$$\sum_{t} \sum_{i \in S_t, j \in R_t} (a_{ij} - 1/2)$$

to be maximized. His algorithm closely follows the K-Means partitioning procedure: starting with a partition R, every cluster S_t is collected to contain the entities $i \in I$ maximally contributing to R_t ($t = 1, ..., T$). Then, the procedure is repeated, this time starting from S_t. These iterations are performed repeatedly until no change is observed in S or R.

A similar algorithm was developed by Govaert 1980 for contingency data (see more details in Section 6.4).

A particular approach to entity-to-variable data bipartitioning was suggested by E. Braverman et al. 1974 based on the so-called method of extremal grouping of the variables (Braverman 1970, Braverman and Muchnik 1983). The method seeks a partition of the set of the variables (data table columns) maximizing the average contribution of the first principal component of each cluster of the variables to the total variance of the cluster. Then (or, in later versions, simultaneously) a partition of the entities by every variable cluster is found which gives an aggregate representation of the data: every row of the data matrix is represented as a sequence of the entity clusters (made by different variable clusters) it belongs to.

Hierarchical Biclustering

In analysis of species-to-site contingency data in ecology, an efficient hierarchical biclustering method (Hill 1979, ter Braak 1986) is employed probably based on the particular feature of the data that, usually, each species tends to inhabit geographically close areas. The method has been implemented in a program involving: (1) step-by-step splitting of the site clusters via ordination of the sites along the first correspondence analysis factor; (2) at each division step, finding the so-called differential species who inhabit, mostly, only one of the splits; (3) recoding the original abundance of the species into the so-called pseudo-species which are nothing but some Boolean categories defined by thresholds. In the original program, the bihierarchy was represented by corresponding rearrangement of both the rows and columns (Hill 1979). Then, ter Braak (1986) suggested representing the results with only one hierarchy, of the sites, supplemented with two kinds of discriminating conceptual descriptions for each cluster: in terms of the differential species and the environmental variables.

A distinctive hierarchical biclustering method is developed in De Boeck and Rosenberg 1988, Rosenberg, Van Mechelen and De Boeck 1996 for Boolean data tables. The method involves arbitrary subsets of the rows or columns ordered by set-theoretic inclusion; the subsets may arbitrarily overlap each other, which generates a most general kind of hierarchy, maybe not as good for representing scientific theories but well adapted to representing real-world knowledge and structures.

Let us consider the method in more detail using data set Tasks (Table 1.8, p. 34) where seven persons are described in terms of their success/failure patterns, which provides us with the following 7×6 Boolean matrix:

$$r = \begin{pmatrix} 0 & 0 & 1 & 0 & 0 & 0 \\ 0 & 0 & 1 & 0 & 0 & 0 \\ 1 & 1 & 0 & 1 & 0 & 0 \\ 1 & 1 & 1 & 0 & 1 & 1 \\ 1 & 1 & 1 & 0 & 1 & 1 \\ 1 & 1 & 1 & 1 & 1 & 1 \\ 1 & 1 & 1 & 1 & 1 & 1 \end{pmatrix}$$

In the matrix $r = (r_{ij})$, every row $i \in I$ defines a corresponding column subset

$W_i = \{j : r_{ij} = 1\}$ and every column $j \in J$ defines a row subset $V_j = \{i : r_{ij} = 1\}$; the subsets are partially ordered by set-theoretic inclusion, which is extended into corresponding row and column partial preorders. That means that $j_1 \preceq j_2$ when $V_{j_1} \subseteq V_{j_2}$; and similar $\preceq$ relation is defined for the rows (with sets W_i).

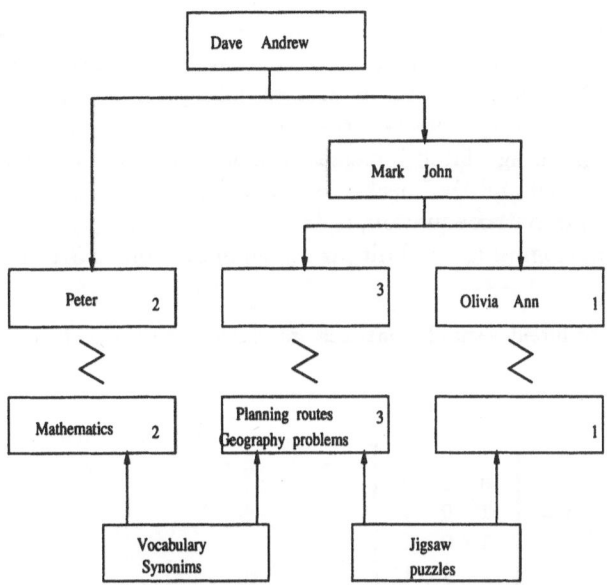

Figure 3.8: Tasks data biclustering: Row cluster hierarchy goes up above zigzags while column cluster hierarchy goes down; the numbers denote the latent variables.

For instance, row set 3-4-5-6-7 corresponding to column 1, obviously, contains overlapping row sets, 3-6-7 and 4-5-6-7, corresponding to columns 4, 5 and 6.

The graphical representation of the partial preorders in Fig.3.8 shows: (a) the hierarchy of the rows (by inclusion among column subsets) corresponding to persons (from top to the middle), (b) the hierarchy of the columns (by inclusion among row subsets) corresponding to tasks (from bottom to the middle), (c) correspondence between classes of the lowest levels in the hierarchies (empty boxes are needed to have only the boxes of the lowest levels connected). The correspondence relation in (c), in fact, states an association between the row and column classes of the equivalence relations corresponding to relation $\preceq$: a given column class is associated with all the classes of the rows which have some of their attributes belonging to that column class, and vice versa, a given row class is associated with all the column classes that are the attributes of that row class.

This association is easily seen in Fig.3.8 by transitivity. For instance, correspondence between row 3 (Peter) and column 4 (Mathematics) makes also association between rows 6/7 (Dave/Andrew) and columns 1/2 (Vocabulary/Synonyms).

Actually, the picture in Fig.3.8 is a graphical representation of the Boolean matrix equation $r = v \odot w$ where $v = (v_{ik})$ and $w = (w_{kj})$ and (i, j)-th entry in $v \odot w$ is equal to $\max_k \min(v_{ik}, w_{kj})$; k can be interpreted as a "latent" Boolean attribute which is meaningful for both the rows and columns. Every latent variable corresponds to a join in Fig.3.8 (which is shown by corresponding numbers in the boxes). The equation means that every observed row- or column-class, actually, is equal to the union of corresponding "latent" classes; this is called "disjunctive model" in Rosenberg, Van Mechelen and De Boeck 1996. The "conjunctive model" of these authors refers to a similar matrix equation, $r_{ij} = \min_k \max(v_{ik}, w_{kj})$. Note that the models are Boolean analogues to the bilinear model in Sections 2.3.1 and 6.1.

In the example considered, the latent variable matrices associated with the picture in Fig.3.8 are as follows:

$$v = \begin{pmatrix} 1 & 0 & 0 \\ 1 & 0 & 0 \\ 0 & 1 & 0 \\ 1 & 0 & 1 \\ 1 & 0 & 1 \\ 1 & 1 & 1 \\ 1 & 1 & 1 \end{pmatrix}$$

and

$$w = \begin{pmatrix} 0 & 0 & 1 & 0 & 0 & 0 \\ 1 & 1 & 0 & 1 & 0 & 0 \\ 1 & 1 & 1 & 0 & 1 & 1 \end{pmatrix}$$

Although in Rosenberg, Van Mechelen and De Boeck 1996 this example was employed for the conjunctive model decomposition only, the factors found with the disjunctive model also can be interpreted in terms of some basic strategies employed by the persons for resolving the tasks: nonverbal (1), nonspatial (2), and nonquantitative (3).

An alternating optimization approach to fit the Boolean matrix equation when some of the entries can be erroneous is mentioned in De Boeck and Rosenberg 1988. The method leads to an original and impressive output and deserves to be further elaborated.

3.2.5 Conceptual Clustering

The roots of conceptual clustering can be traced back to work done in the late fifties
- early sixties by Williams and Lambert 1959 and Sonquist et al. 1973. A new
impetus was done in the eighties by Breiman et al. 1984 who developed a technique
along with a bunch of theoretical results on the consistency of classification trees
obtained for a random sample with the population structure, and by Michalski
and Stepp 1983 who put the concept in the perspective of artificial intelligence as
a basic notion in machine learning.

To present the variety of most common conceptual clustering techniques developed, let us use the following notions:

1. *Learning Task.* There have been four classification learning tasks considered:

   (a) Learning a class (a class $S \subset I$ is given, the classification tree must give
       a decision rule for distinguishing between S and non-S (Quinlan 1986).

   (b) Learning a partition (a partition of the set I is given; the classifica-
       tion tree must give a decision rule for distinguishing between its classes
       Breiman et al. 1984, Fisher 1987).

   (c) Self-learning (the classification tree must represent a conceptual struc-
       ture of the given set of the variables, qualitative or quantitative
       (Williams and Lambert 1959, Sonquist, Baker, and Morgan 1973, Lbov
       1981, Rostovtsev and Mirkin 1985).

   (d) Learning association between two sets of the variables (there are two
       groups of the variables considered, X and Y; the classification tree is
       designed with X-variables to produce clusters that are meaningful with
       regard to Y-variables (Rostovtsev and Mirkin 1985).

   Obviously, the last item, (d), can be considered as a generalization of the
   items (b) (Y consists of a unique categorical variable representing the partition given) and (c) ($Y = X$).

2. *Data Availability.*

   The data for learning a tree can be available: (1) all immediately, or (2) incrementally, entity by entity, which is considered better corresponding to real
   world learning problems (especially when no reprocessing of the previously
   incorporated observations is permitted).

3. *Method for Tree Construction.*

   The major conceptual clustering algorithms construct decision trees from
   top to bottom (the root representing the universe considered) in a divisive
   manner, each time deciding the following problems:

1. Which class (node of the tree) and by which variable to split?

2. When to stop splitting?

3. How to prune/aggregate the tree if it becomes too large?

4. Which class to assign to a terminal node?

Item 4 concerns the first two learning tasks: class or partition learning. The aggregating classes in item 3 is the only way, in this framework, to get more than just conjunctive concepts formed by sequential addition of the splitting characters.

In the case of the incremental clustering, pruning/aggregation decisions can be made while producing the tree; this requires the items 2 and 3 to be substituted by the operation of choosing the best action with regard to an entity x being incorporated in a class. The actions to choose from are as follows: (a) put x into an existing child class; (b) create a new child class containing x; (c) merge two child classes into a new class containing x; (d) split the class into its children, adding x to the best of them (Fisher 1987, Gennari 1989).

To decide which class S to split and by which variable it is to be split, a goodness-of-split criterion must be defined. There is a general way for defining such a criterion based on a measure of dispersion (called impurity in Breiman et al. 1984)) of the variable y in subset S, $\delta(y, S)$. When S is divided in subclasses $\{S_t\}$, the change of the measure, $\Delta(y, \{S_t\}) = \delta(y, S) - \sum_t p_t \delta(y, S_t)$, where p_t is the proportion of the entities of S sent to S_t under the split, can be used as the criterion of goodness-of-split with regard to the variable y. The larger $\Delta(y, \{S_t\})$, the better the split. In Breiman et al. (1984), such a measure is employed in the situation when y is just a pre-given partition of the entities, and $\delta(y, S)$ is the Gini index or qualitative variance of the corresponding partition, equal to $\delta(y, S) = 1 - \sum_u p^2(u)$ where $p(u)$ is the proportion of the entities in S belonging to category u of y, and summation is made by all the categories of y. When y is a quantitative variable, the measure $\delta(y, S)$ can be its variance in S. In this case, the ratio $\Delta(y, \{S_t\})/\delta(y, S)$ has a particular meaning in statistics, being the so-called correlation ratio (squared) of the variable y with regard to the split (see Section 6.1.2).

When learning task (c) or (d) is considered, the changes $\Delta(y, \{S_t\})$ are summed up over all the target variables $y \in Y$ ($Y = X$, in task (c)). In this case, a normalization of change $\Delta(y, \{S_t\})$ for each y is needed to keep the variables equivalent. In Rostovtsev and Mirkin 1985 two normalization methods are considered: by value, as it is done to get the correlation ratio above, and by distribution: in standard statistical assumptions, the change $|S_t|\Delta(y, \{S_t\})$ converges (by distribution) to distribution χ_p^2 with a number p of degrees of freedom, which makes meaningful Fisher's approximation

$\phi = \sqrt{2\chi_p^2} - \sqrt{2p-1}$ of the normal distribution standardized to serve as the standardization formula. When the assumptions are fair, this normalization admits a direct interpretation: $1 - \phi$ is the probability of getting a better split by splitting randomly.

Among the other goodness-of-split criteria considered for categorical variables, the following two should be mentioned:

(a) *Twoing Rule* (Breiman et al. 1984) applied when split of S is made into two subclasses, S_1 and S_2, only:

$$tw(y, S_1, S_2) = \frac{p_1 p_2}{4} [\sum_u |p(u/S_1) - p(u/S_2)|]^2$$

where u are the categories of y.

(b) *Category Utility Function* (Fisher 1987, referring to Gluck and Corter 1985) applied when there is a set of categorical variables Y and the split is made into any number T of subclasses S_t, $t = 1, ..., T$. In our denotations, the criterion is

$$CU(Y, \{S_t\}) = \sum_{y \in Y} [\sum_{u_y} \sum_t \frac{p_{u_y t}^2}{p_t} - \sum_{u_y} p_{u_y}^2]/T$$

which is proportional to the so-called reduction of the error of proportional prediction (see p. 5.1.4).

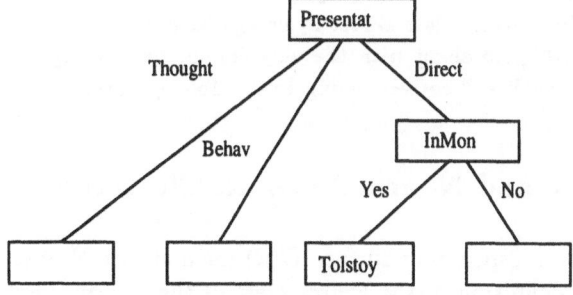

Figure 3.9: A concept tree for L. Tolstoy novel class.

Let us consider the learning task of separating the Leo Tolstoy masterpieces using the two qualitative variables from the Masterpiece data in Table 1.12, p. 40. The goodness-of-split criterion should address the problem as it is, thus relating only to the separated class s, without involving the non-Tolstoy class.

Since no criterion for that has been suggested in the literature, let us consider a criterion averaging the squared differences between the general probability p_s of the class and its probabilities $p(s/t)$ in the decision classes t created: the larger the absolute value of the difference, the clearer presence or absence of s in t:

$$W(\{S_t\}) = \sum_t (p(s/t) - p(s))^2 p(t)/(T - 1)$$

where $p(s)$, $p(t)$, and $p(s/t)$ are (empirical) probabilities of the corresponding events. This measure is consistent with the following assignment rule: a terminal node, t, is labeled by the Tolstoy class mark s if the difference $p(s/t) - p(s)$ is considerably high; otherwise, t is assigned with a no-s label.

In the literature, usually, the value $p(s/t)$ itself is considered a good assignment index (see, for example, Breiman et al 1984) which is fair when $p(s)$ is relatively small. When $p(s)$ is relatively large, values $p(s/t)$ tend to be also large for each t. The absolute probability change $p(s/t) - p(s)$ seems a more flexible measure, in this case (Mirkin 1985).

To decide which of the two variables, InMon or Presentat, has to be used for splitting of the entire set of the masterpieces by its categories, let us calculate the criterion value for each of them:
$W(InMon) = [(2/5 - 1/4)^2 5/8 + (0 - 1/4)^2 3/8]/1 = 0.0375$, and
$W(Presentat) = [(2/3 - 1/4)^2 + (0 - 1/4)^2 1/4 + (0 - 1/4)^2 3/8]/2 = 0.0521$, which implies that splitting by the latter variable is somewhat better. Having now three subsets corresponding to the categories Direct, Behav, and Thought, we can see that only the first of them is of interest for sequential splitting; the other two just contain no Tolstoy novels. Then, we have only the possibility of splitting class Direct by categories of InMon, which produces decision tree presented in Fig.3.9 along with the only conjunctive conceptual cluster "Presentat='Direct' & InMon='Yes' " corresponding to the Tolstoy novel class.

3.2.6　Separating Surface and Neural Network Clustering

To discuss the problem of finding a separating surface $G(x)$ for a subset S of n-dimensional points, one needs a definition of the source class of the discriminant functions G. A quite general formulation based on so-called VC-complexity classes was given by V. Vapnik and A. Chervonenkis in the late sixties (see Vapnik 1982). A simple formulation defines that as the class of linear functions $G(x) = (c, x)$ where c is normal vector to the hyperplane G.

There exists an approach, called the *potential function* method, which is quite general and, at the same time, reducible to linearity. The potential function $\psi(x, y)$

reflects a kind of "influence" of a "prototype point" y upon the variable space point x and, usually, is considered a function of the squared Euclidean distance between x and y such as $\psi(x,y) = 1/(1 + ad^2(x,y))$ or $\psi(x,y) = exp(-ad^2(x,y))$ where a is a positive constant.

The potential discriminant function for a class $S \subseteq I$ is defined then as the average potential with respect to the points of S as the prototype points: $G_S(x) = \sum_{i \in S} \psi(x,y_i)/|S|$. It appears that using in the $G_S(x)$ potential function $\psi(x,y) = exp(-ad^2(x,y))$ with sufficiently large a, the function $G_S(x)$ separates any S from $I - S$ (Andrews 1972).

On the other hand, potential functions depending on x and y through Euclidean distance between them, can be represented as $\psi(x,y) = \sum_p \lambda_p^2 g_p(x)g_p(y)$ where $\{g_p(x)\}$ is a set of the so-called eigen-functions. This allows transforming the classification problem into a so-called "straightening space" based on the transformed variables $z_p = \lambda_p g_p(x)$. In this straightening space, the potential function becomes the scalar product, $\psi(x,y) = (z(x), z(y))$, which makes all the constructions linear.

Obviously, a separating hyperplane exists only for those subsets S whose convex hull is separated from the convex hull of the rest. The first neuron-like learning algorithm, the *perceptron*, was proposed by F. Rosenblatt (see Nilsson 1965) to learn a linear separating surface when S can be linearly separated. The perceptron perceives the entity points coming in sequence, starting from arbitrary c and changing it after every try with the following rule: $c' = c + \theta(c, -y)y$ if $y \in S$ and $c' = c - \theta((c,y))y$ if $y \notin S$ where y is the point perceived and $\theta(x)$ is the threshold neuron output equal to 1 or 0 depending on whether its argument is positive or not. This means that we add to c (respectively, subtract from c) all erroneously classified points from S (respectively, from $I - S$), thus turning c toward a direction between the summary points of S and $I - S$ (the latter point is taken with minus). This guarantees that the method converges.

In a multilayer perceptron, a similar learning idea requires sequential weight changes layer-by-layer starting from the output layer (back-propagating). The *back-propagation* learning process proposed by Rumelhart, Hilton and Wilson 1986 is, actually, a version of the method of steepest descent (a.k.a. hill-climbing) as applied to the square-error criterion $E(c) = \sum_i (x_i - d_i)^2$ where x_i and d_i are actual and ideal (shown by the "teacher") outputs of the neuron network, respectively. Let the output of the p-th neuron in a layer equal $x_{pi}(c) = \theta_p((c_p, y_i))$, where y_i is the input to the neuron from the preceding layer when the input to the network is i-th given point.

Back-propagation

Updating step: change of the weight vector c_p in the neuron is controlled by the equation $\Delta_i c_p = \alpha \delta_{pi} y_i$ where α is the step size factor (usually, constant) and $\delta_{pi} = -\theta_p'((c_p, y_i))(x_i - d_i)$ if this is the output layer, or $\delta_{pi} = -\theta_p'((c_p, y_i)) \sum_q \delta_{qi} c_{qi}$ for a hidden layer where q represents the next (more close to the output) layer's suffix.

As any local optimization method, in a particular computational environment, the back-propagation method can converge to any local optimum or even not converge at all, which does not hamper its great popularity.

The unsupervised clustering problem cannot be put so naturally in the context of separating surfaces since it requires some ad hoc criterion for clustering. In this aspect, the only advantage of this approach over the other optimal clustering techniques is that the theoretical nature of the discriminant functions makes it easy to put in a probabilistic environment (Braverman and Muchnik 1983, Briant 1991).

As to the neural-network unsupervised clustering, it seems currently in quite a beginning phase involving rather arbitrary approaches (see Carpenter and Grossberg 1992, Kamgar-Parsi et al. 1990, Kohonen 1989, 1995, Murtagh 1996, Pham and Bayro-Corrochano 1994).

3.2.7   Probabilistic Clustering

Let us consider the mixture-of-distributions and high-density clustering approaches mentioned above.

Let $y_1, ..., y_N$ be a random sample of N n-dimensional observations from a mixture of densities $f(x) = \sum_{t=1}^{T} p_t f(x, a_t)$ where the unknown parameters are: the mixing weights p_t and parameters a_k defining within-class densities ($a = (\mu, \Sigma)$ when $f(x, a)$ is the normal density function with mean μ and covariance matrix Σ unknown). To estimate the parameters, the maximum likelihood method is applied in each of the following two versions: (1) in the *mixture likelihood* approach (MA), the problem is to find the parameters p_t, a_t, $t = 1, ..., T$, maximizing

$$L = \log\{\prod_{i=1}^{N} \sum_{t=1}^{T} p_t f(y_i, a_t)\};$$

no partitioning of the sample is assumed, formally, which is not the case in the other approach, (2) *classification likelihood* (CA), where supplementarily the sample must be partitioned in T classes maximizing the log-likelihood expressed as

$$L_C = \sum_{i=1}^{N} \sum_{t=1}^{T} z_{it} \log(p_t f(y_i, a_t))$$

where z_{it} are the membership values, $z_{it} = 1$ if y_i is assigned to class t, and $z_{it} = 0$ otherwise.

To deal with the maximization problems, the MA criterion is reformulated as

$$L = \sum_{i=1}^{N} \sum_{t=1}^{T} g_{it} \log p_t + \sum_{i=1}^{N} \sum_{t=1}^{T} g_{it} \log f(y_i, a_t) - \sum_{i=1}^{N} \sum_{t=1}^{T} g_{it} \log g_{it} \qquad (3.8)$$

where g_{it} is the posterior probability that y_i came from the density of class t, defined as

$$g_{it} = \frac{p_t f(y_i, a_t)}{\sum_t p_t f(y_i, a_t)}.$$

The alternating optimization algorithm for these criteria is called EM-algorithm.

EM-algorithm

Starting with any initial values of the parameters, alternating optimization is performed as a sequence of the so-called estimation (E) and maximization (M) steps.

E-step: estimation of the current values g_{it} by the p_t, a_t given.

M-step: g_{it} given, the parameters are found maximizing the log-likelihood function (3.8), which is not difficult for most common density functions.

If, for example, f is the normal density function, then the optimal values of parameters can be found with the following formulas:

$$\mu_t = \sum_{i=1}^{N} g_{it} y_i / g_t, \quad \Sigma_t = \sum_{i=1}^{N} g_{it}(y_i - \mu_t)(y_i - \mu_t)^T / g_t$$

where $g_t = \sum_{i=1}^{N} g_{it}$.

The EM iterations stop when the parameter estimates found do not differ too much from those found at the preceding iteration. If the user needs an assignment of the observations to the classes, the posterior probabilities g_{it} are utilized: i is assigned to that t for which g_{it} is the maximum. Also, the values g_{it}/g_t can be considered as fuzzy membership values.

The problem of estimating the number of clusters, T, still has not found any reasonable solution; a practical approach requires estimating the model for several different T-values to select the one which leads to a larger value of the log-likelihood function.

In the CA approach, some estimates of the crisp membership z_{it} must be put within the computation process since the posterior estimates become biased here: the number of estimated values z_{it} grows when number of the observations increases. An extension of the EM algorithm to this situation was suggested by Celeux and Diebolt (see Celeux and Govaert 1992, Aivazian et al. 1989), adding yet another step dealing with stochastic partitioning of the sample (S-step) between E- and M-steps. This S-step produces a random partitioning of the sample according to the posterior probabilities g_{it}; subsequently, the density parameters could be estimated directly by the sampling classes (p_t is estimated as the proportion of the entities in class t, etc.)

Let us consider now high density clustering and present an algorithm for obtaining unimodal clusters S_t ($t = 1, ..., m$) along with a "ground" cluster S_0 (Kovalenko 1993), which can be considered a theoretically-substantiated version of a well known mode-search clustering algorithm by Wishart 1969. The algorithm involves approximation of the general density function via the k-nearest neighbors method (Devroye and Wagner 1977) applied to the sample. Let $V_k(y)$ be volume of the minimal ball with its center at y containing k sample nearest neighbors of y, and $d_k(y)$ be its radius, that is, the distance $d(y, y_i)$ where y_i is k-th nearest neighbor of y. Then, $p_N(y) = V_k(y)(k/N)$ is the estimate of the density. A subsample S is called a h-level cluster if the maximum of the differences $p_N(y_i) - p_N(y_j)$ ($i, j \in S$) is larger than $h > 0$. The algorithm designs all the h-level unimodal clusters leaving the other entities in the ground cluster S_0 (k is considered fixed; each of the k nearest neighbors of a point will be referred to as its nearest neighbor).

The algorithm starts with renumbering sample entities y_i with regard to $d_k(y_i)$ increase.

Unimodal high-density clustering

An iteration: For every entity i, its nearest neighbors among the preceding $i - 1$ entities are considered. If there are none, y_i starts a new cluster. If the neighbors are in clusters $S_{t_1}, ..., S_{t_l}$ ($t_1 < t_2 < ...t_l$) and all of the clusters are labeled as finished, put entity i in the ground cluster S_0. If some of the clusters are not finished, check which of them are h-level clusters. If none or only one, then merge all the clusters $S_{t_1}, ..., S_{t_l}$ into cluster S_1 and put i into S_1. Otherwise, label the h-level clusters as finished, putting the others (entity i included) in the ground class.

The labeling and checking of h-level clusterness are the model-driven options added to Wishart 1969 algorithm; based on them, consistency of the algorithm can be proved.

3.2.8 Discussion

1. In the survey presented, we focus on two major problems: (1) finding a cluster structure in the data and (2) finding a "theoretical", intensional cluster structure to describe an "empirical", extensional one which is pregiven. The latter subject can be described as also interpreting the extensional cluster structure in terms of the intensional structure.

2. This latter subject usually is not considered to belong in clustering; it is labeled as "supervised learning", or "pattern recognition", in contrast to "unsupervised learning", or "proper clustering". We do not see any contradiction in that. This is the matter of a different perspective we focus on: in pattern recognition, a decision rule must be designed to deal with potential infinity of the set of the instances; in clustering, this is the question of finding an intensional structure which would be a best-fit for a pregiven extensional structure. It should be pointed out, that the clustering perspective as it is stated here, actually, has never been pursued; in particular, no measure of discrepancy between the extensional sets and discriminant functions has been suggested so far (neither in this volume). However, it is only now when the discriminant function is identified as a classification structure, such a perspective can be suggested. Actually, this comment is in line with current data analysis developments. Some authors prefer to apply several known clustering techniques to the same data thus finding both extensional and intensional descriptions of the same structure, which gives a better understanding of what has been really found.

3. The clustering discipline seems yet to have not reached "maturity". Clustering is considered helpful to generate questions and hypotheses when our knowledge of the phenomenon in question is poor, but it is still primarily an art rather than science (Jain and Dubes 1988). "Do so-and-so" recepies prevail without any clear explanation why. For instance, in agglomerative clustering, we calculate a similarity measure between clusters, merge the most similar clusters, update the similarities. This is a rule as simple as it is puzzling. Which similarity measure should be taken for the original entities and for updating? A list of the problems facing a clustering algorithm user includes the following:

   (a) Choice of the parameters in a particular clustering algorithm (between-entity distance measure, between-cluster distance measure, number of clusters, threshold of significance, etc.);

   (b) Evaluation of the extent of correspondence between the clusters and the data;

   (c) Interpretation of the clusters found;

   (d) Choice of a clustering algorithm?

(e) Distinguishing between the results based upon the data and those upon the algorithm;

(f) Comparison of the results obtained with different algorithms;

(g) Processing mixed data; etc.

Mathematical modeling in clustering should address at least some of the issues. A unified framework based on extensions of the bilinear low-rank approximation models to cluster analysis problems, provided in Chapters 4 to 6 and Section 7.6, gives answers to some of the problems listed.

3.3   Interpretation Aids

3.3.1   Visual Display

In distance/similarity based cluster analysis, visual representation of the geometry of the variable space along with the entities represented as space points is considered a major interpretation tool.

Updated reviews of the graphical techniques employed can be found in Young, Faldowski and McFarlane 1993 and Dawkins 1995. In clustering, we can indicate two major graphic designs: box-plot and scatter-plot.

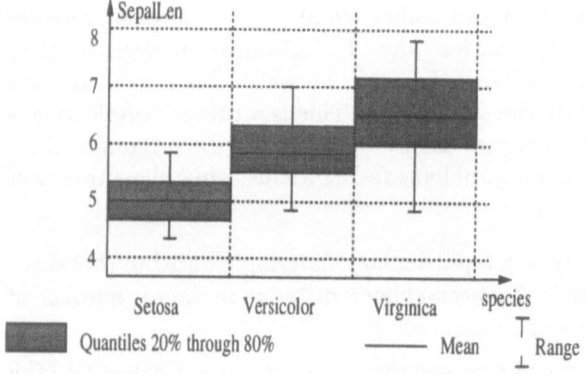

Figure 3.10: Box-plot of three classes of Iris specimens from Table 1.10 by the variable v1 (sepal length); the cluster bodies are presented by quantile boxes; the quantile value can be adjusted by the user.

A *box-plot* is an all-clusters-to-one-variable picture; it shows the distribution of the within cluster ranges of a variable in the entire domain of the variable; the averages and medians can be shown as in Fig. 3.10. A *scatter-plot* is a screen display representation of all the entity points in a plane generated by two variables or by two linear combinations of the variables (currently, many programs include some facilities for 3-dimensional representation).

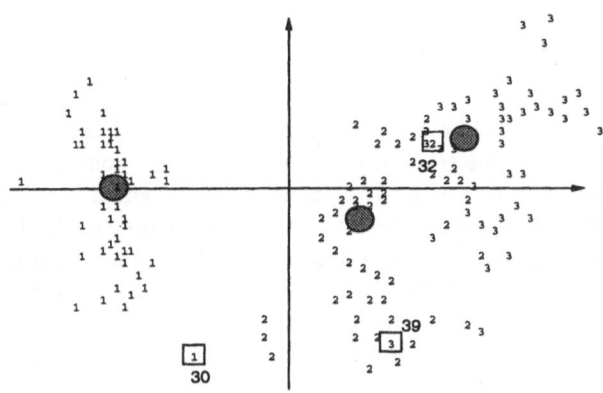

Figure 3.11: Scatter-plot of the Iris specimens in the plane of the first two principal components; the class centroids are presented by grey circles; the most deviant entities (30 in class 1, 32 in class 2, and 39 in class 3) are in the boxes.

Among the linear combinations, the following three kinds are mostly used: (a) principal components for entity-to-variable or entity-to-entity data, (b) correspondence analysis factors for contingency data, (c) canonical correlation components between the data and cluster partition indicator matrices.

These kinds of pictures seem to have little relation to the interpretation problems beyond the situations when the spatial arrangement is employed as a tool for testing clustering algorithms or has another substantive meaning.

3.3.2   Validation

Validation is an activity referring to testing the results of clustering. Examples of the problems under attack: to which extent the clusters found are cohesive and/or isolated, is the number of the clusters determined correctly or not, etc. Although validation and interpretation are not coincident, there are many common features to allow thinking of them as of quite intermixed: for instance, finding a good interpretation is a part of validation; conversely, if the clusters are invalid, the interpretation appears unnecessary. For a theoretical cluster structure, its validity depends on the validity of the corresponding model, which can be addressed, at least theoretically, in the framework of the model, which is the case of probabilistic clustering.

The following concerns only empirical cluster structures, mostly subsets, partitions, and hierarchies. Usually, two different bases for validation are considered: internal and external. Internal validity testing is claimed to be oriented toward comparing the cluster structure and the data the clusters derived from, while the external validation is to be based upon information which was not utilized for producing the clusters (see Jain and Dubes 1988, Chapter 4 and references within). However, in common practice, the "internal" validation is devoted to comparing the cluster structure with any data, internal or external (formation of clusters is just omitted), while the "external" validation refers to relating a cluster structure to another one, given previously. This is why we prefer referring to *data-based* validation, in the former case, and to *cluster-based* validation, in the latter case. Let us discuss these two kinds of validation for each of the three kinds of empirical cluster structures (subsets, partitions and hierarchies) in turn.

Validity of Single Clusters

Cluster-Based Validity

Correspondence between a cluster S and a pre-fixed subset T can be analyzed by comparing these two subsets: the cluster is perfectly valid if $S = T$; the larger the difference between the subsets, the less the cluster validity. There are two kinds of error to be taken in account somehow: the number of elements of S which are out of T (corresponding to error of the first kind in statistics), and the number of elements of T which are out of S (error of the second kind). There exist many indices suggested to measure the degree of correspondence (see Chinchor, Hirshman, and Lewis 1993, Gebhardt 1994 and Section 4.1.2).

Data-Based Validity

The general premise for this kind of validation is that the more cluster structure is unusual, the more it is valid (Jain and Dubes 1988).

When the cluster is found from entity-to-variable data, its validity can be tested by comparing the within-cluster means of the variables with their grand means in all the sample. The larger the differences, at least by some of the variables, the better the cluster. This kind of comparison could be made in the analysis-of-variance style, although the latter is applied usually only for studying a partition, not just a single cluster.

Another reasonable kind of parameter to look at is correlations between the variables. A great difference between the within-cluster correlation pattern and the pattern in the whole entity set can be utilized for both validity testing and producing a meaningful interpretation.

When a cluster is compared with a dissimilarity matrix, its validity is analyzed based on a concept of cluster. For example, when the user considers a subset S being a clump cluster if its internal dissimilarities are smaller than those oriented outward, $d_{ij} < d_{ik}$ for any $i, j \in S$ and $k \in I - S$, the cluster validity is evaluated based on comparison of the set of the triples $(i, j, k) \in S \times S \times (I - S)$ for which the inequality above holds with all the set $S \times S \times (I - S)$. The closer the subsets, the better validity. Or, when a cluster is defined as a clique component in a threshold graph, two quantities can be used: number of the actual within-cluster edges compared with $|S|(|S| - 1)$, number of the edges when S is a clique, and number of the actual outward edges compared with zero (Jain and Dubes 1988).

Validation of a Partition

Partition-Based Validation

This is frequently used for evaluating performance of a clustering algorithm: the data are generated according to a prior "noisy" partition; the more resulting partition resembles the prior one, the better performance. A contingency table having its rows corresponding to clusters found, and its columns, to classes pre-given (frequently called a confusion table) seems a best means for checking the resemblance as it is. There have been many indices of similarity/dissimilarity between partitions suggested in the literature based on the contingency table (see, for example, Goodman and Kruskal 1979, Hubert and Arabie 1985, Jain and Dubes 1988, Agresti 1984); some of them are considered in Section 5.1.4.

Data-Based Validation

The primary problem of estimating the number of clusters has found no satisfactory general solution yet since there is no satisfactory theory of what the clusters are. In our opinion, no general concept of cluster can be suggested; clusters are different depending on the problem, the data measurement, the variables involved, etc., which makes the question irrelevant except for the cases when the data are just another form of a partition. However, there can be some heuristic simplicity/stability considerations employed. For example, sometimes, in indexed trees produced with agglomerative algorithms, the larger classes are much higher than their constituents (see Fig. 3.6); this indicates a distance zone which can be cut to produce more-or-less stable clusters. In another example, when there are several related sets of variables (chemical tests, radiology-based tests, and physiological tests, in medicine), that number of clusters can be considered relevant when the partitions made by the different bases are most similar.

In the context of entity-to-variable data, many general suggestions have been made on testing the hypothesis that two clusters have been generated from the same multidimensional distribution; if the hypothesis is confirmed, the clusters are recommended to merge.

A convenient way to deal with the problem is by defining an index of similarity between the cluster structure and the data table. The earlier developments have been based on the so-called point serial correlation coefficient between the original entity-to-entity distance matrix and the matrix of partition-based distances defined as 0 when the entities belong to the same cluster or 1, otherwise. Although there is no good probabilistic theory for that coefficient, its relative values allow evaluating and comparing cluster structures by their fit into data. In Monte-Carlo study by Milligan 1981 this index (along with that of rank correlation between the same matrices) outperformed many others. More on this index and its relevance to the problem see in Section 5.3.

Validation of a Hierarchy

Hierarchy-Based Validation

The number of between-tree similarity indices suggested in the literature is not as large as the numbers of those for partitions or subsets, just half a dozen. The most respected of them is the so-called cophenetic correlation coefficient which is just the product-moment Pearson correlation coefficient between corresponding ultrametrics (see Sokal and Rohlf 1962, and Section 7.1.9). For this and some other indices, some probabilistic considerations are provided to estimate their statistical significance (see Lapointe and Legendre 1992, Steel and Penny 1993).

Data-Based Validation

Here also, matrix correlation coefficients are employed: correlation between the hierarchy ultrametric and the entity-to-entity distance matrix, as well as its rank correlation twin (see Jain and Dubes, pp.166-168). No sound experience has been accumulated yet: it seems that the theory of hierarchic clustering structures is taking its very first steps.

Let us make several concluding remarks:

First, in spite of some theoretical developments made on validation as described above, the most convincing validation technique is a kind of experimental testing which is called cross-validation. Cross-validation consists of systematic replicated clustering for various subsamples of the data and comparing the results. For example, you may be quite confident in your results if you get the same number of clusters along with almost the same centroids, having received similar results for many randomly chosen subsamples.

Second, so far, the only statistical hypothesis tested against the cluster structure found is that of completely random sources of data. Since every reasonable algorithm finds a "nonrandom" substructure in the data set, the hypothesis usually fails; moreover, such testing should be considered a testing of randomness of the data rather than of the clustering results. It would be fair to expect some more restrictive hypotheses suggested for testing against the cluster structure found.

Third, the matrix correlation concept (considered by Daniels 1944, Sokal and Rohlf 1962, Mantel 1967, etc.) seems to be a universal tool for measuring resemblance between different kinds of cluster structures and data sets. In the sequel, this opinion will be supported with some more examples of the indices that can be considered as the matrix correlation coefficients (including most known statistical coefficients such as chi-square contingency coefficient or correlation ratio). In the present author's opinion, the coefficient can be adjusted for all possible validation problems (for two-way data), which should be considered a direction for future developments.

3.3.3 Interpretation as Achieving the Clustering Goals

Current clustering literature provides no general analysis of the interpretation problem except perhaps the monograph by Romesburg 1984 where many interesting interpretation issues are raised. For example, in his elegant account of using clustering for scientific discovery (slightly modified here), Romesburg suggests that any discovery involves the following three sequential steps: (1) Asking a question on a phenomenon observed; (2) Creating a hypothesis answering the question; (3) Testing the hypothesis. If the hypothesis fails, another one (or more) is formulated based on the knowledge acquired in testing. The cluster analysis can be considered a framework to put the steps in a restricted setting. The question (1) here has the following form: why are the clusters such? Or, just why did these two entities occur in the same cluster? Or, in different clusters? For example, Romesburg 1984, p. 41-52, considers a set of mammals characterized by their dental formulas involving the number of teeth of each kind (incisor, canine, premolar, molar) in each of the jaws. It turns out, the mole and the pig have identical dental formulas while the walrus' formula is very different from the other carnivores. A hypothesis arises: the dental formula is related to the animal's diet. To test, we need a clustering of the set by diet; it appears, it is very different from the former one, which leads to rejecting the hypothesis. A new hypothesis says that the animal diet is related to their tooth morphology (shapes, not kinds, of the teeth) which requires a new data set, etc. This shows how clustering can help in pursuing a purpose.

To return to clustering as an activity on its own, let us consider interpretation problems in the framework of the main goals of clustering as a classification activity. In Section 1.2.4, the goals have been identified as analyzing the structure of phenomena and relating different aspects of those to each other.

The structure of a phenomenon is represented, first of all, by its essential parts and their relationship, and this is, basically, what clusters are about. Of course, the "essence" of the parts represented by the clusters is predetermined by selection of the variables and entities in the data; their relationship is reflected in the cluster structure to that extent which can be caught with a particular clustering structure involved. It should be pointed out that methods involved in finding a cluster structure have nothing to do with its interpretation. It is quite appropriate that the structure can be generated by the user just by a guess — still a data set must be involved in the process to guarantee that the user may not be completely arbitrary in his efforts; it is only the real contents of the data which provide its intensional description(s) and, thus, interpretation.

Basically, each of the three levels, (1) classification structure itself, (2) a single class, and (3) an entity, may lead to an interpretation breakthrough. Just a few examples for each of the levels:

(1) Woese 1981 cluster-analyzed a dissimilarity matrix between nucleotide sequences for 16 species of bacteria. He found out that, in addition to two well established classes of bacteria, prokaryotes and eukaryotes, a third cluster emerged which has been recognized, afterwards, as another bacteria class, archaebacteria.

The manual worker class, presented with seven occupations in Table 1.22, usually is divided into two quality subclasses (upper and lower); however, clustering the mobility table suggests a different structure (manufacturing/other), see Breiger 1981 and p. 281.

(2) Since the mid-sixties, a high stroke mortality rate has been documented in an ill-defined geographical region in the southeastern United States (a Stroke Belt), though the cause(s) of that, in spite of intense research, is still unknown (Howard et al. 1995). An even more important issue, in medicine, is defining a "regular" cluster consisting of those individuals who have not suffered, in a given time period, any illness.

(3) The Darwinian issue about the origin of humankind can be well advanced if we know which cluster of primates the homo *sapiens* species belongs to (Li and Graur 1991; see also Table 1.5, p. 30, and corresponding discussion).

Actually, every reliable fact on a cluster structure yields a series of questions about explaining and employing it.

Let us now discuss, in brief, the possibilities of relating different aspects of phenomena to each other via clustering. An intensional description of a classification structure in terms of a variable space can be considered an elementary interpretation unit. Combining them may produce a great deal of advancement in understanding phenomenon. Let us consider a couple of examples.

(A) **One classification, two or more variable spaces.** This is the most frequent case.

In the Planets data set (Table 1.3, p. 27), we have two planet clusters defined in terms of the distance to the sun. Amazingly, there are a bunch of variables having no logical connection to the distance, but still following the partition very closely: the matter is solid in one cluster and liquid (or mixed) in the other; the number of moons with a planet is not higher than 2 in the first cluster while it is larger than 7 in the second; etc. There is a strong relationship between the distance to Sun and the 5 variables in the data set, though there is no answer yet why it is so. The ninth planet, Pluto, discovered just in this century, does not fit into this regularity: all the five parameters fall in the first cluster, which makes some believe that this is not a genuine planet.

In the Digits data set (Table 1.9, p. 36), there is a Confusion cluster, 1-4-7 (see Table 5.5, p. 260), which is distinguished by the fact that its numerals have no lower segment (e7) in their styled representation (Fig 1.9). This, along with the features of the other Confusion clusters, makes a hypothesis on the plausible psychological mechanism related to digit perception as based on the absence/presence of just a few particular segments in the pictures.

In a social psychology research, three predefined groups of respondents gave quite different frequency profiles in their answers to the question concerning most special things at home. It is sufficient to say that the most special (on average), for children, was Stereo set; for parents, Furniture; and, for grandparents, Photos. However, with the respondents' comments about their choices analyzed, it was found that the real meanings of the things were not so different: they related to the respondents' feelings about important memories and relationship (Csikszentmihalyi and Rochberg-Halton 1981).

In a most profound problem of the history of humankind, where there is no consensus of scientists yet reached, bringing together three different aspects (archaeological, genetic, and linguistical) has served well to establish a plausible classification of the nations (Cavalli-Sforza et al. 1988).

(B) **Two classifications, different variable spaces.**

In Braverman et al. 1974, a data table (thirty statistical indicators for some sixty developing nations) was partitioned in two variable groups interpreted as (1) level of economic development (national income per capita was the most loading

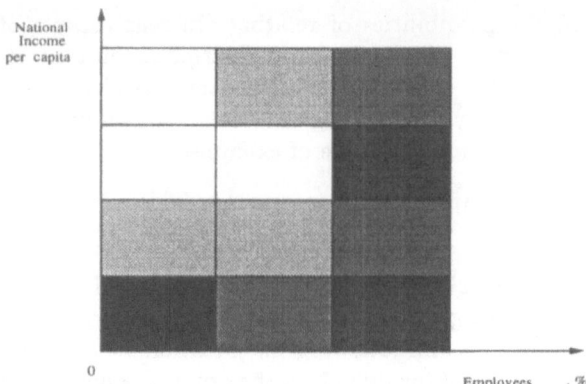

Figure 3.12: Two-factor classification of the developing countries; the darker cell, the larger the number of countries within.

variable), (2) level of market relations in the economy (employee proportion was the closest variable). Then, the set of the countries was clusterized by each of the "factors" found (see qualitative picture of the cross-classification in Fig.3.12). It can be seen that there is no simultaneous change of the factors as it would seem natural to expect. In contrast, the level of market relations increases when the level of economic development is small; the latter increases only after the former becomes high. Such a structural picture much resembles a shift in the optimal trajectory of economic development proven for a linear economy model; this could be considered empirical evidence supporting the model (claimed by many to have no empirical meaning).

(C) Many classifications, different variable spaces.

In the end of the Soviet era, a data set was collected about 47 statistical indicators concerning the following nine aspects of the socio-economic status of each of 57 regional units in Moscow region: (1) Natural reproduction of population (NRP), (2) Migration (M), (3) Demography (D), (4) Labor and work (LW), (5) Industrial and agricultural production (IAP), (6) Economics of services (ES), (7) Social infrastructure (SI), (8) Housing (H), and (9) Environment (E). The data related to 1979 and 1988. The goal was to reveal associations between the variables for using them in controlling the demographic processes in the region. Unfortunately, no quantitative relations among the variables were found (with dozens of tries involving various regression and factor analyses computations). Then, yet another attempt was made employing a structural approach: clustering of the set of units was made by each of the aspects above, along with subsequent analysis of their cross-classifications (Mirkin and Panfilova 1991). It turned out that the

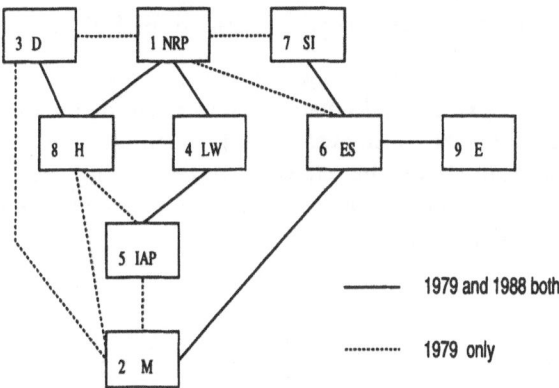

Figure 3.13: Structures of association between nine aspects of socio-economic status in Moscow region, 1979 and 1988.

variables measured, however good as statistical descriptors, were of little help in demography policy making. In the current context, the graph representation of the largest correlations between the vertices representing the aspects in Fig.3.13 is of interest as a "second-layer" structure derived from comparing the clustering results ("first-layer" structures). The graph for 1988 appears to be a part of 1979 graph; dashed edges present those valid for 1979 only. It shows that, in 1979, it was a natural structure of interrelation between industrial, economical and social subsystems; for example, the birth/death process directly correlated with demography, social infrastructure, the service economy, which was lost in 1988, leaving only physiological correlations with work and housing. The natural structure was almost destroyed by 1988, reflecting a devastated society and economy and a great necessity for change (in the near future, beginning 1990, the birth/death balance became negative in the region).

These examples present some different layers of structural relations from a huge variety of possible schemes. Much research in classification analysis is currently in simultaneously finding "empirical" and "theoretical" structures by different aspects adjusted with each other through conceptual clustering techniques (see, for example, Kubat, Pfurtscheller and Flotzinger 1994 or Fisher et al. 1993).

3.3.4 Discussion

Unlike the validation topic, the interpretation subject still has not received too much attention from a theoretical side. However, the validation techniques developed also seem kind of a preliminary study rather than a matured discipline.

However, the definition of clustering given here suggests a framework for theoretical thinking on interpretation. An extensional cluster structure supplemented with an intensional one is suggested as the "elementary unit" of interpretation. Practical advancement in interpretation should be expected in developing methods for finding such elementary units: simultaneous revealing of extensional and intensional cluster structures for the same data set.

Chapter 4

Single Cluster Clustering

FEATURES

- Various approaches to comparing subsets are discussed.

- Two approaches to direct single cluster clustering are described: seriation and moving center separation, which are reinterpreted as locally optimal algorithms for particular (mainly approximational) criteria.

- A moving center algorithm is based on a novel concept of reference point: the cluster size depends on its distance from the reference point.

- Five single cluster structures are considered in detail:

  ◇ Principal cluster as related to both seriation and moving center;

  ◇ Ideal fuzzy type cluster as modeling "ideal type" concept;

  ◇ Additive cluster as related to the average link seriation;

  ◇ Star cluster as a kind of cluster in a "non-geometrical" environment;

  ◇ Box cluster as a pair of interconnected subsets.

- Approximation framework is shown quite convenient in both extending the algorithms to multi cluster clustering (overlapping permitted) and interpreting.

4.1   Subset as a Cluster Structure

4.1.1   Presentation of Subsets

Let us recall that the base entity set is denoted as I and its cardinality, $N = |I|$. A subset $S \subseteq I$ is a group picked out of the "universe" I, for instance, a committee or just the set of all individuals of a given age. As is well known, there are three major forms for presenting subsets: a) Enumeration, when the subset is given by the list of its elements, $S = \{i_1, ..., i_m\}$; b) Indicator: a Boolean vector, $s = (s_i)$, $i \in I$, where $s_i = 1$ if $i \in I$ and $s_i = 0$ otherwise; c) Intensional predicate $P(i)$ defined for $i \in I$, which is true if and only if $i \in S$. The indicator function can be considered a binary variable assigned to S questioning "Whether an $i \in I$ belongs to S" and assigning 1 if Yes, and 0 if No.

For example, a subset of the entities in the Masterpiece data in Table 1.12, (given by their numbers) can be presented as: a) $S = \{1, 2, 3\}$ or, in a more convenient notation, $S = 1 - 2 - 3$; b) $s = (1, 1, 1, 0, 0, 0, 0, 0)$; c) "Works by A. Pushkin" or "Works with LenSent < 16".

Another useful form for presenting subsets: via binary relation $\sigma = S \times S$ which is the Cartesian square of S, that is, the set of all ordered pairs (i, j) with both $i, j \in S$. The relation σ can be presented by Boolean matrix, $s = ss^T$, having $s_i s_j$ as its (i, j)-th element (s is the indicator of S). Obviously, matrix s has a specific structure of ones filling in a "square" corresponding to S. This matrix will be referred to as the *set indicator matrix*.

The indicator matrix presentation of S is, actually, a similarity matrix with all the within similarities equal 1 and all the outer ones, 0. Yet, in some situations, the intensity of the mutual within similarities in the indicator matrix can be different from 1. Matrix λs has its elements 0 and λ; it can be interpreted as an S-cluster with weight intensity λ. This weighting can be extended to the outer similarities shift from zero to another value (Mirkin 1987b).

Representing subsets S along with their standard (centroid) points $c = (c_1, ... c_n)$ will be done through the matrix format, sc^T, whose elements are $s_i c_k$, $i \in I, k = 1, ..., n$, which can be referred to as a type-cluster structure; the rows of such a matrix are equal to c or zero depending on whether corresponding i's belonging to S.

A set of weighted clusters, $S_W = \{S_w, \lambda_w\}$, $w \in W$, can be presented as an additive structure, that is, a matrix $\sum_{w \in W} \lambda_w s_w s_w^T$ with its entries, $\sum_{w \in W} \lambda_w s_{iw} s_{jw}$, $i, j \in I$. However, this requires recovering the original clusters from the additive structure, which has been proven possible for some kinds of sets of clusters, like (weak) hierarchy. Analogously, type-clusters S_w, c_w, $w \in W$ can be presented by a summary matrix $\sum_{w \in W} s_w c_w^T$ with its entries equal to

$\sum_{w \in W} s_{iw} c_{wk}$, $i \in I$, $k = 1, ..., n$.

4.1.2 Comparison of the Subsets

For any two subsets, $S, T \subseteq I$, their comparison can be made with a so-called four-fold table (see Table 4.1) containing the numbers of elements in corresponding subset intersections (see Fig. 4.1).

Set	T	$\bar{T}$	Total
S	a	b	a+b
$\bar{S}$	c	d	c+d
Total	a+c	b+d	$a + b + c + d = N$

Table 4.1: Four-fold table.

This is a statistical representation of the information on interrelation between S and T as subsets of I: a refers to the common elements while b and c to the elements presented in only one of the sets; d is the number of "outer" elements in I with regard to S and T.

When weight coefficients are assigned to the entities, the total weights are counted as the entries of the four-fold table. Actually, the four-fold table is nothing but a 2×2 contingency table for the binary variables associated with the subsets.

The contents of the table can be presented in a vector form as $[S, T](I) = (a, b, c, d)$.

We can distinguish between (a) structural and (b) relational approaches in comparing subsets, which can be introduced with the following example.

In a survey of 7000 jury trials, the judges were asked about the decisions they would have made if they had been on the jurors' bench. From their responses, it was found that they chose the same verdict as jurors in 78% of the cases. The remaining 22% (about 1540) of the judges had a different opinion: they would have freed 210 defendants who were convicted by the juries, and convicted 1330 defendants who were acquitted by the juries. This was claimed to be a serious discrepancy.

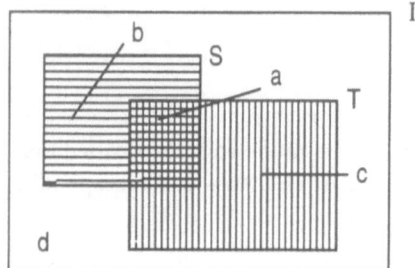

Figure 4.1: General pattern of overlapping sets S and T as parts of I.

Let us analyze the story in terms of the four-fold table. The information can be put in Table 4.2.

Jury Judge	Freed	Convicted	Total
Freed	a	b=210	a+b
Convicted	c=1330	d	c+d
Total	a+c	b+d	7000

Table 4.2: The data on the judges' opinions on the jury's verdicts reached (Parade Magazine, New York, July 30, 1995, p.12).

In her presentation (Parade Magazine, New York, July 30, 1995, p.12), the author does not bother herself to supply yet another figure needed to fill in all the table since it is not necessary for the final conclusion that there are $b + c = 1540 = 22\%$ of contradicting opinions, and c supplies the major part of it. It is only the extent of coincidence between the judge's and jury's Freed verdict, which matters. Based on this data, we can conclude that the juries are more likely to make acquittals than the judges since $a + c = a + 1330$ (juries' acquittals) is larger than $a + b = a + 210$ (judges' acquittals). However, this leaves another question unanswered. Does the juries' verdict affects the judges' opinion (in the poll) or not? To address the issue, we need to compare some proportions, not just the subsets themselves. For example, if the proportion, $a/(a+c)$, of the defendants to be freed by the judges' opinion among those acquitted by the juries is larger than that of the defendants freed by the judges among the all defendants, $(a+b)/N$, that would mean that a positive influence does exist. If the proportions coincide, there is no bias in the judges' opinions related to the jury's verdict.

However, having no information about the key value a in the example, we cannot derive anything about the bias. If, for instance, $a = 3000$, the quotient of proportion, $aN/(a+b)(a+c)$, is larger than 1, and if a=30, it is less than 1; in the former case, the judges are more likely to free the defendants freed by juries, while it is opposite, in the latter case.

The structural approach involves comparison of the subsets as they are while the relational approach concerns their interdependence as measured by comparing proportions. This is why, usually, the structural measures are employed for comparing entities (characterized by some binary variables) while (binary) variables are compared with the relational approach. In the former case, S and T are sets of the attributes that are present at the entities compared. In the latter case, S and T relate to the entity sets attributed by the binary variables compared.

Structural Approach

The most important structural indices are:

1. *Match coefficient*
$$s(S,T) = \frac{a+d}{a+b+c+d},$$

2. *Mismatch coefficient*
$$\delta(S,T) = \frac{b+c}{a+b+c+d},$$

3. *Jaccard mismatch coefficient*
$$e(S,T) = \frac{b+c}{a+b+c},$$

4. *Jaccard match coefficient*
$$r(S,T) = \frac{a}{a+b+c},$$

Obviously, the coefficients s, δ and e, r complement each other to unity. The Jaccard coefficient does not involve the d observations included neither in S nor in T, thus referring only to those showing up in the sets.

The following properties of the four-fold table are relevant to the structural approach.

Statement 4.1. *Equality $b = 0$ holds if and only if $S \subseteq T$, and $c = 0$ if and only if $T \subseteq S$; $b = c = 0$ means $S = T$; $a = 0$ means $S \cap T = \emptyset$.*

Proof: Obvious.   □

Thus, presence of a zero in the table witnesses a set-theoretic inclusion, which might be useful in deriving implication rules about variables.

Statement 4.2. *The numerator of the coefficient above, $d(S, T) = b + c$, is equal to the city-block distance between corresponding indicator vectors s, t and to the squared Euclidean distance between them.*

Proof: The statement follows from the equality $|s_i - t_i| = (s_i - t_i)^2$ which holds for every $i \in I$ since values s_i, t_i can be 0 or 1 only.   □

In other words, $d(S, T)$ is the cardinality of the symmetric difference $S \triangle T = (S - T) \cup (T - S)$, which means that

$$d(S, T) = |S \triangle T| = |S| + |T| - 2|S \cap T|. \tag{4.1}$$

It should be noted that the city-block distance between Boolean vectors is frequently referred to as Hamming distance. Yet another interpretation of the distance: it is the minimum number of elementary changes (1 to 0 or 0 to 1, which means adding or excluding an element) necessary to transform s into t.

To realize how different the symmetric-difference metric space is from our habitual Euclidean geometry, let us consider a concept of *betweenness*. Set R will be referred to as being between sets S and T if it makes equality in the triangle inequality: $d(S, T) = d(S, R) + d(R, T)$, $R, S, T \subseteq I$.

Statement 4.3. *Set R is between S and T if and only if $S \cap T \subseteq R \subseteq S \cup T$, or, equivalently, $s_i t_i \leq r_i \leq s_i + t_i$.*

Proof: Let $d(S, T) = d(S, R) + d(R, T)$; that is, $\sum_{i \in I}(|s_i - r_i| + |r_i - t_i| - |s_i - t_i|) = 0$, which can hold only if $|s_i - r_i| + |r_i - t_i| - |s_i - t_i| = 0$ for every $i \in I$, since $|s_i - r_i| + |r_i - t_i| \geq |s_i - t_i|$ by the properties of the absolute value. Thus, $s_i = t_i$ implies $r_i = s_i = t_i$.   □

In terms of the indicator vectors, s and t, considered as vertices of the N-dimensional Boolean cube, all the vectors of the interval between $s \wedge t$ and $s \vee t$ (where $s_i \wedge t_i = \min(s_i, t_i)$ and $s_i \vee t_i = \max(s_i, t_i)$) are between s and t (and nothing else).

Thus, the mismatch coefficient is a metric since its denominator, N, is just a constant. Amazingly, the Jaccard mismatch coefficient also is a distance.

Statement 4.4. *The quantity*

$$e(S,T) = \frac{b+c}{a+b+c} = 1 - \frac{|S \cap T|}{|S \cup T|}$$

is a distance satisfying all metric axioms.

Proof: To prove the triangle inequality: $e(S,T) \le e(S,R) + e(R,T)$ for any $S,T,R \subseteq I$, let us introduce corresponding notation, as follows: $[S,T](R) = (a_1, b_1, c_1, d_1)$ and $[S,T](\bar{R}) = (a_2, b_2, c_2, d_2)$. In these symbols, $e(S,T) = 1 - (a_1 + a_2)/(a_1 + a_2 + b_1 + b_2 + c_1 + c_2)$, $e(S,R) = 1 - (a_1 + b_1)/(a_1 + b_1 + a_2 + b_2 + c_1 + d_1)$, and $e(T,R) = 1 - (a_1 + c_1)/(a_1 + c_1 + a_2 + c_2 + b_1 + d_1)$. Evidently, for $0 < \alpha \le \beta$ and $\gamma \ge 0$,

$$\frac{\alpha}{\beta} \le \frac{\alpha + \gamma}{\beta + \gamma},$$

which can be verified directly. Let us add c_2 as γ to the fraction parts in $e(S,R)$ while b_2 is added to $e(T,R)$. This makes both of the denominators equal to each other, which leads to the following:

$$e(S,R) + e(R,T) \ge 2 - (2a_1 + b_1 + b_2 + c_1 + c_2)/(a_1 + a_2 + b_1 + b_2 + c_1 + c_2 + d_1).$$

The right part of the inequality, obviously, is equal to $(2a_2 + b_1 + b_2 + c_1 + c_2 + 2d_1)/(a_1 + a_2 + b_1 + b_2 + c_1 + c_2 + d_1)$. Subtracting d_1 from both numerator and denominator in the last expression, we have

$$e(S,R) + e(R,T) \ge (2a_2 + b_1 + b_2 + c_1 + c_2 + d_1)/(a_1 + a_2 + b_1 + b_2 + c_1 + c_2)$$

$$\ge (b_1 + b_2 + c_1 + c_2)/(a_1 + a_2 + b_1 + b_2 + c_1 + c_2) = e(S,T),$$

which proves the statement. □

It is not difficult to show that $R = S \cup T$ satisfies equality $e(S,T) = e(S,R) + e(R,T)$; all the other R, in general, do not fit into that.

Relational Approach

Conditional probability $p(T/S) = a/(a+b)$ is a key concept in the relational approach as a measure of the predictive capability of S with regard to T: it shows the rate of correct predictions of T when a randomly selected entity is observed to belong to S (prediction rule: $S \to T$). Respectively, $b/(a+b)$ shows the proportion of errors occurring when this prediction rule is used.

However, as it was shown in Section 1.6.1, the conditional probability alone is not sufficient for comparison. "Eating cucumbers is the major cause of death: 100% of the dead ate cucumbers". This somewhat somber joke pinpoints the necessity to consider what occurs in another part of I (which is $\bar{S}$) to make a proper conclusion.

In statistics, two approaches have been developed for taking into account both, S and $\bar{S}$.

The first relates to testing a statistical hypothesis. In this context, S presents a hypothesis on T, which involves two of the prediction rules: $S \rightarrow T$ and $\bar{S} \rightarrow \bar{T}$. Respectively, two kinds of error can occur: the first related to errors when T is predicted (based on an observation from S), the second when errors occur when $\bar{T}$ is predicted (based on an observation from $\bar{S}$). The first is $E1 = p(\bar{T}/S) = b/(a+b)$, called an error of the first kind while the second, $E2 = p(T/\bar{S}) = c/(c+d)$, is called an error of the second kind. Note, the two errors are based on different parts of the four-fold table, which implies they may be quite different.

An ideal situation occurs when both of the errors are zero: this is the case when $S = T$. A situation is not bad when one of the errors is zero, while the other is small. That means that one of the two prediction rules works correctly; $S \rightarrow T$, if E1=0, or $\bar{S} \rightarrow \bar{T}$, if E2=0; and still the other rule allows for correct negation in most cases, in contrast to the example of cucumber-eating above, where E1=0 while E2=1.

The other approach, which we just discussed in the trial example, is to relate both S and $\bar{S}$ to T only (without involving $\bar{T}$) based on comparison of the conditional probability, $P(T/S) = a/(a + b)$, and the unconditional probability of T, $p(T) = (a + c)/N$. When these two probabilities coincide, the subsets S and T are referred to as *statistically independent*. In statistics, the difference between the probabilities is expressed, traditionally, using highly symmetrical expressions such as difference, $aN - (a + b)(a + c)$, or quotient, $Q = aN/[(a + b)(a + c)]$.

The quotient, $Q = P(T/S)/P(T) = P(S \cap T)/[P(S)P(T)]$, has also a predictive meaning: it shows the change of probability of T when S is taken into account. It is called *odds ratio* (see p. 54) and widely used as a measure of interdependence of S and T; its inverse logarithm, $I(S, T) = -\log Q$ is called *mutual information*.

There can be two other prediction-based measures indicated: the *absolute probability change* (APC), $\Delta(T/S) = P(T/S) - P(T)$, and *relative probability change* (RPC), $\Phi(T/S) = (P(T/S) - P(T))/P(T) = Q - 1$ (considered in Chapter 1 in the framework of aggregable data analysis, see Section 1.6). Each of these indices can be interpreted as an estimate of improvement in predicting, for a randomly selected entity, whether it belongs to T, when it becomes known that the entity is from S, not just from all I.

Obviously,

$$\Delta(T/S) = \frac{ad - bc}{(a + b)N}, \quad \Phi(T/S) = \frac{ad - bc}{(a + b)(a + c)} \qquad (4.2)$$

A remark: Δ seems more relevant when $P(T)$ is large as in the example considered; Φ seems better fitting in the situations when $P(T)$ is small. For example, in a district nearby the Chernobyl nuclear reactor in Belarus, the rate of thyroid cancer (T event) has reached some 101 cases per million children after the accident at 1986 (S event); before, it was one case per million children. The difference is 0.0001 but the ratio is 100 to show that the risk of getting thyroid cancer has increased 10,000%!

Yet a different approach should be presented, which can be referred to as a geometrical approach.

Geometrical approach

This approach is based on the binary indicator presentation of the subsets; the indicators considered as the variables lead to measures of covariance or correlation between them. For instance, the noncentral product-moment correlation coefficient between vectors $s = (s_i), t = (t_i)$, $i \in I$, equals (by definition)

$$r_0 = \sum_i s_i t_i / (\sum_i s_i \sum_i t_i)^{1/2} = \frac{a}{\sqrt{(a+b)(a+c)}}$$

while the Pearson's product-moment correlation coefficient is

$$r = \frac{ad-bc}{\sqrt{(a+b)(a+c)(b+d)(c+d)}}.$$

The latter coefficient is quite popular; it can be interpreted operationally through differences between the conditional probabilities: $a_1 = P(S/T) - P(S/\bar{T})$ and $a_2 = P(T/S) - P(T/\bar{S})$ which are, also, the regression coefficients of the indicator vectors, s by t and t by s, respectively. It turns out, the absolute value of the Pearson product-moment correlation coefficient is the geometric mean of these values, $|r| = \sqrt{a_1 a_2}$. Similar interpretation can be provided in terms of the relative change of probability coefficients, $\Phi(T/S)$ and $\Phi(\bar{T}/\bar{S})$, since $r^2 = \Phi(T/S)\Phi(\bar{T}/\bar{S})$.

4.1.3   Discussion

1. The subset is an important classification structure pertaining to an "elementary" classification unit, a class.

2. Mathematical models of subsets involve both extensional and intensional meanings (the enumerated subset and logical predicate); moreover, the indicator concept as a vector space element will be employed in the approximation clustering models in Chapters 4 to 6.

3. Correspondence between subsets can be evaluated in different frameworks; among them, the structural, relational and geometrical approaches are distinguished.

4.2   Seriation: Heuristics and Criteria

4.2.1   One-by-One Seriation

Let, for every subset $S \subset I$ and every $i \in \bar{S} = I - S$, a similarity measure $a(i, S)$ (or, a dissimilarity measure, $d(i, S)$), between i and S, be known. Such a measure could be called a similarity (or dissimilarity) *linkage* between i and S. A set of linkages can be defined based on an entity-to-entity similarity matrix given, $A = (a_{ij})$, $i, j \in I$, as follows.

1. *Single linkage* or *Nearest neighbor*

$$sl(i, S) = \max_{j \in S} a_{ij};$$

2. *Summary linkage*

$$sul(i, S) = \sum_{j \in S} a_{ij};$$

3. *Average linkage* or *Average neighbor*

$$al(i, S) = \sum_{j \in S} a_{ij}/|S|;$$

4. *Threshold linkage*

$$l_\pi(i, S) = \sum_{j \in S}(a_{ij} - \pi) = \sum_{j \in S} a_{ij} - \pi|S|;$$

where π is a fixed threshold value.

In the situation when the elements of S are ordered in a series, $S = \{i_1, ..., i_{|S|}\}$, yet another measure might be useful,

5. *Chain linkage* or *Last neighbor*:

$$ll(i, S) = a_{ii_{|S|}}.$$

Similar linkage functions can be defined in terms of dissimilarities; in this latter case, the minimum must be taken in the single linkage. Linkage measures can be defined also for different kinds of data. For example, for an entity-to-variable matrix, $Y = (y_{ik})$, $i \in I, k \in K$, a similarity linkage measure $a(i, S)$ can be defined as the scalar product of the row $y_i = (y_{ik})$, $k \in K$, and the center of gravity $c = (c_k)$ of S-rows y_j, $j \in S$. Yet this particular measure can be reformulated as the average linkage above. Indeed, since $c = \sum_{j \in S} y_j / |S|$, the scalar product equals

$$a(i, S) = (y_i, c) = \sum_{j \in S} (y_i, y_j) / |S| = al(i, S) \qquad (4.3)$$

for $a_{ij} = (y_i, y_j)$, $i, j \in I$.

Yet another, dissimilarity, linkage function defined in terms of the entity-to-variable data:

$$ml(i, S) = \sum_{k \in K} \min_{j \in S} |y_{ik} - y_{jk}| \qquad (4.4)$$

which is an example of "holistic" linkage which is not reducible to pair-wise dissimilarities between entities.

One might think also of a situation when such a linkage measure arises just as a kind of primary data, which seems possible in technical applications connected to VLSI or image processing.

For any linkage measure, a one-by-one adding procedure for seriation of the entities in I can be defined as follows.

One-by-One Seriation
The procedure starts with $S = \{i_0\}$ where $i_0 \in I$ is an arbitrary entity. General step: given S, find $i^* \in I - S$ maximizing similarity (minimizing dissimilarity) linkage measure $s(i, S)$ with regard to all $i \in I - S$ and join i^* as the last element in S seriated. Unless $S = I$, the general step is repeated.

The output of the process is I seriated along with the sequence of linkage measure values between every entity and set of the preceding elements in the series. Obviously, number of basic comparison operations here is $O(N^2)$ since, at every step, S is compared with all the remaining elements (average number of them is $N/2$), and the number of steps is $O(N)$. Obviously, the result may depend on the starting point i_0. Comparing single linkage seriation with the algorithm for finding a minimum (maximum) spanning tree (MST) (see p. 88), we can see no essential differences between them: just single linkage seriates the entities without drawing the MST tree itself, though the minimal distance values kept are exactly the MST distances.

To simplify the computations, it would be nice to have the linkages $d(i, S)$ or $a(i, S)$ calculated at every step not just from scratch but updating the linkage values of the previous step. That means that we should be able to calculate linkages between i and $S + j$ (where j is the entity added to S at the step considered) for $i \notin S + j$ using linkages between i and S and the pair-wise linkage between i and j in a Lance-Williams formula fashion (see Section 3.2.3):

$$a(i, S + j) = \alpha_S a(i, S) + \alpha_j a(i, j) + \beta a(j, S) + \gamma |a(i, S) - a(i, j)| + const \qquad (4.5)$$

where α_S, α_j, β, γ, *const* are coefficients to be specified according to the particular linkage method used. Although when S is fixed the *const* does not change the order of the values and, thus, does not affect the seriation process, it does change the values of the linkage measure (across different Ss) and is included to take into account the threshold linkage formula. Table 4.3 contains the coefficients corresponding to the five formulas above.

Linkage Method	α_S	α_j	β	γ	*const*
Single	1/2	1/2	0	-1/2	0
Summary	1	1	0	0	0
Average	$|S|/(|S|+1)$	$1/(|S|+1)$	0	0	0
Threshold	1	1	0	0	$-\pi$
Chain	0	1	0	0	0

Table 4.3: Coefficients of the modified Lance-Williams formula corresponding to the similarity linkage measures considered; when dissimilarity linkage is assumed, the sign of γ in the first row must be reverted.

Although β is zero in the table, we prefer keeping it since it is not so in some other methods.

In clustering, seriation has no independent meaning: this is just a means to find a "suitable" cluster with a "suitable" cut of the series found; after a cut has been done, the initial fragment of the series forms the resulting cluster.

4.2.2   Seriation as Local Search

The seriation strategy can be put in a theoretical framework in the following way. Let $f(S)$ $(S \subset I)$ be a numerical function (on the set of subsets of I) to be maximized (or minimized).

A local search algorithm for maximizing $f(S)$ is defined by a neighborhood, $N(S)$, which must be assigned to any feasible solution S. Then the local search algorithm starts from an initial S and repeatedly performs finding a best solution in $N(S)$ along with subsequent substitution of the initial subset S by the new-found S. The search ends when the new-found solution is not better than S found at the previous iteration.

A particular neighborhood associated with seriation is defined as this: $N(S)$ consists of the subsets $S + i$ for all $i \in I - S$. With this neighborhood, the local search iteration consists of adding to S that element $i \in I - S$ which maximizes the increment $\delta(i, S) = f(S + i) - f(S)$, which is, basically, an iteration of the seriation procedure with $\delta(i, S)$ utilized as a similarity linkage measure. To make the resemblance tight, let us start the computation from a singleton $S = \{i_0\}$.

> However, there is an important difference between the local search and seriation procedures: the local search algorithm may seriate not all the entities; it must stop when the increment $\delta(i, S)$ becomes negative.

Is it possible to find set functions behind the linkage methods considered? Can a set function be specified for a given linkage in such a way that the linkage formula coincides with $\delta(i, S)$ for the set function? In general, the answer is no. However, for the collection of particular linkages above, the following set functions (defined in terms of a similarity matrix $A = (a_{ij})$) can be considered, to some extent, as the optimized criteria:

1. $SL(S) = \max_{i \in I - S} \max_{j \in S} a_{ij}$;

2. $SUL(S) = \sum_{i \in S} \sum_{j \in S} a_{ij}$;

3. $AL(S) = \sum_{i \in S} \sum_{j \in S} a_{ij} / \nu(S)$

4. $L(\pi, S) = \sum_{i \in S} \sum_{j \in S} (a_{ij} - \pi) = \sum_{i \in S} \sum_{j \in S} a_{ij} - \pi |S| \nu(S)$

   where π is a fixed threshold value;

5. $LL(S) = \sum_k a_{i_k i_{k+1}}$

   where S is considered seriated: $S = \{i_1, i_2, ..., i_{|S|}\}$.

In these formulas, $\nu(S)$ equals $|S|$ if the diagonal entries a_{ii}, $i \in I$, are present and $\nu(S) = |S| - 1$ if not.

Three of the functions, $SUL(S)$, $L(\pi, S)$, and $LL(S)$, match the problem above perfectly: the increment $\delta(i, S)$ for each of them equals $sul(i, S)$, $l_\pi(i, S)$, and $ll(i, S)$, respectively. This implies that when $A = (a_{ij})$ is nonnegative, functions $SUL(S)$ and $LL(S)$ suggest, actually, no stopping rule, collecting all the entities in the "universal" optimal cluster. In contrast, $L(\pi, S)$ does suggest a stopping rule: the seriation must stop when $l_\pi(i, S)$ becomes negative, which highly depends on the threshold value π.

Also, it can be easily seen that, due to the symmetric form of criteria $SUL(S)$, $L(\pi, S)$ and $AL(S)$, the matrix $A = (a_{ij})$ can be considered symmetric for each of them; otherwise, its entries must be changed to $(a_{ij} + a_{ji})/2$, which does not change the optimal solution.

Although the single linkage function, $sl(i, S)$, cannot be presented as an increment of SL, still sl-based seriation steps are equivalent to local search iterations for minimizing $SL(S)$. The local search stopping rule requires ending the process when the respective value of SL goes down (for dissimilarities) or up (for similarities).

Let us consider AL criterion in more detail since it is related with some further analysis. The average neighbor seriation is connected with optimizing criterion $AL(S)$ in the following sense. The increment of $AL(S)$ equals:

$$\delta AL(i, S) = \frac{\nu a_{ii} + 2|S|al(i, S) - AL(S)}{\nu(S) + 1} \qquad (4.6)$$

where ν equals 1 or 0 depending, respectively, on presence or absence of the diagonal entries in $A = (a_{ij})$; the same condition defines $\nu(S)$ as $|S|$ or $|S| - 1$, respectively.

Thus, the seriation process due to the local search algorithm applied to $AL(S)$ is controlled by yet another linkage function $Al(i, S) = \nu a_{ii} + 2|S|al(i, S)$. Therefore, AL-based local search seriation coincides with that produced by the average linkage if the diagonal similarities are constant or absent. The stopping rule, turning $\delta AL(i, S)$ negative, basically, depends on comparison between $2|S|al(i, S)$ and $AL(S)$ (in (4.6)).

Increment $\delta AL(i, S)$ itself can be considered a linkage. Recalculation of this in the seriation procedure (when S becomes $S + j$) can be made in terms of a Lance-Williams formula (4.5) where all the coefficients (except γ) are nonzero. We leave the task of identifying the coefficients to the reader, while suggesting another way of using recalculated values, as follows.

Local Search for AL(S)
At every iteration, the values $Al(i, S) = \nu a_{ii} + 2|S|al(i, S)$ $(i \in I - S)$ are calculated and their maximum $Al(i^*, S)$ is found. If $Al(i^*, S) > AL(S)$ then i^* is added to S; if not, the process stops, S is the resulting cluster. To start a new iteration, all the values are recalculated:
$al(i, S) \Leftarrow (|S|al(i, S) + a_{ii^*})/(\nu(S) + 1)$
$AL(S) \Leftarrow (|S|AL(S) + Al(i^*, S))/(\nu(S) + 1)$
$\nu(S) \Leftarrow \nu(S) + 1$.

4.2.3   Clusterness of the Optimal Clusters

Description of the linkage-based clustering in terms of the corresponding optimization criteria may be considered useful for better understanding of what kind of clusters should be expected with each of the criteria. Such an understanding could be reached with a thorough consideration of specially crafted examples, as it is done by Sneath and Sokal 1973, Spaeth 1985, but the general analysis may provide somewhat more general disclosure. Moreover, having a criterion may provide a better computational outfit than just one-by-one seriation procedure. Let us review the criteria in sequence.

Single Linkage Criterion

Criterion SL has been considered both for clustering and nonclustering purposes. Nonclustering applications are connected with operations research "bottleneck" problems as, for instance, the problem of location of $|S|$ noxious industrial facilities and $N - |S|$ residential facilities in the given sites $i \in I$ in order to maximize the minimum distance between a noxious and a residential facility (Hsu and Nemhauser 1979). Analysis of the problem in the clustering framework can be found in Delattre and Hansen 1980, Muchnik and Zaks 1989, Hansen and Jaumard 1993. The basic observation made by Delattre and Hansen 1980 is that, for every nonempty $S \subset I$, $SL(S)$ is equal to the weight of an edge in every minimum spanning tree (MST) (the dissimilarity setting of the data and criterion is assumed here). This means, that an optimal cluster is found by cutting an MST by an edge of the maximum weight. This provides us with a simple algorithm for optimizing $SL(S)$ and allows describing an optimal cluster as follows.

Let π be the weight of the MST edge cut, then set S is disconnected with its complement, $I - S$, in the π-threshold graph defined by the dissimilarity matrix analyzed since all the other edges between S and $I - S$ have greater weights (by the definition of MST). On the other hand, S is a component in the threshold graph since all its vertices are connected by the MST edges, at least. Thus, the single linkage optimal clusters are just components of a threshold graph, which shows that they may have a complicated spatial shape (see comment on p. 142).

Summary Linkage Criterion

Criterion $SUL(S)$, when the data entries are nonnegative, gives a trivial solution in both cases: all set I if maximized (when similarities are considered) or any singleton $\{i\}$ if minimized (when dissimilarities are treated). Fixing the number of the entities in S, seemingly inappropriate in the cluster analysis framework, is considered convenient in operations research. Maximizing $SUL(S)$ with $|S|$ fixed is an NP-complete problem since if an algorithm for the problem is developed, it can be used for answering the question whether a clique of a fixed size in a graph exists (which is NP-complete). Indeed, let us consider the adjacency matrix of the graph as a similarity matrix, maximize $SUL(S)$ for this graph and see the value of $SUL(S)$: since it is equal to the number of the edges in the subgraph on the vertex set S, it shows whether S is a clique or not. If not, there is no clique of the fixed size in the graph. Minimizing $SUL(S)$ with $|S|$ fixed is an NP-complete problem, also.

Min Cut Criterion

It is convenient to discuss here a criterion which can be minimized in polynomial time (its maximization is a hard problem), though it relates, actually, to splitting I into two parts S and $I-S$ and, in this case, is equivalent to a "uniform partitioning" criterion $SU(0, S)$ (5.18), p. 256. It is the so-called min cut criterion, a most popular concept in combinatorial optimization. Let

$$CUT(S) = \sum_{i \in S} \sum_{j \in I-S} a_{ij},$$

the overall similarity between the cluster S and its surrounding $\bar{S} = I - S$.

The problem of finding a minimum of $CUT(S)$ can be put in the context of the so-called network flow theory developed by Ford and Fulkerson 1962 which is described in every textbook on graph theory or combinatorial optimization (see, for example, Papadimitriou and Steiglitz 1982 or Bondy and Murty 1976). This problem involves a nonnegative similarity (called capacity) matrix $A = (a_{ij})$ considered as a graph along with the edge capacities a_{ij}. Two entities (vertices) are fixed and called poles; one is considered the source of a flow restricted by the edge capacities into the other, called the sink. The problem of finding a maximum flow between the source i_1 and sink i_2 is well studied and has a relatively simple solution leading to a subset S minimizing $CUT(S)$ with regard to all subsets $S \subset I$ containing i_1 and not containing i_2. This itself can be utilized for clustering with criterion $CUT(S)$ since, frequently, the user has prior knowledge of a most appropriate entity (i_1) and a most inappropriate one (i_2).

However, this can be employed also for minimizing $CUT(S)$ without any constraints. As it can be easily shown, the values $cut(i_1, i_2)$ of the minimum cut $CUT(S)$ in the restricted problem, satisfy the ultrametric inequality (see (7.1) in Section 7.3) and, thus, have not more than $N - 1$ different values, which can be found effectively (Ford and Fulkerson 1962) to resolve the min cut problem.

Summary Threshold Linkage

When a_{ij} may fall on either side of zero, the problem of maximizing $SUL(S)$ can be considered as the problem of maximizing $L(S, \pi)$ for nonnegative a'_{ij} since, obviously, the original entries a_{ij} can be considered as $a'_{ij} - \pi$ with $\pi \geq |min_{i,j \in I} a_{ij}|$ and, thus, all $a'_{ij} = a_{ij} + \pi$ nonnegative.

The problem of maximizing $L(\pi, S)$ is NP-complete for some π. To prove that, let us consider $A = (a_{ij})$ to be the adjacency matrix of a graph (with no loops, thus, with no a_{ii} given). Let $\pi = (N-1)/N$. Evidently, for any $S \subset I$, the value $L(\pi, S)$ is positive if and only if S is a clique. Then, the value is maximum when the clique contains the maximum number of vertices, which implies that the problem of maximizing $L(\pi, S)$ is NP-complete, in this case.

Yet a necessary optimality condition gives a cluster meaning to the optimal subset. Subset $S \subset I$ will be referred to as a π-*cluster* if, for every $i' \in S$ and $i'' \notin S$, $al(i'', S) \leq \pi \leq al(i', S)$. In this definition, $al(i, S)$ is defined, in the standard way, for all $i \in I$.

Statement 4.5. *A subset S maximizing $L(\pi, S)$ is a π-cluster.*

Proof: If S is optimal, then $l_\pi(i, S)$ is nonnegative for every $i \in S$ and nonpositive for every $i \notin S$. The proof follows from the fact that $l_\pi(i, S) = |S|(al(i, S) - \pi)$. $\square$

Since the threshold linkage seriation algorithm involves only this kind of manipulation with $l_\pi(i, S)$, the proof can be applied to the cluster found with this algorithm (do not forget, it stops when $l_\pi(i, S)$ becomes negative!), which implies that the clusters found with the threshold linkage algorithm satisfy a part of the defining inequality: for every $i'' \notin S$, $al(i'', S) \leq \pi$. As to the elements i' within the cluster S, some of them may also have low linkages $al(i', S) \leq \pi$ because all the later added elements may have their individual similarities to i' less than π. To get a π-cluster, excluding the entities from S must be permitted, which will be done in the next section.

Statement 4.5. suggests an interpretation of the threshold linkage clusters in terms of the average similarities within and out of them with regard to the threshold π. An explicit meaning of threshold π also becomes clear: in contrast to the "hard" threshold utilized in threshold graphs, this threshold is "soft"; it is applied to the average, not individual, similarities.

When the threshold increases, the number of the entities in an optimal cluster may only decrease.

Statement 4.6. *Let $\pi_1 > \pi_2$, and S_t be a solution to $L(\pi_t, S)$ optimization problem $(t = 1, 2)$. Then $|S_1| \leq |S_2|$.*

Proof: By definition, $L(\pi, S) = L(0, S) - \pi|S|\nu(S)$. The fact that S_1 is optimal for π_1 and S_2 for π_2 implies:

$$L(0, S_1) - L(0, S_2) \geq \pi_1(|S_1|\nu(S_1) - |S_2|\nu(S_2)),$$

$$L(0, S_1) - L(0, S_2) \leq \pi_2(|S_1|\nu(S_1) - |S_2|\nu(S_2)).$$

These inequalities combined lead to $(\pi_2 - \pi_1)(|S_1|\nu(S_1) - |S_2|\nu(S_2)) \geq 0$ which implies that $|S_1|\nu(S_1) \leq |S_2|\nu(S_2)$. Inequality $|S_1| \leq |S_2|$ holds since either of the functions $f(x) = x^2$ or $f(x) = x(x - 1)$ is strictly monotone for $x \geq 1$. $\square$

Average Linkage Criterion

This criterion emerges in the approximation framework, Sections 4.4 and 6.1.

The intuitive meaning of the criterion can be highlighted when it is rewrit-ten as this: $AL(S) = a(S)|S|$ where $a(S)$ is the average similarity, $a(S) = \sum_{i,j \in S} a_{ij}/|S|\nu(S)$. The criterion is the product of the average similarity a_{ij} within S and the number of entities in S, which provides a compromise between these two mutually contradicting criteria: the smaller the $|S|$, the greater the $a(S)$. If, for example, S is supposed to have only two elements, the criterion will be maximized by the maximum of a_{ij}'s. With the number of elements increased, one can only decrease the average $a(S)$.

When similarities are nonnegative, the problem of maximizing criterion $AL(S)$ is complicated, yet can be resolved with a polynomial-time algorithm. In graph theory, set-function $AL(S)$ is known as the *density* function; in a graph $G = (V, E)$ having vertex set $V = I$ and weights on the edges ij equal to a_{ij}, $AL(S)$ is equal to the total summed edge weight within S divided by the number of vertices in S. The problem of maximizing $AL(S)$, in this context, is known as the maximum density subgraph problem. It turns out, the problem can be resolved as a sequence of max-flow-min-cut problems for a network associated with the problem. Let us describe a method from Gallo, Grigoriadis, and Tarjan 1989 involving a four-layer network, the layers being consecutively: a source so, set E, set I, and a sink si. The source is connected with every edge ij by an arc (so, ij) of capacity a_{ij}; every edge ij is connected with its ends, $i, j \in I$, by arcs of "infinite" (as great as necessary) capacity; and every entity $i \in I$ is connected with the sink by an arc (i, si) of a capacity $\lambda > 0$. The vertices $i \in I$ belonging to the subset cut which contains sink si, form a subset S maximizing the criterion $SUL(S) - \lambda|S|$ and, thus, satisfying the inequality $AL(S) \geq \lambda$ when the criterion value is not negative. Indeed, the min cut value in the network equals $\lambda|S| + \sum_{ij \notin S \times S} a_{ij} = \sum_{i,j \in I} a_{ij} - (SUL(S) - \lambda|S|)$. The crucial fact is that the optimal cut set S can only lose some of its vertices when λ is increased, thus providing us with a simple method for finding the AL-optimal S which is the last nonempty cut set found while λ is increasing. Fig. 4.2 illustrates this: (b) represents the network corresponding to graph (a); with $\lambda = 2$, cut set $S = 2 - 3 - 4$ is the maximum density subgraph corresponding to $\lambda = AL(S) = 3$.

Unfortunately, this method cannot be extended to the case of arbitrary a_{ij}. Moreover, the problem becomes hard in the general case.

The criterion can be interpreted also in terms of the Boolean indicator func-tion $s = (s_i)$ where $s_i = 1$ if $i \in S$ and $s_i = 0$ when $i \notin S$. Obviously, $AL(S) = s^T As/s^T s$ where $A = (a_{ij})$. This is the so-called Raleigh quotient, having the maximum eigenvalue of A as its maximum with regard to arbitrary s in the case when no Boolean restriction on s is imposed. As it is well known, if A is nonnegative, its eigenvector corresponding to the maximum eigenvalue has all its coordinates nonnegative. This makes one suggest that there must be a correspon-dence between the components of the globally optimal solution (the eigen-vector) and the solution in the restricted problem with Boolean s. However, even if such a

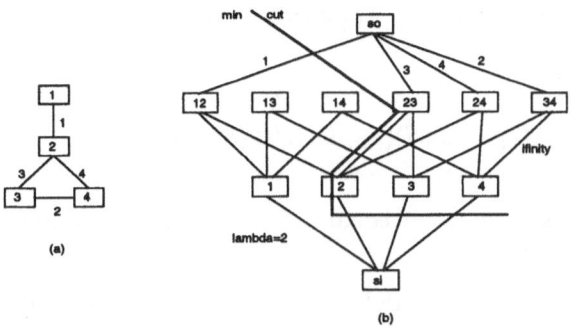

Figure 4.2: The four-layer network (b) for finding a maximum density subgraph in graph (a); for $\lambda = 2$, the maximal flow saturates edges joining si with 2, 3, and 4, and so with 1, which defines min cut shown by the bold line.

correspondence exists, it is far from straightforward. For example, there is no correspondence between the largest components of the eigen-vector and the non-zero components in the optimal Boolean s: the first eigen-vector for the 20-vertex graph in Fig. 2.8 has its maximum value corresponding to vertex 5 which, obviously, does not belong to the maximum density subgraph, the clique $1 - 2 - 3 - 4$.

Still the optimal clusters satisfy a clusterness condition. Let us refer to a subset $S \subset I$ as a *strict cluster* if it is an $a(S)/2$-cluster where $a(S)$ is the average similarity, defined as $a(S) = \sum_{i \in S} \sum_{j \in S} a_{ij} / \nu(S)|S|$. This means that, for every $i' \in S$ and $i'' \in I - S$,

$$al(i', S) \geq \frac{a(S)}{2} \geq al(i'', S).$$

The threshold value $\pi = a(S)/2$ here is not constant: it is high for a "dense" cluster and small for a "sparse" one. What seems important is that the cluster is separated from the outsiders with a barrier: each of them has its average linkage (similarity) to the cluster not larger than half of the average similarity within. The proof will follow from Statement 4.7. characterizing AL-optimal clusters.

For an $S \subseteq I$, let us set $z_i = 1$ for $i \in S$ and $z_i = -1$ for $i \in I - S$. Obviously, vector $z = (z_i)$ can be expressed with S's indicator s as $z = 2s - 1$. For any other $T \subset I$, let z_T be the vector obtained from z by making all the components z_i, for $i \notin T$, zero. Let dal denote the vector consisting of deviations $al(i, S) - a(S)/2$ as its components.

Statement 4.7. *Subset S is a global maximizer of $AL(S)$ if and only if, for any $T \subset I$,*

$$(z_T, dal) \geq z_T^T A_T / 2\nu(S) \tag{4.7}$$

Proof: Obviously, S is optimal if, for any T, $AL(S\Delta T) - AL(S) \leq 0$. On the other hand,

$$AL(S\Delta T) - AL(S) = (z_T^T A z_T - 2\nu(S)(z_T, dal))/(\nu(S) - (z_T, u))$$

where u is the vector having all its components equal to 1, which proves the statement. □

This statement shows that the differences $dal_i = al(i, S) - a(S)/2$ play a key role in the problem of optimizing $AL(S)$. When similarity matrix A is positively definite, the right part of inequality (4.7) can be used to set bounds in a version of the branch-and-bound method for optimizing $AL(S)$ (Mirkin 1987a). The strict clusterness of the optimal clusters follows from (4.7) when T consists of a single entity.

4.2.4   Seriation with Returns

Considered as a clustering algorithm, the seriation procedure has a drawback: every particular entity, once being caught in the sequence, can never be relocated, even when it has low similarities to the later added elements.

After the optimization criteria have been introduced, such a drawback can be easily overcome. To allow exclusion of the elements in any step of the seriation process, the algorithm is modified by extending the neighborhood system.

Let, for any $S \subset I$, its neighborhood $N(S)$ consist of all the subsets differing from S by an entity $i \in I$ being added to or removed from S. The local search techniques can be formulated for all criteria above based on this modification.

For the sake of the brevity, we consider here the modified algorithm only for criterion $AL(S)$. Its increment equals

$$\delta AL(i, S) = \frac{\nu s_{ii} + 2z_i|S|al(i, S) - z_i AL(S)}{\nu(S) + z_i} \tag{4.8}$$

where $z_i = 1$ if i has been added to S or $z_i = -1$ if i has been removed from S. Thus, the only difference between this formula and that in (4.6) is change of the sign in some terms. This allows the modified algorithm being formulated analogously.

Local Search with Return for AL(S)

Every iteration, values $Al(i, S) = \nu s_{ii} + 2z_i|S|al(i, S)$ $(i \in I)$ are calculated and their maximum $Al(i^*, S)$ is found. If $Al(i^*, S) > z_{i^*} AL(S)$ then i^* is added to or removed from S by changing the sign of z_{i^*}; if not, the process stops, S is the resulting cluster. To start the next iteration, all the values participating in the formulas are recalculated:

$al(i, S) \Leftarrow (\nu(S)al(i, S) + z_{i^*} s_{ii^*})/(\nu(S) + z_{i^*})$

$AL(S) \Leftarrow (|S|AL(S) + z_{i^*} Al(i^*, S))/(\nu(S) + z_{i^*})$

$\nu(S) \Leftarrow \nu(S) + z_{i^*}.$

Statement 4.8. *The cluster found with the modified local search algorithm is a strict cluster.*

Proof: The stopping criterion involves the numerator of (4.8): $\nu s_{ii} + 2z_i|S|al(i, S) - z_i AL(S) \leq 0$, for any $i \in I$. Thus, the cluster found satisfies the inequality: $z_i(al(i, S) - AL(S)/2|S|) \leq 0$ for any $i \in I$. This completes the proof since $AL(S)/2|S| = a(S)/2$. $\qquad\square$

Analogously, it can be proven that the cluster optimizing $L(\pi, S)$ with the modified local search method admitting returns is a π-cluster.

4.2.5 A Class of Globally Optimized Criteria

The seriation techniques may be considered as another formulation for the so-called greedy algorithms (Papadimitriou and Steiglitz 1982, Helman, Moret, Shapiro 1993). These kinds of algorithm have been studied from the following point of view: given a class of set functions (usually, linear set functions that can be represented as linear combinations of the function values at the individual entities), characterize the class of feasible subsets to guarantee optimality of the seriation technique results. In cluster analysis, the structure of the set of all feasible subsets, usually, is simple; just the set of all subsets (perhaps containing a pre-fixed bunch of the entities, which, basically, does not change anything since the fixed elements do not vary). However, set functions here, usually, are not linear. Nevertheless, the single linkage function has a nice property of monotonicity which can be extended to a class of dissimilarity linkage measures; the related class of nonlinear set functions can be globally optimized with a version of the seriation algorithm.

Let us refer to a dissimilarity linkage function $d(i, S)$, $S \subset I, i \in I - S$, as a *monotone linkage* if $d(i, S) \geq d(i, T)$ whenever $S \subset T$ (for all $i \in I - T$). Two of the linkage functions considered, $sl(i, S)$ and $ml(i, S)$, are monotone.

A set function $M_d(S)$ can be defined based on a linkage function $d(i, S)$:

$$M_d(S) = \min_{i \in I-S} d(i, S). \tag{4.9}$$

Following terminology of Delattre and Hansen 1980, $M_d(S)$ can be referred to as the *minimum split function* for linkage $d(i, S)$. The function measures the minimum linkage between S, as a whole, and $I - S$ as set of the "individual" entities.

A set function $F(S)$ ($\emptyset \subset S \subset I$) is referred to as *quasi-convex* if it satisfies the following condition: For any overlapping S_1 and S_2 ($S_1 \cap S_2 \neq \emptyset$),

$$F(S_1 \cap S_2) \geq \min(F(S_1), F(S_2)) \tag{4.10}$$

Statement 4.9. *The minimum split function for a monotone linkage is quasi-convex.*

Proof: Let $F(S) = \min_{i \in I-S} d(i, S)$ for some monotone linkage $d(i, S)$ and S_1, S_2 be overlapping subsets of I. Let $F(S_1 \cap S_2) = d(i, S_1 \cap S_2)$ while $F(S_1) = d(j, S_1)$ and $F(S_2) = d(k, S_2)$. By definition of F, i does not belong to S_1 or S_2, say, $i \notin S_1$. Then, $d(i, S_1) \geq F(S_1) = d(j, S_1)$ and $F(S_1 \cap S_2) = d(i, S_1 \cap S_2) \geq d(i, S_1)$ since $d(i, S)$ is a monotone linkage, which proves that F is quasi-convex. $\square$

Let us define now the *maximum join* linkage function d_F for any set function $F(S)$:

$$d_F(i, S) = \max_{S \subseteq T \subseteq I-i} F(T) \tag{4.11}$$

Statement 4.10. *The maximum join linkage function $d_F(i, S)$ is monotone if F is quasi-convex.*

Proof: Obvious, since any increase of S makes the set of maximized values in the definition of d_F smaller. $\square$

It appears that in the setting defined by conditions of quasi-convexity and monotonicity, the functions d_F and M_d are dual, that is, for any quasi-convex set function $F : \mathcal{P}(I) \rightarrow R$, the minimum split function of its maximum join linkage coincides with F. Vice versa, for any monotone linkage $d : I \times \mathcal{P}(I) \rightarrow R$, the maximum join linkage of its minimum split function coincides with d.

Statement 4.11. *For any quasi-convex set function $F : \mathcal{P}(I) \rightarrow R$, the minimum split function of its maximum join linkage coincides with F.*

Proof: Let S_i be a maximizer of $F(T)$ with regard to all T, satisfying the condition $S \subseteq T \subseteq I - i$ for $i \notin S$ so that $d_F(i, S) = F(S_i)$. The minimum split function for d_F, by definition, is equal to $M(S) = \min_{i \notin S} F(S_i)$. Thus, $M(S) \leq F(\cap_{i \notin S} S_i)$, due to quasi-convexity of $F(S)$. But $\cap_{i \notin S} S_i = S$ since $S \subseteq S_i$ and $i \notin S_i$, for every $i \notin S$, which implies $M(S) \leq F(S)$. On the other hand, $F(S_i) \geq F(S)$, $i \notin S$, since S belongs to the set of feasible subsets in the definition of S_i as a maximizer of F; this implies that $M(S) \geq F(S)$, which proves the statement. □

The duality proven is asymmetric from the algorithmic point of view: it is quite easy to construct a minimum split function M_d based on a linkage $d(i, S)$ while determining the maximum join linkage d_F by $F : \mathcal{P}(I) \to R$ may be an NP-hard problem: the former task involves enumerating the elements $i \in I - S$ while the latter task requires maximizing a set function $F(T)$. This implies that it would be more appropriate to consider monotone linkage as a tool for defining a quasi-convex set function rather than, conversely, quasi-convex set function as a tool for representing the monotone linkage.

Let us consider a quasi-convex set function F represented through a monotone linkage function $d(i, S)$ by equation (4.9), $F = M_d$. Let us refer to a series, $(i_1, ..., i_N)$, as a *d-series* if it is obtained with the algorithm of one-by-one seriation applied to $d(i, S)$; that is, $d(s_{k+1}, S_k) = \max_{i \in I - S_k} d(i, S_k)$ where $S_k = \{i_1, ..., i_k\}$ is a starting subset of the series and $k = 1, 2, ... N - 1$. A subset S will be referred to as a *d-cluster* if S is a maximizer of $F(S)$ with regard to all starting sets, $S_k = \{i_1, ..., i_k\}$, in a *d*-series, $(i_1, ..., i_N)$. Obviously, finding a *d*-cluster can be done simultaneously with designing a *d*-series: this is just any initial fragment whose linkage with the subsequent element of the *d*-series is maximum.

Statement 4.12. *Any maximizer of F includes a d-cluster which is a maximizer of F, also.*

Proof: Let S^* be a maximizer of F and p_i be a *d*-series starting from an $i \in S^*$. Then, let S^* not be a starting set, which means that there are some elements between i and the last element of S^* in p_i, that do not belong to S^*; let i^* be the first of them. Let us prove that set T_i of the elements preceding i^* in p_i is a maximizer of F. Indeed, $d(i^*, S^*) \geq F(S^*)$ due to equality (4.9) applied to $S = S^*$. On the other hand, $d(i^*, T_i) \geq d(i^*, S^*)$ since d is a monotone linkage. Thus, T_i is a maximizer of F. Since T_i is a starting set of p_i, it is a *d*-cluster, which proves the statement. □

Statement 4.13. *If $S_1, S_2 \subset I$ are overlapping maximizers of a quasi-convex set function $F(S)$, then $S_1 \cap S_2$ is also a maximizer of $F(S)$.*

Proof: Obvious from definition. □

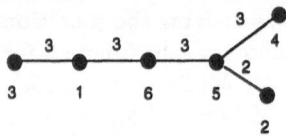

Figure 4.3: A minimum spanning tree for matrix D.

These two statements imply that the minimal (by inclusion) maximizers of a quasi-convex function $F(S)$ defined by a monotone linkage d are not overlapping and they can be found as initial fragments (d-clusters) of some d-series. To do that, a d-series p_i must be determined starting from each $i \in I$. Then, in each of the d-series p_i ($i \in I$), the minimal d-cluster is found as the first starting fragment S_k having maximum $F(S_k) = d(i_{k+1}, S_k)$ over $k = 1, ..., N - 1$. Among the d-clusters, only those maximizing F are left. The structure of the maximizers of a quasi-convex set function is described as follows.

Statement 4.14. *Each maximizer of a quasi-convex set function M_d is union of its minimal maximizers that are d-clusters.*

Proof: Indeed, if S^* is a maximizer of $M_d(S)$, than, for each $i \in S^*$, there is a minimal d-cluster containing i, as it follows from the proof of Statement 4.12.. $\square$

The (minimal) maximizers may not cover all the entities, thus leaving some of them unclustered (ground cluster).

Let us consider set $I = \{1, 2, 3, 4, 5, 6\}$ of the rows of a 6×7 Boolean matrix X:

$$
X = \begin{array}{c|ccccccc}
1 & 0 & 1 & 1 & 0 & 0 & 1 & 1 \\
2 & 1 & 0 & 1 & 0 & 1 & 1 & 0 \\
3 & 0 & 1 & 1 & 1 & 1 & 0 & 1 \\
4 & 1 & 0 & 0 & 1 & 1 & 0 & 0 \\
5 & 1 & 0 & 1 & 1 & 0 & 1 & 0 \\
6 & 0 & 1 & 0 & 1 & 0 & 1 & 0 \\
\end{array}
$$

The matrix of row-to-row Hamming distances (numbers of noncoinciding components) is this:

$$
D = \begin{pmatrix}
0 & 4 & 3 & 7 & 4 & 3 \\
4 & 0 & 5 & 3 & 2 & 5 \\
3 & 5 & 0 & 4 & 5 & 4 \\
7 & 3 & 4 & 0 & 3 & 4 \\
4 & 2 & 5 & 3 & 0 & 3 \\
3 & 5 & 4 & 4 & 3 & 0 \\
\end{pmatrix}
$$

A D-based MST is presented in Fig. 4.3. It can be seen from it that the following five subsets are minimal maximizers of the minimum split single linkage function $SL(S)$: $\{1\}$, $\{2,5\}$, $\{3\}$, $\{4\}$, $\{6\}$, all corresponding to maximum value $SL(S) = 3$. They form a partition of I, which always holds for SL since $SL(S) = SL(I - S)$, implying that all the entities must be covered by SL maximizers.

The situation is slightly different for the minimum split of ml; its minimal maximizers are $\{1\}$, $\{3\}$, $\{4\}$, and $\{6\}$ while none of the elements 2 or 5 belongs to a maximizer of $M_{ml}(S)$. Indeed, let us take a look at six ml-series starting from each of the entities: 1(3)3(3)2(0)5(1)4(0)6, 2(2)5(2)4(2)6(1)1(0)3, 3(3)1(3)2(0)5(1)4(0)6, 4(3)2(1)5(2)6(1)1(0)3, 5(2)2(2)(2)6(1)1(0)3, 6(3)1(2)3(2)2(0)4(0)5. The value $ml(i_{k+1}, S_k)$ is put in parentheses between every starting interval S_k seriated and i_{k+1} ($k = 1, ..., 6$). It can be seen that the maximum value 3 separates each of the four singletons indicated while it never occurs in the series starting with 2 or 5.

The contents of this section is based on a paper by Kempner, Mirkin, and Muchnik 1995, preceded by Mullat 1976 and Zaks and Muchnik 1989.

4.2.6 Discussion

1. Seriation, as a process of iterative ordering a finite set of objects with one-by-one addition of the objects to the order, is an important mental and mathematical operation. Its relation to single cluster clustering can be easily seen with a particular concept of linkage introduced as an entity-to-subset similarity measure.

2. An attempt is made to explicitly describe well-known concepts of single, average, etc. linkages as applied to single cluster clustering rather than to traditional hierarchical clustering.

3. For some linkage functions, set functions being optimized by the seriation procedure (as a local search algorithm) have been indicated. These criteria may be employed for at least two purposes: (a) developing more efficient algorithms, (b) theoretical study of properties of the clusters found with the criteria, both of which are outlined here.

4. The linkage function as a mathematical object has been studied in Russia for almost two decades; some results of the study have been adapted here to allow a complete description of quasi-convex set functions as represented with monotone linkage functions and globally optimized by a simple seriation-like algorithm. No polynomial algorithm is possible for maximizing quasi-convex functions in the oracle-defined form.

4.3   Moving Center

4.3.1   Constant Radius Method

Methods related to the method of moving centers (K-Means) for partitioning present another class of the single cluster clustering methods. They are based on two concepts: (1) a $c(S)$, centroid or standard point or prototype defined for any subset $S \subseteq I$ and (2) $d(i, c(S))$, the distance between entities $i \in I$ and the centroids. These concepts can be defined in various ways (see Section 3.2.2).

To accomplish single cluster clustering, the separative strategy framework is employed in such a way that the cluster sought is just dropped out of the entity set using a pre-defined cluster-size parameter. When the cluster-size parameter is a fixed radius (threshold) $R > 0$ value, here is a separative moving center algorithm:

Constant Radius Moving Center
Starting with arbitrary c, the procedure iterates the following two steps:
1) (Updating the cluster) define the cluster as the ball of radius R around c: $S = \{i : d(i, c) \leq R\}$;
2) (Updating the center) define $c = c(S)$.
Stopping rule: compare the newly found cluster S with that found at the previous iteration; stop if the clusters coincide, else go to 1).

It is not obvious that the algorithm converges, that is, that the stopping rule is satisfied once; and, really, non-convergence may occur when the concept of centroid has no relation to that of the distance $d(i, c)$. Let us say that a centroid concept, $c(S)$, *corresponds* to a dissimilarity measure, d, if $c(S)$ minimizes $\sum_{i \in S} d(i, c)$. For example, the gravity center (average point) corresponds to the squared Euclidean distance $d^2(y_i, c)$ since the minimum of $\sum_{i \in S} d^2(y_i, c)$ is reached when $c = \sum_{i \in S} y_{ik}/|S|$. Analogously, the median vector corresponds to the city-block distance.

It turns out, that when a centroid concept corresponds to a distance measure, the algorithm above can be described as the alternating minimization algorithm for a clustering criterion. To introduce a convenient criterion, let us add a particular distinct point ∞ to I, with all the distances $d(i, \infty)$ equal to the radius R. Let us define:

$$D(c, S) = \sum_{i \in S} d(i, c) + \sum_{i \in I - S} d(i, \infty) \qquad (4.12)$$

to be minimized by both kinds of the variables (one related to c, the other to S). As it has been described already, p. 125, the alternating minimization technique iteratively minimizes the criterion over one group of the variables with the other fixed, using the newly found values for the next iteration.

When c is fixed, minimizing $D(c, S)$ with regard to S is easily achieved: all the is with $d(i, c) < R$ must be put in S, and all the is with $d(i, c) > R$ must be kept out of S while assigning of those i having $d(i, c) = R$ to S or not does not matter. When S is fixed, c is taken to minimize $D(c, S)$, or, equivalently, $\sum_{i \in S} d(i, c)$, since the second term in (4.12) does not depend on c. Obviously, the optimal $c = c(S)$ when the centroid concept corresponds to the distance. We have proven the following statement.

Statement 4.15. *The moving center method (with the radius fixed) is equivalent to the method of alternating minimization applied to criterion $D(c, S)$ in (4.12) when the centroid concept corresponds to the distance measure.*

Corollary 4.1. *The moving center method (with the radius fixed) converges if the center element concept corresponds to the distance measure.*

Proof: Indeed, the alternating minimization decreases the criterion value at each iteration, and the number of the feasible clusters S is finite. □

In Russia, the constant-radius moving center method was suggested by Elkina and Zagoruiko 1966 as the basic part in their separative clustering procedure FOREL (from FORmal ELement, also "trout", in Russian). The original method employs the gravity center and Euclidean distance concepts; and it starts from the center of gravity of the entire entity set. After a cluster is found, the procedure is applied repeatedly to the rest of the entity set. Usually, the algorithm finds a few large clusters; the rest yields to very small clusters that are considered a "swamp" to be excluded from the substantive analysis. Several computations are usually made, with different values of R: the greater R, the smaller the number of "significant" clusters. Convergence of the original algorithm was proved by Blekher and Kelbert 1978 using a form of function (4.12).

4.3.2   Reference Point Method

Yet another moving center method for the entity-to-variable data was developed by the author (Mirkin 1987a, Mirkin and Yeremin 1991), in the approximation clustering context, with a different kind of cluster-size parameter to be specified by the user. This parameter, referred to as the *reference point $a = (a_k)$*, is a particular point in the variable space, which is considered a "center" to look from it at the other entity points. For the sake of simplicity, the origin of the space, 0, will be shifted into the reference point, which is achieved by subtracting a from all the row-points y_i. To make the method more flexible, let us define yet another parameter, a *comparison scale factor*, $\alpha > 0$, to control the cluster size in a "soft" manner. Usually $\alpha = 1$.

Reference-Point-Based Moving Center
Start with an initial standard point c.
1. Updating of the cluster: Define cluster S of points y_i around the center c as $S = \{i : d(y_i, c) \leq \alpha d(y_i, 0)\}$.
2. Updating of the standard point: Compute $c = c(S)$.
3. Stop condition: Compare S with that at the previous iteration. If there is no difference, the process ends: S and $c(S)$ are the result. Else go to Step 1.

This algorithm follows the scheme of the constant radius moving center method, although there are some differences: the size of the cluster, in this method, depends not on the scale factor only, but, mainly, on the reference point location! Indeed, the cluster size is proportional to the distance between c and 0; the less the distance, the less the cluster radius.

This is a violation of the longstanding tradition in clustering, which pertains to homogeneity of the variable space and requires the size of the clusters to depend only on the inter-entity distances, not on their centroid's location. However, the reference point is a meaningful parameter. For example, a moving robotic device should classify the elements of the environment according to its location: the greater the distance, the greater the clusters, since differentiation among the nearest matters more for the robot's moving and acting. Moreover, the homogeneity tradition, actually, does not hold in traditional clustering algorithms also, for example, when they involve scalar-product based similarity measures. The row-to-row scalar product, $(y_i, y_j) = \sum_k y_{ik} y_{jk}$, refers to the angle between the lines joining 0 and each of y_i, y_j, which depends much on location of 0. The connection with the scalar product can be seen quite unequivocally:

Statement 4.16. *In the Euclidean space, updating of the cluster with $\alpha = 1$ in Step 1 of the reference-point-based moving center method is equivalent to the decision rule based on the following discriminant function $f(x)$ (x belongs to S if and only if $f(x) > 0$):*

$$f(x) = (c, x) - (c, c)/2 > 0 \tag{4.13}$$

Proof: The generic inequality in Step 1 does not change if the Euclidean distances are squared:

$$(y_i, y_i) - 2(y_i, c) + (c, c) \leq (y_i, y_i) - 2(y_i, 0) + (0, 0),$$

After obvious reductions, we get (4.13) proven.   □

The reference point, 0, is involved in (4.13) implicitly, through definition of the scalar product.

As can be expected, the reference-point-based method also turns out an alternating optimization algorithm for a particular criterion:

$$D(c, S) = \sum_{i \in S} d(i, c) + \alpha \sum_{i \in I-S} d(i, 0) \qquad (4.14)$$

Statement 4.17. *The reference-point-based moving center method is equivalent to the method of alternating minimization applied to criterion $D(c, S)$ in (4.14) when the standard point concept corresponds to the distance measure.*

Proof: The proof is completely analogous to that in the Statement 4.15.. □

In practical computations, the reference-point-based algorithm shows good results when it starts from the entity-point which is farthest from the reference point. This seems to correspond to the process of typology starting from the most deviate types. A mathematical substantiation of this rule will be presented in Section 4.4.1.

4.3.3   Discussion

1. The moving center method (K-Means) is a clustering method used widely for partitioning. However, there are some versions of the method utilized for finding clusters in the separative strategy framework. At any step, only one cluster is found, which allows us to consider this kind of algorithm as, primarily, single cluster clustering. Separation of a cluster involves a cluster-size parameter as the radius or reference point.

2. The underlying idea is that the clusters are extracted one by one from the "main body" of the entities, which is especially easy to see in the second of the algorithms presented. In this algorithm, the "main body" is considered as resting around a "reference point" while the cluster extracted must have its standard point as far from that as possible, which may be considered a model of typology making based on extracting the "extreme" types. This method has a distinctive feature that the size of the cluster found explicitly depends on its distance from the reference point, which may allow the user to use the "reference point" as an interpretable parameter of the algorithm.

3. Both of the methods presented can be explained in terms of the square-error criterion, which allows proof of their convergence. Moreover, the reference-point-based algorithm, actually, pertains to the principal cluster analysis emerging in the context of approximation clustering (as described in the next section).

4.4   Approximation: Column-Conditional Data

4.4.1   Principal Cluster

Let $Y = (y_{ik})$, $i \in I$, $k \in K$, be an entity-to-variable data matrix. A type-cluster can be represented with its standard point $c = (c_k)$, $k \in K$, and indicator function $s = (s_i)$, $i \in I$ (both of them may be unknown). Let us define a bilinear model connecting the data and cluster with each other:

$$y_{ik} = c_k s_i + e_{ik}, \; i \in I, \; k \in K, \tag{4.15}$$

where e_{ik} are some residuals whose values show how the cluster structure fits into the data. The equations (4.15), basically, mean that the rows of Y are of two different types: a row i resembles c when $s_i = 1$, and it has all its entries small when $s_i = 0$.

To get the model fitted, let us minimize the residuals by c_k or/and s_i with the least-squares criterion:

$$L^2(c, s) = \sum_{i \in I} \sum_{k \in K} (y_{ik} - c_k s_i)^2 \tag{4.16}$$

A minimizing type-cluster structure is referred to as a *principal cluster* because of the analogy between this type cluster and principal component analysis: a solution to the problem (4.16) with no Boolean restriction gives the principal component score vector s and factor loads c corresponding to the maximum singular value of matrix Y (see Section 2.3.1).

The criterion can be rewritten in terms of the subset $S = \{i : s_i = 1\}$ corresponding to s:

$$L^2(c, S) = \sum_{i \in S} \sum_{k \in K} (y_{ik} - c_k)^2 + \sum_{i \in I-S} \sum_{k \in K} y_{ik}^2 = \sum_{i \in S} d^2(y_i, c) + \sum_{i \in I-S} d^2(y_i, 0) \tag{4.17}$$

where d^2 is the Euclidean distance squared.

Criterion (4.17), obviously, is a particular instance of the criterion (4.14) above. This implies the following statement.

Statement 4.18. *Alternating minimization of the least-squares principal cluster criterion (4.16) is equivalent to the reference-point-based moving center method when its parameters are specified as follows: 1) the standard point is the gravity center, $c(S) = \sum_{i \in S} y_{ik}/|S|$; 2) dissimilarity $d(i, c)$ is the Euclidean distance squared; 3) the reference point is in the origin, 4) comparison scale factor $\alpha = 1$.*

The proof follows from the fact that the gravity center, as a standard point concept, corresponds to the Euclidean distance squared (see p. 194).

Another form of the criterion (4.16) allows finding a match to it among the seriation criteria. Let us represent the criterion in matrix form: $L^2 = Tr[(Y - sc^T)^T(Y - sc^T)]$. Putting there the optimal $c = Y^Ts/s^Ts$ (for s fixed), we have

$$L^2 = Tr(Y^TY) - sYY^Ts/s^Ts$$

leading to decomposition of the square scatter of the data $Tr(Y^TY) = \sum_{i,k} y_{ik}^2$ in the "explained" term, sYY^Ts/s^Ts, and the "unexplained" one, $L^2 = Tr(E^TE)$, where $E = (e_{ik})$:

$$Tr(Y^TY) = sYY^Ts/s^Ts + Tr(E^TE) \qquad (4.18)$$

Matrix $B = YY^T$ is a $N \times N$ entity-to-entity similarity matrix having its entries equal to the row-to-row scalar products $b_{ij} = (y_i, y_j)$. Let us denote the average similarity within a subset $S \subseteq I$ as $b(S) = \sum_{i,j \in S} b_{ij}/|S||S|$. Then (4.18) implies that the principal cluster is a Boolean maximizer of the set function

$$g(S) = sYY^Ts/s^Ts = \frac{1}{|S|}\sum_{i,j\in S} b_{ij} = |S|b(S) \qquad (4.19)$$

which is, actually, the Average linkage criterion $AL(S)$ (applied to the similarity matrix B). Two one-by-one local search algorithms for this criterion have been presented p. 182 and 189, as well as the property that the output is a strict cluster. The local search seriation starts with $S = \emptyset$; and the first i to be added is a maximizer of $g(\{i\}) = b_{ii} = (y_i, y_i) = d^2(y_i, 0)$. This means that the starting point of the seriation process here must be among those which are the most distant from 0. Let us put it in a formal way.

Statement 4.19. *The principal cluster is a maximizer of the average linkage criterion, $AL(S)$, applied to matrix $B = YY^T$; $AL(S)$ is the cluster contribution to the square data scatter. The local search algorithm for principal clustering starts in a row-point which is farthest from the origin.*

This shows how close to each other, actually, the two single clustering algorithms are, the reference-point-based moving center algorithm and the local search (seriation) algorithm for the average linkage, with the setting described in the statements above.

Yet another point should be underscored concerning the interpretation aids yielded by the principal cluster criterion. The standard point $c = c(S)$ is a traditional aid which can be supplemented by the relative importance weights of various elements of the cluster structure as measured by their contributions:

Corollary 4.2. *Contribution $g(S)$ (4.19) of the principal cluster (S, c) to the data scatter equals $|S|(c, c)$ while the variable k and the entity i contributions to that are $c_k^2|S|$ and (y_i, c), respectively.*

Proof: The proof follows from (4.18), (4.19), and the fact that $g(S) = |S|(c, c) = \sum_{i \in S}(y_i, c)$. □

Amazingly, contributions of the entities, (y_i, c), can be negative since they are related to the cosine of the angle between y_i and c (at the origin): when the angle is obtuse!

Let us consider some examples. Applied to the Iris data set (square-scatter standardized), the algorithm above found a principal cluster containing 26 specimens, all from class 3, with its relative contribution to the data scatter equal to 22.4%. Applied repeatedly to the rest (124 entities), a 49-element cluster was found (all from class 1, except for specimen number 30 which is somewhat apart from the main body (see the plot in Figure 3.11), its relative contribution to the data scatter, 41.6%. Though the method is first supposed to find the maximally contributing cluster, it is not the case here, due to its local nature.

Table 4.4 presents the cluster centroids along with the relative contributions of the variables (which are proportional to the standardized centroid coordinates squared)

Cluster	Variable	$v1$	$v2$	$v3$	$v4$
	Original scale	7.00	3.16	5.87	2.17
1	Standardized scale	1.40	0.23	1.20	1.28
	Contribution, %	38.42	1.08	28.30	32.20
	Original scale	5.02	3.45	1.47	0.24
2	Standardized scale	-1.00	0.90	-1.30	-1.25
	Contribution, %	19.68	16.10	33.28	30.94

Table 4.4: Two principal clusters found for Iris data set, represented by their centroid values and variable contributions.

It can be seen, from the table, that $v1$ and $v4$ are the most contributing variables for cluster 1, while $v3$ and $v4$ are those, for cluster 2.

Applied to the Disorders data, Table 1.11, p. 39 (centered but not normed since all the variables are measured in the same 7-rank scale), the algorithm produces class 1, as it is, as the principal cluster accounting for 22.3% of the square data scatter (which is the maximum contribution, in this case). Among 17 variables, four account for 66.5% of the total contribution: w13 (19.8), w9 (18.3), w5 (16.7), and w8 (11.7) (the figures in parentheses are the relative contributions of the variables).

Let us look at the values of the most contributing variables in the cluster: w13 is 6 for seven and 5 for four of the patients while w9 is 6 for all eleven of them. Can we employ the observation for intensional description of the cluster, say, as the subset where w13 is 5 or 6? Yes, we can, though it will not be completely satisfactory since there is a patient, number 32, who does not belong in the cluster, though he satisfies the description. In this aspect, variable w9 provides a better intensional description: w9=6 for all the patients from the cluster and only for them.

How did it occur that a better variable had a lesser contribution? Because the contribution is based on the statistical average concept which may not follow the intensional aspect as closely as necessary. The contribution is proportional to the difference (squared) between the within-cluster and the grand means of the variable; the difference for w9, 3.39, is smaller than that for w13, 3.53, though w13 is more confusing, in terms of the intensional description. However, "in average", the difference between the means reflects the intensional difference.

There can be other criteria employed to fit the model (4.15). Among them, least moduli, $L_1 = \sum_{i,k} |e_{ik}|$, and least maximum (Chebyshev), $L_\infty = \max_{i,k} |e_{ik}|$, are. Let us consider them in sequence.

Least Moduli Fitting

$$L_1 = \sum_{i \in I} \sum_{k \in K} |y_{ik} - c_k s_i| = \sum_{i \in S} d_{cb}(y_i, c) + \sum_{i \in I-S} d_{cb}(y_i, 0) \qquad (4.20)$$

where d_{cb} is the city-block metric. This formulation suggests use of the Reference-point-based moving center method with 0 as the reference point, the median as the standard point concept, and $\alpha = 1$, and the alternating minimization technique can be applied for minimizing the criterion.

Yet another formulation of the criterion:

$$L_1 = \sum_{i,k} |y_{ik}| - 2 \sum_{i \in S} ([y_i, c] - [c, c]/2) \qquad (4.21)$$

where c is the median vector of the variables in S; and $[,]$ is the l_1-scalar product defined on p. 82. Obviously, $[c, c] = \sum_k |c_k|$. Again, we have here another form of the criterion, $g_1(S) = 2 \sum_{i \in S} ([y_i, c] - [c, c]/2)$, the cluster contribution to the L_1-scatter of the data to be maximized. The local search algorithm (seriation[with returns]) can be easily adjusted to this particular criterion, which is left to the reader (see Mirkin 1990).

Due to the additive form of $g_1(S)$, contributions of the single variables and/or entities to the total data 1-scatter can be easily extracted, to be used as the interpretation aids.

Applied to the Iris data set, the least-moduli principal cluster algorithm produced a 25-element cluster, being a subset of the principal cluster found with the least-squares criterion (within class 3). The second cluster found coincides with class 1. The first cluster contributes 13.2% to the least-moduli scatter while the second, 39.0% (which, again, illustrates the local nature of the algorithm).

Least Maximum Fitting

This criterion,

$$L_\infty(c, S) = \max_{i,k} |y_{ik} - c_k s_i| = \max\{\max_{i \in S} d_\infty(y_i, c), \max_{i \in I-S} d_\infty(y_i, 0)\},$$

where d_∞ is Chebyshev distance, cannot be presented as the difference between the data scatter and its explained part. The alternating minimization technique, however, can be applied.

> **Least Maximum Principal Clustering**
> An iteration: S given, the optimal c can be taken as the midrange point of the variables (within S only). When c is given, S and $I - S$ are defined as the point subsets belonging in two equal-size n-dimensional cubes having c and 0 as their centers.

The original, still uninvestigated, combinatorial problem here is to find two nonoverlapping cubes of the same minimum size (one with a mandatory center in 0) to contain all the entities.

Applied to the Iris data set (infinity-scatter standardized), the least-maximum based algorithm produced a 19-element "principal" cluster within class 1 while the second cluster (applied to the remaining 131 specimens) gave a 31-element cluster containing the rest of class 1.

4.4.2 Ideal Type Fuzzy Clustering

The concept of fuzzy set has been introduced (by L. Zadeh 1965) to describe an indefiniteness in assigning entities to a subset (see, for example, Zimmerman 1991, Diamond and Kloeden 1994). A vector $f = (f_i)$, $i \in I$, is referred to as a *fuzzy set* or *membership function* if $0 \le f_i \le 1$, for every $i \in I$. Value f_i is interpreted as the degree of membership of i in f. A fuzzy set becomes a traditional "hard" set when all the membership degrees equal 1 (belongs) or 0 (does not). A pair (c, f) is called a *fuzzy cluster* if f is an N-dimensional membership function and c is a standard point in the variable space.

The approximation framework (developed by Mirkin and Satarov 1990) suggests a bilinear model for revealing an underlying fuzzy cluster structure, which is much like that of the principal cluster (4.15):

$$y_{ik} = c_k f_i + e_{ik}, \quad i \in I, \ k \in K, \tag{4.22}$$

Fitting the model with the least-squares criterion

$$L(c, f)^2 = \sum_{i \in I} \sum_{k \in K} (y_{ik} - c_k f_i)^2 \tag{4.23}$$

with regard to arbitrary c_k and nonnegative $f_i \leq 1$ is a non-convex mathematical programming problem (belonging to the so-called semidefinite programming).

If no constraints are placed on f_i, the problem is just the problem of finding the singular vectors c^*, f^* corresponding to the maximum singular value of matrix Y and satisfying the equations:

$$c_k^* = (y_k, f^*)/(f^*, f^*), \quad f_i^* = (y_i, c^*)/(c^*, c^*)$$

where y_k is k-th column of Y, $k \in K$. The constraints yield a modified locally-optimal solution:

$$c_k^* = (y_k, f^*)/(f^*, f^*), \quad f_i^* = \begin{cases} 0 & for \quad g_i \leq 0 \\ g_i & for \quad 0 < g_i \leq 1 \\ 1 & for \quad g_i > 1 \end{cases} \tag{4.24}$$

where $g_i = (y_i, c^*)/(c^*, c^*)$.

It turns out, g_i never exceeds 1 in locally optimal solutions, and, moreover, any pair (c^*, f^*) satisfying the equalities above with $g_i \leq 1$ for all $i \in I$ is a locally optimal solution (Mirkin and Satarov 1990). This can be put in geometrical terms. Let $H(c) = \{x \in R^n : (x, c) = 0\}$ be the hyperplane passing through the origin 0 with normal vector c. Then, f_i^* equals the relative length of the projection of y_i on c^* for all y_i located at the same side of $H(c^*)$ as c^*, and $f_i^* = 0$ for all the y_is at the other side. No point y_i exceeds point c^* in the aspect that all of them are located on the same side of the hyperplane $H(c^*) - (c^*, c^*)$ which is parallel to $H(c^*)$ and passes through c^*. This latter statement means that, in contrast to the hard cluster case, the standard point here is by no means an "average" of the given entities; on the contrary, the "bilinear" standard point is an extreme of Y's row points, modeling therefore the "ideal type" concept. The extreme, actually, can be as distant from the origin as possible. More accurately, if (c^*, f^*) is a locally optimal minimizer of (4.23) then $(\gamma c^*, f^*/\gamma)$ also will be a minimizer, for any $\gamma > 1$. This is an obvious implication from the bilinearity of criterion (4.23) along with the fact that the criterion equals the same value for this new pair.

This shows the extent of similarity between the problem considered and that of finding the singular vectors. Singular vector is defined up to a direction only; it may take any move both ways in the axis while the ideal standard point may have only one-way moves.

To find a locally optimal point in an appropriate location, the alternating minimization algorithm based on iterative application of equations (4.24) should be employed.

Alternating Minimization for Ideal Type

Start with $c = y_i$ where y_i is the farthest row-point from the space origin. Then, the corresponding f is computed with the second equation in (4.24) and, using this f, a new c is computed by the first equation. The iterations are repeated until the newly found vectors f and c coincide (up to a pre-fixed error) with those found on the previous iteration.

The algorithm converges to a locally optimal solution. Indeed, at each step the criterion decreases, which guarantees a limit point since the values of (4.23) are bounded from below by 0. The limit point satisfies all the conditions and, thus, is a locally optimal solution.

For the Iris data set, the ideal type fuzzy cluster found with the algorithm above is presented in Table 4.5.

Variable	$v1$	$v2$	$v3$	$v4$
Original scale	7.83	2.96	7.67	2.88
Standardized scale	2.40	-0.22	2.22	2.21
Contribution, %	36.88	0.31	31.55	31.26

Table 4.5: Characteristics of the ideal fuzzy cluster found for the Iris data set.

The ideal values for three of the variables are exaggerated; they are higher than the averages by 2.2-2.4 standard deviations. Moreover, they are higher than any real values of the variables: just compare 7.83 (ideal) with 7.7 (maximum real), 7.67 (ideal) with 6.9 (maximum real), and 2.88 (ideal) with 2.5 (maximum real). The membership function is zero for all the entities in class 1 and it is rather high for the entities in class 2. Table 4.6 presents all the entities (enumerated in the order of Table 1.23 in p. 57) with the membership value larger than 0.2.

Entity	Membership	Entity	Membership	Entity	Membership
51	0.215	108	0.740	129	0.332
56	0.324	109	0.393	130	0.553
57	0.307	110	0.741	131	0.482
63	0.354	111	0.488	132	0.818
65	0.247	112	0.369	133	0.257
68	0.232	113	0.240	134	0.389
70	0.343	114	0.508	135	0.721
78	0.223	115	0.232	136	0.248
80	0.290	116	0.446	137	0.428
81	0.209	117	0.242	138	0.333
82	0.202	118	0.300	140	0.551
83	0.250	119	0.227	141	0.576
88	0.254	120	0.540	142	0.550
94	0.228	121	0.270	143	0.350
97	0.232	122	0.495	144	0.376
101	0.500	123	0.530	145	0.475
102	0.476	124	0.354	146	0.593
103	0.666	125	0.242	147	0.453
104	0.745	126	0.253	148	0.204
105	0.493	127	0.212	149	0.446
106	0.617	128	0.428	150	0.375
107	0.725				

Table 4.6: Membership function's values exceeding 0.2 for the ideal fuzzy cluster found for the Iris data set.

4.4.3 Discussion

1. Approximation clustering is based on considering the data table as a point in a multidimensional space, which is approximated in a subset of the space related to cluster structure of a specific kind. This approach is substantiated here by indicating the cluster structure criteria and properties that lead to

interpreting the approximate clusters in terms of the standard variable space.

2. In the approximation framework, some particular indices arise, as the contribution weight of the cluster to the data scatter, of the variable to the cluster,
   or of the entity to the cluster, and these indices can be employed as the
   interpretation aids.

3. Principal cluster clustering is a method closely following the principal component analysis methodology except for the factor here required to be Boolean.
   With the least-squares criterion, it underlies two of the single cluster clustering methods: the reference-point-based moving center and AL-seriation
   clustering, which was not obvious at all.

4. Ideal type fuzzy clustering is another cluster-wise derivative of the principal
   component analysis. In contrast to principal clustering which is in line with
   traditional clustering, the ideal type method differs from the traditional fuzzy
   clustering algorithms, especially in the aspect that the standard points here
   tend to be extremes, not the centers of the clusters, thus modeling the ideal
   type logical concept.

4.5   Approximation:
Comparable/Aggregable Data

4.5.1   Additive Clusters

Let $A = (a_{ij})$, $i, j \in I$, be a given similarity or association matrix and $\lambda s = (\lambda s_i s_j)$
be a weighted set indicator matrix which means that $s = (s_i)$ is the indicator of
a $S \subseteq I$ along with its intensity weight λ. When A can be considered as a noisy
information on λs, the following model seems appropriate:

$$a_{ij} = \lambda s_i s_j + e_{ij} \qquad (4.25)$$

where e_{ij} are the residuals to be minimized. Usually, matrix A must be centered
(thus having zero as its grand mean) to make the model look fair.

The least-squares criterion for fitting the model:

$$L^2(\lambda, s) = \sum_{i,j \in I} (a_{ij} - \lambda s_i s_j)^2 \qquad (4.26)$$

is to be minimized with regard to unknown Boolean $s = (s_i)$ and, perhaps, real
λ (in some problems, λ may be predefined). When no diagonal similarities a_{ii}
are specified, $i \neq j$ in all the summations by pairs i, j. When λ is not subject to
change, the criterion can be presented as

$$L^2(\lambda, s) = \sum_{i,j\in I} a_{ij}^2 - 2\lambda \sum_{i,j\in I} (a_{ij} - \lambda/2)s_i s_j$$

which implies that, for $\lambda > 0$ (which is assumed for the sake of simplicity), the problem in (4.26) is equivalent to the following:

$$\max L(\lambda/2, s) = \sum_{i,j\in I} (a_{ij} - \lambda/2)s_i s_j = \sum_{i,j} a_{ij}s_i s_j - \lambda/2 \sum_{i,j} s_i s_j \qquad (4.27)$$

which is just the Summary threshold linkage criterion (see p. 184). The criterion in (4.27) shows that the intensity weight λ is just the similarity-threshold value doubled.

Let us now turn to the case when λ is not pre-fixed and may be adjusted based on the least-squares criterion. There are two optimizing options available here.

The first option is based on the representation of the criterion as a function of two variables, S and λ, made above to allow using the alternating optimization technique.

> **Alternating Optimization for Additive Cluster**
> Each iteration includes: first, finding a (locally) optimal S for $L(\pi, S)$ with $\pi = \lambda/2$; second, determining the optimal $\lambda = \lambda(S)$, for fixed S, by the formula below. The process ends when no change of the cluster occurs.

The other option is based on another form of the criterion, as follows.

For any fixed S, optimal λ can be determined (by making derivative of $L^2(\lambda, s)$ by λ equal to zero) as the average of the similarities within S:

$$\lambda(S) = a(S) = \sum_{i,j\in I} a_{ij}s_i s_j / \sum_{i,j\in I} s_i s_j$$

The value of L^2 in (4.26) with the $\lambda = \lambda(S)$ substituted becomes:

$$L^2(\lambda, s) = \sum_{i,j} a_{ij}^2 - \left(\sum_{i,j} a_{ij}s_i s_j\right)^2 / \sum_{i,j} s_i s_j \qquad (4.28)$$

Since the first item in the right part is constant (just the square scatter of the similarity coefficients), minimizing L^2 is equivalent to maximizing the second item which is the Average linkage criterion squared, $AL^2(S)$. Thus, the other option is just maximizing this criterion with local search techniques described in Section 4.2.2.

Obviously, the problem in (4.28) is equivalent to the problem in (4.27) when λ in (4.27) equals $\lambda(S) = a(S)$.

Both of the criteria found already have been considered in Section 4.2.2 as the Summary threshold linkage (with a particular threshold, $\pi = \lambda/2$) and Average linkage. All the material about them, thus, remains valid: the seriation (local search) algorithms and clusterness properties. Some things left untouched are:

(1) both of the criteria present a contribution of the cluster to the square scatter of the similarity data, which can be employed to judge how important the cluster is (in its relation to the data, not just in a personal opinion);

(2) since the function $AL(S)$ here is squared, the optimal solution may correspond to the situation when $AL(S)$ itself is negative as well as $L(\pi, S)$ and $a(S)$. Since the similarity matrix A normally is centered, that means that such a subset consists of the most disassociated entities and should be called anti-cluster. However, using local search algorithms allows us have the sign of $a(S)$ we wish, either positive or negative: just the initial extremal similarity has to be selected from only positive or only negative values;

(3) in the local search procedure, change of the squared criterion when an entity is added/removed may behave slightly differently than that of the original $AL(S)$ (a complete account of this is done in Mirkin 1990);

(4) when $A = YY^T$ where Y is an entity-to-variable matrix, the additive cluster criterion is just the principal cluster criterion squared, which implies that the optimizing clusters must be the same, in this case. Yet there is a primary interpretation of the cluster contribution which relates to the relevance of the cluster to the model (4.25).

Let us consider the matrix

$$
A = \begin{pmatrix}
2 & 32.5 & & & & & & \\
3 & 2.5 & 2.0 & & & & & \\
4 & 7.0 & 9.0 & 6.5 & & & & \\
5 & 44.0 & 29.5 & 2.0 & 9.5 & & & \\
6 & 12.5 & 15.0 & 1.5 & 3.5 & 15.5 & & \\
7 & 4.0 & 2.0 & 5.0 & 5.0 & 6.0 & 3.0 & \\
8 & 14.0 & 7.5 & 3.0 & 6.5 & 10.5 & 7.0 & 5.0
\end{pmatrix}
$$

which is a symmetrized version of the Switching data table 1.23, p. 57 (the diagonal removed). The local search algorithm starts with the maximum similarity $a_{51} = 44$ thus defining initial $S = \{1, 5\}$. Obviously, it is entity 2 which has the maximum of the mean similarity with S, $a(2, S) = (32.5 + 29.5)/2 = 31$, which is larger than half of $\lambda(S) = 44$. This implies that 2 must be put within S leading to $S = \{1, 2, 5\}$. Now entity 6 has maximum mean similarity with S, $a(6, S) = (12.5 + 15.0 + 15.5)/3 = 14.33$ which is smaller than half the average within similarity $\lambda(S) = (44 + 29.5 + 32.5)/3 = 35.33$. This finishes the process with $S = \{1, 2, 5\}$. The same result can be obtained by preliminarily subtracting the mean similarity $a = 9.67$ from all the entries. In this latter case, the average within similarity also will be less, $\lambda(S) = 25.66$. The contribution of the cluster to the variance is 71.67%, which is quite a high value. The cluster $S = \{1, 2, 5\}$ has been revealed in many other studies; it comprises Coke and the other cola drinks (Pepsi and 7-Up) (see, for instance, Arabie et al. 1988, De Sarbo 1982).

Additive Cluster with Constant Noise

The simplest modification of the model (4.25) to allow shifting the threshold level of the entire similarity data within the model is as follows:

$$a_{ij} = \lambda s_i s_j + \mu + e_{ij} \tag{4.29}$$

where both λ and μ may be found by minimizing the residuals or fixed preliminarily or anything else.

When both λ and μ are fixed (and $\lambda > 0$), the least-squares fitting problem is equivalent to the problem of maximization of Summary threshold criterion $L(\pi, S)$ with $\pi = \lambda/2 + \mu$.

When both λ and μ are adjusted due to the least squares fitting criterion, the following equality can be easily proved as an analogue to that in the theory of linear regression, p. 65:

$$L^2(S) = \sum_{i,k} e_{ik}^2 = N\nu(N)\{\sigma^2(A) - \lambda^2\sigma^2(ss^T)\} \tag{4.30}$$

This shows that the last item (in the curled bracket), to be maximized, represents the cluster's contribution to the data variance scatter.

The variance of the matrix indicator function ss^T equals $\sigma^2(ss^T) = |S|\nu(S)[N\nu(N) - |S|\nu(S)]/N^2\nu(N)^2$ which is maximum when $|S| = N/2$.

The optimal μ is the average association "out" of S (with regard to all $(i, j) \notin S \times S$) while optimal $\lambda + \mu$ is the average similarity $a(S)$ within; the optimal λ is an index of "contrast" between the averages within and out of S.

The local search seriation (with returns) strategy is applicable to produce a cluster which is a π-cluster. However, the threshold $\pi = \lambda/2 + \mu$ here is half the sum of the average similarities within and out of S. The condition that $al(i, S)$ is greater/lesser than this threshold has nothing to do with cohesion of the cluster, in contrast to the model with no noise, which undermines potential use of the "enriched" model with noise. The empirical results obtained with this criterion support this conclusion (see Mirkin 1987b).

However, in the particular problem of clustering by the matrix A above, the modified algorithm leads to the same cluster $S = \{1, 2, 5\}$. Indeed, the average out-similarity is equal to 6.6, for this S; added to the within similarity, 35.33, this gives threshold $\pi = (6.6 + 35.33)/2 = 20.96$ which is larger than any outer similarity while lesser than any similarity within, which guarantees that this S is the optimum.

Multi-way Similarity Data Clustering

Let us consider a 3-way 2-mode similarity data set, which is the case when several similarity matrices $A_k = (a_{ij,k})$, $k \in K$, are available where $k \in K$ may be related to different groups (in psychology or sociology) or time periods (in marketing or international comparisons). Following to Carroll and Arabie 1983 who developed their model for the case when there are many clusters to find, the single cluster model can be presented as this:

$$a_{ij,k} = \lambda_k s_i s_j + e_{ij,k} \tag{4.31}$$

assuming, thus, that there is the same cluster set S for all $k \in K$ while all the differences among A_k, $k \in K$, may be explained by the differences among the intensity weights λ_k.

It is not difficult to show that the least-squares fitting criterion is equivalent to the criterion of maximizing the contribution of the cluster to the square scatter of the data (which is the case with 2-way data, also). The contribution is equal to

$$\sum_{k \in K} AL_k(S)^2$$

where $AL_k(S)^2 = (\sum_{i,j \in I} a_{ij,k})^2/|S|\nu(S)$. This shows that, actually, the criterion may be maximized by a local search algorithm based on the sum of linkage functions $Al_k(i, S)$ defined for each of the matrices A_k.

The alternating optimization technique here involves recalculation of $\lambda_k = a_k(S)$ for every S, and, then, analysis of matrix $A = \sum_{k \in K} \lambda_k(A_k - \lambda_k/2)$ since the least-squares fitting of the model (4.31) with λ_k fixed is nothing but optimizing $SUM(S)$ by matrix A (Mirkin 1990). This operation is repeated iteratively until the cluster does not change anymore. Convergence obviously follows from the fact that the criterion decreases at each iteration while there is only a finite number of the clusters feasible.

4.5.2   Star Clustering

Sometimes the association data are "non-geometrical", that is, the fact that each of two entities is close to a third one does not imply that those two elements are similar to each other. It can happen, for example, in protein sequence fragments comparisons. The traditional concept of a cluster as a subset of mutually similar objects does not meet this peculiarity. In such a situation another concept of cluster could be used considering cluster as a subset of a "star" structure: each of the cluster elements must be close to the single "standard" element which is considered the center of the star cluster, while mutual similarities between some or all of the "ray" elements could be small. The center of the star can be considered its representative.

Let us refer to a Boolean matrix $r = (r_{ij})$, $i, j \in F$, as an *i-star* or *star with center i* if all the elements of r are equal to zero except for some elements r_{ij} in its i-th row. Topologically, an i-star corresponds to a star graph with arcs connecting center i to the set $S(i)$ of the elements j with $r_{ij} = 1$. Subset $S = \{i\} \cup S(i)$ of the star vertices will be considered as the corresponding cluster. A set of stars $r^1, ..., r^m$ (their centers may be different) along with a set of positive intensity weights $\lambda_1, ..., \lambda_m$ will be referred to as a *star structure*. A star structure will be considered as a model of the similarity matrix if it minimizes the value $\sum_{i,j \in I} e_{ij}^2$ of the squares of the residuals in the following equations:

$$a_{ij} = \sum_{t=1}^{m} \lambda_t r_{ij}^t + e_{ij}$$

where a_{ij} are given and λ_t, r_{ij}^t are sought.

Let us restrict ourselves to the case when only one star is sought ($m = 1$). Obviously, in this case, the problem is to minimize criterion $\sum_{i,j \in I}(a_{ij} - \lambda r_{ij})^2$ with regard to arbitrary λ and r.

The problem can be expressed with one of the following two criteria to maximize:

$$l_{\lambda/2}(i, S(i)) = \sum_{j \in S(i)} (a_{ij} - \lambda/2) = \sum_{i,j \in S(i)} a_{ij} - \lambda/2|S(i)|, \qquad (4.32)$$

when λ is fixed, or

$$g(i, S(i)) = (\sum_{j \in S(i)} a_{ij})^2 / |S(i)| = \lambda^2(r)|S(i)| \qquad (4.33)$$

when λ is optimal, thus defined as the average similarity in the star $\{i, S(i)\}$ presented by Boolean matrix r:

$$\lambda(r) = \sum_{i,j} a_{ij} r_{ij} / (|r| - 1),$$

$|r| = |S(i)| + 1$ is the number of elements (vertices) in the star.

When the cardinality of the star is pre-fixed, both of the criteria are equivalent to:

$$f(i, S(i)) = \sum_{j \in S(i)} a_{ij} \qquad (4.34)$$

All the three criteria satisfy the following property.

Statement 4.20. *If j' belongs to an optimal i-star, and $a_{ij'} < a_{ij''}$, then j'' also belongs to the optimal star.*

Proof. Let $\{i, S(i)\}$ be an optimal i-star, and, on the contrary, $j'' \notin S(i)$. In such a case, let us substitute j' by j'' in $S(i)$. Obviously, each of the criteria (4.32)-(4.34) will be increased, which contradicts to the optimality of the star. $\qquad\square$

This property yields a set of simple algorithms to find an optimal i-star for any $i \in I$ by one of the criteria considered.

Star-clustering for (4.34)
When the number m of elements in star is pre-fixed, it is necessary to sum up, for any $i \in I$, the $m - 1$ greatest similarities in i-th row of A. The maximum of the totals determines solution: the corresponding i and $m - 1$ indices of the greatest elements in i-th row.

The algorithm for criterion (4.32) is almost as simple: for any $i \in I$, the sum of positive $a_{ij} - \lambda/2$ is considered only (with corresponding j forming optimal $S(i)$), and then again the maximum of the sums defines the optimal star.

The most difficult of these problems (still quite simple) is to maximize criterion (4.33). This problem is resolved with multiple seriation techniques as follows.

Maximizing (4.33)
For any $i \in I$, sort I by decreasing a_{ij}, and then, for any initial segment of the series, calculate (4.33): the maximum of these values gives the optimal i-stars when it occurs. Then, a best of these i-stars is taken as the solution.

4.5.3   Box Clustering

Two-mode clustering is applied when association between the rows and columns in a comparable data matrix $B = (b_{ij})$, $i \in I$, $j \in J$, is analyzed. A box cluster is represented by two Boolean vectors, $v = (v_i)$, $i \in I$ and $w = (w_j)$, $j \in J$, and an intensity weight λ to generate the box cluster matrix λvw^T of dimension $|I| \times |J|$. This matrix corresponds to the Cartesian product $V \times W$ where V, W are subsets for which v, w are respective indicator vectors.

The least-squares approximation criterion for box clustering is as follows:

$$L^2 = \sum_{i \in I, j \in J} (b_{ij} - \lambda v_i w_j)^2. \tag{4.35}$$

This criterion is very similar to that in (4.25) of additive clustering; the difference is that products $v_i w_j$ here involve components of different vectors while the additive clustering employs the same vector. Thus, a great part of the properties of the box- and additive cluster models will be alike.

For any λ, criterion (4.35) clearly can be rewritten as follows:

$$AB(V, W) = \sum_{i \in V} \sum_{j \in W} (b_{ij} - \lambda)^2 + \sum_{(i,j) \notin V \times W} b_{ij}^2. \tag{4.36}$$

Consider its increment when a row $k \notin V$ is added to V:

$$\Delta = AB(V \cup \{k\}, W) - AB(V, W) = \sum_{j \in W} (b_{kj} - \lambda)^2 - \sum_{j \in W} b_{kj}^2. \tag{4.37}$$

This value can be either negative or positive depending on the closeness to λ or 0 of the subset of row k corresponding to W, which necessitates respective inclusion or rejection of k with regard to V.

Another form of the criterion,

$$AB(V, W) = \sum_{i \in I, j \in J} b_{ij}^2 + \sum_{i \in V} \sum_{j \in W} [(b_{ij} - \lambda)^2 - b_{ij}^2]$$

(with the contents of the brackets in the last term transformed using the elementary formula $a^2 - b^2 = (a - b)(a + b)$), presents the criterion as the square scatter of the data minus

$$g(V, W, \lambda) = \sum_{i \in V} \sum_{j \in W} \lambda(2b_{ij} - \lambda). \tag{4.38}$$

Thus, to minimize (4.36), criterion (4.38) must be maximized. Based on the latter criterion, let us offer another interpretation of the optimality condition based on the change of sign of (4.37) from negative to positive when $V \times W$ is optimal. Indeed, the change in (4.38) when $k \in I$ is added to V (leaving W invariant) equals:

$$\Delta_g(V, W, k) = \sum_{j \in W} \lambda(2b_{kj} - \lambda) = 2\lambda l_{\lambda/2}(k, W) \tag{4.39}$$

where $l_{\lambda/2}$ is the threshold linkage function.

Thus, for positive λ, criterion (4.38) is actually decreased when the average linkage $al(k, W)$ value is less than $\pi = \lambda/2$.

When the optimal value of λ is used, for a given box cluster $V \times W$, it equals the average internal proximity

$$\lambda = b(V, W) = \sum_{i \in V} \sum_{j \in W} b_{ij} / |V||W|. \qquad (4.40)$$

With the optimal λ substituted into $g(V, W, \lambda)$ (4.38), the criterion become equal to

$$g(V, W) = (\sum_{i \in V} \sum_{j \in W} b_{ij})^2 / |V||W| = b^2(V, W)|V||W|. \qquad (4.41)$$

where $b(V, W)$ is the average value of $b_{ij}, i \in V, \ j \in W$. This form of criterion (4.38) does not involve λ (which can be determined afterward from formula (4.40)) and can be easily adjusted to the case when the optimal λ is negative (corresponding to the most disassociated elements as in the anti-cluster concept).

The local search algorithm (with returns) based on the neighborhood defined by adding/removing an arbitrary row/column, actually, copies the local search algorithm for criterion $AL(S)$:

Local Search Box Clustering
1. (Start) Find a pair $(i, j) \in I \times J$ maximizing b_{ij} (when only positive values of λ are sought) or criterion (4.41) which equals b_{ij}^2 for singleton boxes $\{i\} \times \{j\}$ (when λ is permitted to be negative) and set $V = \{i\}, W = \{j\}$, and $\lambda = b_{ij}$ for the corresponding i, j.
2. (Iterative Step) For any row $k \in I$ and for any column $l \in J$, calculate the change in the criterion (4.38) (having λ equal to $b(V, W)$ given in (4.40)) or constant) caused by adding k to V if $k \notin V$, or removing k from V if $k \in V$, and similarly acting for l and W, and find the maximum of those changes. If the change is positive, add/remove the row or column to/from the box cluster, and repeat Step 2 from the beginning. If not, end.

Applied to the Disorders data set, the algorithm found a box comprising rows 1 to 11 (class 1 pre-given) and columns 1,2,3,5,9,13 corresponding to the variables which are rated highly at class 1. Three of the columns (variables), 5, 9, and 13, have been found important in the preceding analysis of the data with principal clustering; yet the other "important" variable, in that analysis, column 8, does not belong to the box because its values are quite low at class 1: note, all the entries are considered comparable here, in contrast to the case of principal clustering. The contribution of the box to the data scatter is 10.8% while its intensity is 2.33 (after the grand mean has been subtracted).

4.5.4 Approximation Clustering for the Aggregable Data

A specific approximation clustering strategy emerges for the aggregable data based on the two features considered in Section 2.3.3: (1) it is transformed RCP data $q_{ij} = p_{ij}/p_{i+}p_{+j} - 1$ to be approximated rather than the original data; (2) it is the weighted least-squares criterion employed rather than the common unweighted one. To be more definite, let us consider approximation Box clustering for two-mode contingency data. First, take the box cluster model

$$q_{ij} = v_i w_j + e_{ij}$$

applied to the RCP matrix $Q = (q_{ij})$ defined by contingency table P. Second, consider the following weighted least-squares criterion instead of (4.35):

$$\Phi^2(\lambda, V, W) = \sum_{i \in I, j \in J} p_{i+}p_{+j}(q_{ij} - \lambda v_i w_j)^2, \qquad (4.42)$$

For the sake of simplicity, consider only the case when λ minimizes (4.42) for any fixed V and W. It is not difficult to derive (by setting the derivative of $\Phi^2(\lambda, V, W)$ with respect to λ equal to zero) that the optimal λ has the same RCP meaning, this time for V and W. That is, the optimal λ equals:

$$\lambda(V, W) = q_{VW} = \frac{p_{VW} - p_V p_W}{p_V p_W}, \qquad (4.43)$$

where the aggregate frequencies are defined as usual:

$$p_{VW} = \sum_{i \in V} \sum_{j \in W} p_{ij}, \quad p_V = \sum_{i \in V} p_{i+}, \quad p_W = \sum_{j \in J} p_{+j}.$$

This observation can be considered as a legitimization of using the weighted least-squares criterion in any kind of analyses of the aggregable data (correspondence analysis included), though it has arisen somewhat unexpectedly in the context of clustering.

Substituting this value of λ into (4.42), the criterion could be expressed as follows:

$$\Phi^2(V, W) = \sum_{i \in I, j \in J} p_{i+}p_{+j}q_{ij}^2 - \lambda^2(V, W)p_V p_W. \qquad (4.44)$$

The latter form of the criterion shows that its final term must be maximized to find an optimal box. By substituting expression (4.43) for the optimal λ into that term, the criterion can be written:

$$f(V, W) = \left(\sum_{i \in V} \sum_{j \in W} p_{i+}p_{+j}q_{ij}\right)^2 / (p_V p_W). \qquad (4.45)$$

Now a local search algorithm (with returns) can be formulated analogously to the local search Box clustering algorithm described above.

Aggregable Box Clustering
Start with $V = \{i\}$ and $W = \{j\}$ corresponding to maximum $f(\{i\}, \{j\}) = p_{i+}p_{+j}q_{ij}^2$ by $i \in I, j \in J$. Then, at any step, that one row i or column j is added to/removed from V or W, respectively, which maximizes the increment of $f(V, W)$ with respect to all $i \in I$ and $j \in J$. The process is finished when the maximum increment is not positive.

Although the neighborhood and the algorithm above are rather simple, the RCP values within (V, W) found deviate highly from the others.

Statement 4.21. *For any row i or column j outside the found cluster box $V \times W$, the absolute values of relative changes $RCP(V/j)=RCP(j/V)$ and $RCP(W/i)=RCP(i/W)$ are not larger than half the absolute value of the relative "internal" change $RCP(V/W)=RCP(W/V)$.*

Proof: Let us consider the increment value, $D_i = f(V + i, W) - f(V, W)$, for any box $V \times W$ and row $i \notin V$. After simple transformations, we have

$$D_i = [p_V F^2(V + i, W) - (p_V + p_{i+})F^2(V, W)]/[p_V(p_V + p_{i+})p_W].$$

Then,

$$D_i = (1/(p_W(p_V + p_{i+})))[2F(V, W)F(i, W) + F^2(i, W) - (p_{i+}/p_V)F^2(V, W)].$$

since

$$F^2(V + i, W) = F^2(V, W) + F(V, W)F(i, W) + F^2(i, W)$$

For a box (V, W) found by the algorithm, $D_i \leq 0$ for any $i \notin V$. Thus,

$$2F(V, W)F(i, W) + F^2(i, W) \leq (p_{i+}/p_V)F^2(V, W)$$

If $F(V, W)$ is positive, dividing the inequality by $F(V, W)$ leads to:

$$2F(i, W) + F^2(i, W)/F(V, W) \leq (p_{i+}/p_V)F(V, W),$$

which implies

$$2F(i, W) \leq (p_{i+}/p_V)F(V, W),$$

since the second term above is positive. Dividing the last inequality by p_W, the required inequality, $2q_{iW} \leq q_{VW}$, is obtained. The other cases (negative $F(V, W)$ and/or $j \notin W$) are considered analogously, which ends the proof.   □

The statement proven ensures that the data fragment corresponding to the box found reflects a pattern which is quite deviant from the average RCP.

4.5.5 Discussion

1. An additive cluster is a concept related to "classical" clusters. The other two structures considered relate to more distinctive cluster concepts: (a) the star is a subset of the entities which are connected to a "central" entity while they may be mutually disconnected from each other; (b) the box is a pair of mutually related subsets. The concept of star seems to have been never considered before explicitly.

2. The concept of the anti-cluster emerges naturally in the approximation framework as a subset, the entities in which are mutually related with a negative intensity. The concept fits especially smoothly into the aggregable data clustering framework where the negativity means just a decrease of the observation flow in the cluster in comparison with the average behavior.

3. The approximation criteria relate to interesting combinatorial optimization problems: the principal and additive clustering, to the so-called maximum density subgraph problem (which is polynomial when the similarities are positive), the star cluster criterion seems a new easy-to-optimize combinatorial criterion, and many approximation criteria such as those for box or least-maximum principal cluster seem to have been never considered before.

4. The local search (with returns) approximation algorithms employed are interpreted in terms of direct clustering. The approximation framework allows for automatic (not expert-driven) specification of some of the parameters of the algorithms, though a significant freedom in choice of some of them still remains (as in centering and standardizing the data). In this aspect, approximation algorithms for the aggregable data, where all the algorithm parameters are specified according to the model-driven considerations, look quite impressive.

4.6 Multi Cluster Approximation

4.6.1 Specifying SEFIT Procedure

All the single cluster approximation clustering methods discussed (principal, ideal type, additive, star, and box clustering) can be extended to the situation with a multitude of clusters assumed.

To unify the discussion, let us consider the data table as a vector $y \in R^l$ where $y = (y_u)$, $u \in U$, and $l = |U|$. Obviously, the cluster models listed can be embedded in that framework when $U = I \times K$ and $u = (i, k)$, for principal and ideal type clustering; $U = I \times I$ and $u = (i, j) \in I \times I$, for additive and star clustering;

and $U = I \times J$ and $u = (i, j) \in I \times J$, for box clustering. Analogously, let us consider that the clusters underlying the data can be represented by the vectors $z \in R^l$ belonging to a particular subset $D \subset R^l$ of admissible vectors. The subset D consists of the matrix-format $N \times n$ "vectors" $sc^T = (s_i c_k)$ where $c \in R^n$ and s is a Boolean or probabilistic N-dimensional vector in principal or fuzzy ideal type clustering, respectively; matrix-format $N \times N$ "vectors" $ss^T = (s_i s_j)$ where s is an arbitrary N-dimensional Boolean vector, in additive clustering; Boolean $N \times N$ matrices $r = (r_{ij})$ having non-zero elements in one row only, in star clustering; and Boolean matrices $vw = (v_i w_j)$ where $v \in R^{|I|}$ and $w \in R^{|J|}$ are Boolean vectors, in box-clustering. Thus, the m clusters of a given type can be presented as vectors $z_1, ..., z_m$ in the set $D \subset R^l$ comprising the cluster type under consideration.

An additive multi-clustering model, in this general setting, can be formulated as

$$y = \sum_{t=1}^{m} \mu_t z_t + e \qquad (4.46)$$

where e is the residual vector which should be minimized according to a criterion of form $L_p(e) = \sum_u |e_u|^p$, $p > 0$ (remember that $p = 2, 1$, and ∞ correspond to the least-squares, least-moduli, and least-maximum criterion, respectively). The weighted least-squares criterion for the aggregable data also fits into this formulation since the weights, actually, can be put within the criterion.

The problem of fitting the model (minimizing the criterion with regard to z_t, μ_t sought) much depends on what kind of information on the solution to be found is known a priori.

Let us consider that nothing is known, except for the set D. In this situation, we suggest using the greedy approximation procedure SEFIT from Section 2.3.4 to fit the model.

The procedure consists of the iterations involving two major steps: (1) single cluster clustering, (2) subtracting the solution found from the data. Only the least-squares criterion will be discussed here.

SEFIT for Additive Multi Clustering
Iteration: (1) For given y_t, minimize the least-squares criterion $||y_t - \mu z||^2$ with regard to arbitrary $\mu \in R$ and $z \in D$ and set $\mu_t = \mu^*$ and $z_t = z^*$ where μ^*, z^* are the minimizers found.
(2) Calculate the residual vector $y_{t+1} = y_t - \mu_t z_t$.
Initial setting: $t = 1$, $y_1 = y$. After an iteration is completed, a stop-condition is checked; if it does not hold, t is increased by unity and the next iteration begins.

Performing the procedure does not require anything new: the optimizing step (1) is nothing but single (principal, ideal type, additive, star, or box) cluster clustering as discussed in the preceding sections of this chapter.

However, there are two points to be made on the results of SEFIT:

1. The square scatter of the data is additively decomposed with the solutions found:

$$(y, y) = \sum_{t=1}^{m} \mu_t^2(z_t, z_t) + (e, e) \tag{4.47}$$

Let us point out that the decomposition holds for any set of clusters z_t, provided that mu_t are optimal; it is based on the consecutive character of SEFIT rather than on correlations among $z_1, ..., z_m$ (see Statement 2.1. on p. 102).

The decomposition allows for accounting for the contribution of every single cluster (along with a consequent more detailed decomposition of the contribution possible) which can be employed in the stopping condition as based on the contribution accumulated or just on a single contribution declined too much (see Section 2.3.4 for detail). On the other hand, the contributions are a major interpreting aid, which will be seen with the examples below, Section 4.6.2, and in Section 6.3. Yet another feature of the contributions $\mu_t^2(z_t, z_t)$ is that each of them is the criterion maximized at step (1) of SEFIT.

2. The SEFIT procedure cannot guarantee, in general, that the solutions found minimize (e, e) globally; moreover, in clustering, it almost never is an optimal solution. This generates two kinds of questions: (1) Does SEFIT really lead to extracting all the important clusters, in a general situation? (2) What kind of a cluster structure is optimally fit with SEFIT?

The answer to the first question, in the present context, is provided by Statement 2.2. in Section 2.3.4: it is yes, since all the sets D and the local search algorithms presented satisfy Condition E, which guarantees that $(e, e) \to 0$ when $m \to \infty$.

As to the second question, there is no answer to that yet. However, since the contributions, $\mu_t^2(z_t, z_t)$, in fact, are the single cluster clustering criteria, the SEFIT procedure should be used only in the situations when the cluster contributions are very different; if the contributions of some of the clusters are equal to each other in the "underlying" cluster structure, the SEFIT procedure could mix the clusters and reveal the picture in a wrong way!

The procedure SEFIT can be modified for better adjusting the cluster structure found to the data. Here are three versions involving modifications of step (2), residuation of the data, in SEFIT:

(1) Recalculating the intensity weights μ_t.

At every iteration t, after step (1), vector y is linearly decomposed due to equation $y = a_1 z_1 + ... + a_t z_t + e$ minimizing (e, e) (least-squares-based linear regression problem), which is known as the orthogonal projection operation, $a = (Z_t^T Z_t)^{-1} Z_t^T y$, where Z_t is $l \times t$ matrix with $z_1, ..., z_t$ being its columns. Then, at step (2), the residual vector y_{t+1} is defined as $y_{t+1} = y - Z_t a$.

(2) Reiterating.

This procedure is applied when all m cluster vectors $z_1, ..., z_m$ are found already (or given somehow). For every $t = 1, ..., m$, y_t is defined as $y_t = y - \sum_{s \neq t} \mu_s z_s$ to search for a better vector $\mu_t z_t$ at step (1). This reiteration is repeated until it stabilizes.

(3) Reiterating on "cleaned" data.

After m cluster vectors $z_1, ..., z_m$ are found, the data vector y is "cleaned" with $\sum_t \mu_t z_t$ substituted instead (that is, $y - e$ is considered as y for the next stage). Then, the reiterating procedure is applied (the cleaning can be conducted after every major reiterating step).

It can be proved that the result found at any step in version (1) is linearly independent from the preceding cluster vectors z_t, which leads to a finite decomposition of y (Mirkin 1990).

In the author's several experiments with real-world comparable data, the original SEFIT outperformed the modified versions.

The approximation clustering methods discussed, along with SEFIT-based clustering algorithms, were introduced by the author starting in the mid-seventies (see his reviews in Mirkin 1987b, 1990 and 1994). The additive cluster clustering model have been developed in Shepard and Arabie 1979, Arabie and Carroll 1980, and Carroll and Arabie 1983, however, the algorithms presented in these papers seem somewhat more heuristic; in particular, the two concepts, cluster intensity λ and its contribution weight, $\lambda^2 |S|$, could not be distinguished within the approaches developed in those papers. Box clustering methods are developed in Mirkin, Arabie, and Hubert 1995 (see also Mirkin 1995b); the model itself appeared in some earlier publications (Mirkin and Rostovtsev 1978, DeSarbo 1982).

4.6.2   Examples

Points

The principal clustering algorithm SEFIT-extended to cover 99% of the data variance and applied to the Points data set, standardized (Table 2.1, p. 62), produced 6 clusters as presented in Table 4.7.

The original data table is still not exhausted. This is why the additive representation

No	Cluster	y_1	y_2	Contribution, %
1	5, 6	1.34	0.97	45.4
2	1,3,5	-0.45	-1.05	32.6
3	1,2	-0.85	0.04	12.0
4	5,6	0.22	0.53	5.4
5	2,4,6	-0.24	0.21	2.6
6	1,3,5	0.21	-0.19	2.0

Table 4.7: Principal clusters found with the sequential fitting method for the square-scatter standardized Points data set.

of the data entries by the clusters holds up to minor errors such as, for example, that for the row 1, (-1.07, -1.21): (-0.45, -1.05)+(-0.85, 0.04)+(0.21, -0.19)=(-1.09, -1.20) as supplied by clusters 2, 3 and 6 containing entity 1.

Functions

The SEFIT-wise additive single clustering algorithm has been applied to the data set Functions, Table 1.13 in p. 42, which presents pair-wise similarity rates between nine elementary functions (the average similarity, 2.69, has been subtracted from all the entries preliminarily; the diagonal is not considered). The results are shown in Table 4.8.

Cluster	Intensity	Contribution, %
e^x, lnx	4.31	16.6
$1/x$, $1/x^2$	4.31	16.6
$x^2, x^3, \sqrt{x}, \sqrt[3]{x}$	1.97	20.9
$x^2, |x|$	2.31	4.8
x^2, x^3	1.33	1.6
$lnx, \sqrt{x}, \sqrt[3]{x}$	0.98	2.6

Table 4.8: Additive clusters found with sequential fitting method for data set Functions.

The six clusters presented count for 63.1% of the similarity data variance. The process of clustering has been stopped because these take into account all the positive similarities (after the mean was subtracted); the seventh cluster had negative intensity weight and included all but two functions, which is not compatible with the substantive problem. On the other hand, individual contributions became quite small, thus implying that the new clusters found might reflect just data noise.

Kinship

This example involves a three-way two-mode sorting 15 kinship terms data (made by six groups of respondents), Tables 1.14 and 1.15, pp. 44, 45. The semantically related groups of the terms may be investigated with the additive clustering method based on equation (4.31). The results (compared to those published in Arabie, Carroll, and De Sarbo 1987 as found with INDCLUS algorithm described in Carroll and Arabie 1983) are put in Table 4.9.

The author did the computation in 1992 with a group of students in École Nationale Supérieure des Télécommunications, Paris, and we were certain that the data table was a similarity index. The dissimilarity nature of the data was recognized only because the program found the clusters with all the negative intensity weights (which demonstrates that the least-squares approximation method, basically, is invariant in its relation to the orientation in measuring proximities).

The results found with the sequential fitting strategy are quite similar to those reported in Arabie, Carroll, and De Sarbo 1987, though some findings, as the nuclear family cluster or the contributions found, seem better fit into the interpretation.

Behavior

The sequential fitting strategy involving the box algorithm, applied to the Behavior data set in Table 1.18, p. 49, produced the following results (see Table 4.10).

The results shown in Table 4.10 demonstrate again that the greedy local search procedure generally does not lead to the global maximum of the cluster contribution (just compare contribution of box 3, 2.8%, with that of box 4, 5.1%, which would have been reversed if the method gave the optimal solution).

In general, the boxes in Table 4.10 seem to be reasonable both according to their content and the coverage of the raw proximities. For example, in the first of the boxes, the proximities in submatrix $X(V_1, W_1)$ are much higher than the other values of the corresponding columns, Kiss, Eat, Laugh, with the only two exceptions occurring in the column Talk: the proximity 3.95 to Job Interview (see Table 1.4) is taken into account in another box (the third), and the low proximity 0.47 to Movies, which is still positive and much greater than all the other proximities in the row Movies (excluding the column Cry, taken into account in Box 6). There is one entry in the first box cluster, Bus/Kiss, which has a rather small proximity value, -0.24 (similarly, the proximity for the entry Class/Sleep in the third cluster equals -0.91). Although it may seem unnatural to have that entry in the first cluster, the presence of Bus/Kiss can be explained by the column

Clusters by INDCLUS	Interpretation	Clusters by SEFIT	Contrib. %
brother, father, grandfather, grandson, nephew, son, uncle	Male relatives, excluding cousin	brother, father, grandfather, grandson, nephew, son, uncle	8.01
aunt, daughter, granddaughter, grandmother, mother, niece, sister	Female relatives, excluding cousin	aunt, daughter granddaughter, grandmother mother, niece, sister	8.25
aunt, cousin, nephew, niece, uncle	Collateral relatives	aunt, cousin nephew, niece, uncle	9.01
brother, daughter, father, mother, sister, son	Nuclear family	daughter, father, mother, son	9.11
		brother, sister	3.94
granddaughter, grandfather, grandmother, grandson	Direct ancestors and descendants 2 generations removed	granddaughter, grandfather, grandmother, grandson	8.81
		aunt, uncle nephew, niece	2.02 1.89

Table 4.9: The results of two different approaches to three-way additive clustering of the Kinship data (the left column and interpretation are from Arabie, Carroll, and De Sarbo 1987, while the other two columns on the right are found with the algorithm above).

Kiss being connected to the other rows of the cluster more tightly than it is disconnected from Bus, so the exclusion of either Bus or Kiss from Cluster 1 will decrease the value of the criterion maximized in (4.41). It appears that our least-squares estimation strategy is heavily dependent on the value of the threshold ($\lambda(V, W)/2$, in this case) and sometimes allows the inclusion of marginal proximity values in the best-fitting solution. In support of this explanation, we note that for the data in Table 4.10, the inclusion of Kiss in Box 1 gives a better fit compared to its exclusion. Also the Bus/Kiss proximity value of -0.24 in Table 1.4 is still considerably greater than any of the proximities between Kiss and the row-items not included in the first box. Similar explanations could be offered for other

Box	Rows	Columns	λ	Contr. %
1	Date, Bus, Park, Sidewalk, Family dinner, Elevator, Own room, Dorm lounge, Bar, Movie, Football game	Talk, Kiss, Eat, Laugh	2.68	26.5
2	Class, Bus, Park, Own room, Dorm lounge	Write, Sleep, Read	2.60	8.5
3	Class, Date, Job interview, Bar, Park, Restroom, Own room, Football game	Talk, Laugh	1.46	2.8
4	Park, Own room	Run, Mumble, Read, Belch, Argue, Jump, Cry, Shout	1.96	5.1
5	Football game	Jump, Shout	3.02	1.5
6	Movie, Own room	Cry	2.09	0.7

Table 4.10: Additive boxes found with the sequential fitting strategy for the Behavior data set.

Box	Columns	Rows	RCP, %	Contrib., %
1	ASAF, IFAA	PER	79.5	34.5
2	EUAM, IFEA	PER	-46.0	20.8
3	ASAF, IFAA	POL, ECO	-40.5	9.9
4	IFEA	OTH, POL	46.1	9.7
5	EUAM	POL, MIL, ECO, MTO	18.5	9.3
6	IFEA, ASAF, IFAA, IFI	MIL, MTO	-17.5	5.5

Table 4.11: Box cluster structure of the Worries data set.

possible anomalies in the obtained solution (e.g., the Class/Sleep entry in Box 3 of Table 4.10). More on the example can be found in Mirkin, Arabie, and Hubert 1995.

Worries Applied to the contingency data in Table 1.6, p. 31, the aggregable box clustering algorithm produces 6 clusters; the total contribution of the clusters in the initial value Φ^2 equals some 90 % (see Table 4.11).

The next clusters are not shown in Table 4.11 because their contributions are too small. The content of Table 4.11 corresponds to the usual joint display given by the

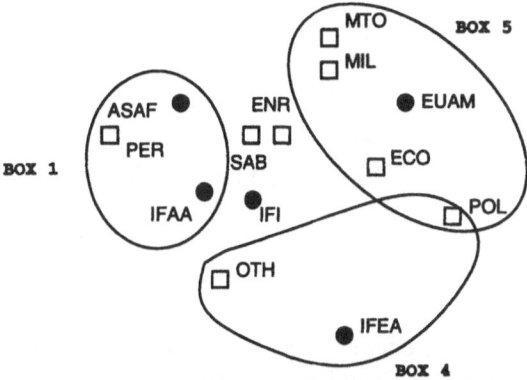

Figure 4.4: Positive RCP boxes in the correspondence analysis factors plane.

first two correspondence analysis factors (see Fig.4.4 where the columns and the rows are presented by the circles and the squares, respectively).

All the boxes with positive aggregate RCP values (clusters 1, 4, and 5) correspond to the continuous fragments of the display (shown on Fig.4.4); boxes with the negative RCP values are associated with distant parts of the picture. Recalling the meaning of the values as the relative probability changes, the interpretation of this phenomenon becomes obvious. The box clusters reflect either positive or negative correlations between columns and rows: for example, row PER correlates positively with columns ASAF and IFAA, but row PER negatively relates to columns EUAM and IFEA: the individuals coming from Asia and Africa have much higher worries about their personal economic status than those from Europe or America.

Data analysts are used to interpreting the results of correspondence analysis by presenting the rows and columns as points on the joint display with this kind of connected fragments distinguished intuitively. Here, the fragments are revealed by a formal procedure based on RCP values that have a clear interpretation. Statement 4.21. ensures that the fragments reflect those parts of the data table where RCP values are most deviant from the average pattern. Thus, the box clustering method could be used as an aid to interpret the display obtained with the traditional correspondence analysis, as well as a complementary instrument to it. Interestingly, the anti-clusters (clusters having the negative intensity weight) here have quite ordinary interpretation: they just show a pattern of behavior which

No	Cluster	Intensity	Contrib., %
1	Coke, 7-Up, Pepsi	25.65	71.7
2	Coke, Pepsi, Sprite	5.77	3.6
3	All eight	-3.37	11.5
4	7-Up, Sprite	8.69	2.7
5	Coke, Pepsi, Fresca	6.05	4.0

Table 4.12: Additive cluster structure of the Switching data set.

occurs rarer than in average. Thus, it can be claimed that the fragments obtained with aggregable box clustering present either "compact" or "marginal" parts of the visual correspondence analysis plot, depending on the sign of the aggregate RCP value. Plus corresponds to the common way for interpreting the display via the visually distinct "clusters" of the row and column points. Minus, to fragments that relate to the margins of the plot and are hardly visible in the display; still interpretation of those fragments seems rather clear since they correspond to relatively unconnected events. As a complementary tool, box clustering avoids the controversial issue of simultaneously presenting both the rows and columns in the same display (discussed in Section 2.3.3), since box clustering is based on all the original data, not on a few factors.

Switching

Continuing to extract the clusters from symmetrized and centered matrix A for the Switching data begun in Section 4.5.1 (see p. 57), we obtain a few clusters having contributions greater than 1% as shown in Table 4.12. Basically, all the clusters overlap the first one (note, cluster 3 appeared just because all the "significant" positive entries were subtracted so that the algorithm "decided" to add 3.37 to all the entries), thus, relating to the opposition diet — non-diet (there have been other normalizations utilized with similar results found, see, for example, Arabie et al. 1988, Eckes and Orlik 1993).

Let us try now the aggregable box clustering approach adapted to the data aggregability. The result is not too promising: the box clustering applied ad hoc to the matrix produced only one non-trivial box (that is, a box that was not a dyad of the same brand with itself) consisting of three diet cola drinks: Tab, Like, and Diet Pepsi, both as the row and the column clusters. Although this cluster was repeatedly found in the other author's analyses, too, the main principle of organization among the drinks was generated by the contrasts of cola versus non-cola and diet versus non-diet. The use of box clustering techniques allows us to display yet another aspect of the data.

Brand switching data typically have large values for the elements in the principal diagonal, corresponding to brand-loyal consumers. Colombo and Morrison

1989 have argued that the influence of the diagonal entries should be mitigated when emphasizing the information in the off-diagonal entries. Let us consider a type of statistical independence hypothesis (as in Colombo and Morrison 1989) that replaces the observed diagonal values with those of the so-called potential switchers (with the hard-core loyals removed as a kind of different people). The hypothesis suggests that any nondiagonal entry p_{ij} $(i \neq j)$ can be expressed as $p_{ij} = \alpha f_i g_j$ where f_i is the probability of switching from brand i, g_j is the probability of switching to brand j, and α is the probability of potential switching behavior (applied to both loyal and nonloyal purchasers). After the unknown values of f_i, g_j, and α are obtained (with least-squares techniques, for example), the proportion of the potential switchers for any given principal diagonal entry p_{ii} is estimated as $\beta = (\alpha - \sum_{i \neq j} p_{ij}) / \sum p_{ii}$.

In the example, these proportions βp_{ii} are as follows: 35, 14, 1, 1, 25, 4,2, and 3. When these values substitute for the original diagonal entries in Table 1.23, p. 57, the algorithm works a little bit differently: it recognizes that there are three important boxes, accounting for 52.7% of the total sum of weighted squares of the data. These boxes are: Diet Pepsi x Tab (RCP=490%, contribution 24.6%), Tab x Like (RCP=358%, contribution= 19.05%), and Like x Tab, Diet Pepsi (RCP=185%, contribution=9.05%). The same three drinks found here are contained in the non-trivial box obtained using the original data, but the boxes now give more interpretable information: obviously, the consumers do not like any of these drinks, and keep changing the selection as if ever hoping for an acceptable diet entry to appear.

4.6.3   Discussion

1. In the multi-clustering problems, the approximation single cluster clustering can be utilized along the line of the sequential fitting SEFIT strategy. This strategy involves single cluster clustering reiterated with the residual data defined, at any step, just by subtracting the cluster structure found from the data table.

2. The data scatter can be decomposed into the sum of contributions corresponding to the clusters found (even when no condition of mutual orthogonality holds), which may be utilized in interpreting the clusters. However, such a strategy may be employed only in the case when the contributions carried by the different clusters are quite different, so that the first cluster's contribution to the data scatter is larger than that of the second cluster, etc.

3. Still no general theory has been developed on particular additive structures to be recovered, nor on the structures that can be entirely decomposed with SEFIT clustering. However, a theory of additive decomposition of dissimilarities by two-cluster partitions (splits) will be presented in Section 7.4 (which is related also to decomposition of the similarities by weak hierarchies).

Chapter 5

Partition: Square Data Table

FEATURES

- Forms of representing and comparing partitions are reviewed.

- Mathematical analysis of some of the agglomerative clustering axioms is presented.

- Approximation clustering methods for aggregating square data tables are suggested along with associated mathematical theories:

  ◇ Uniform partitioning as based on a "soft" similarity threshold;

  ◇ Structured partitioning (along with the structure of between-class associations);

  ◇ Aggregation of mobility and other aggregable interaction data as based on chi-squared criterion and underlying substantive modeling.

5.1 Partition Structures

5.1.1 Representation

Set Terms

A set of subsets $S=\{S_1, ..., S_m\}$ is called a partition if and only if every element $i \in I$ belongs to one and only one of the subsets, $S_1, ..., S_m$, called classes; that is, S is a partition if $I = \cup_{t=1}^{m} S_t$, and $S_l \cap S_t = \emptyset$ for $l \neq t$. Let us denote the cardinality of S_t by $N_t = |S_t|$; the set $(S)=(N_1, ..., N_m)$ of the class cardinalities is referred to as the (cardinality) *distribution* of S. Sometimes, the proportions (called also frequencies) $p_t = N_t/N$ of the classes will be considered as the distribution elements rather than the cardinalities themselves.

The partition is a basic model for classifications in the sciences and logics.

In the data Masterpieces, we have two partitions of the eight novels considered: one, by the writers, $S = \{1 - 2 - 3, 4 - 5 - 6, 7 - 8\}$, attributing the classes to A. Pushkin, T. Dostoevsky, and L. Tolstoy, respectively, and the second, by the variable Presentat, $R = \{1 - 7 - 8, 2 - 3, 4 - 5 - 6\}$, attributing the classes to its categories, Direct, Behav, and Thought, respectively. Their distributions are $(S) = (3, 3, 2)$ and $(R) = (3, 2, 3)$.

Nominal Scale

A partition can be considered as another representation of the nominal scale. Each nominal scale variable is defined up to any possible one-to-one recoding of its values, which means that the variable, really, is defined only up to the classes corresponding to its different values.

However, the nominal scale variables are not confined to the partitions only, they relate to substantive theory of the phenomenon in question; yet this aspect of the nominal variable concept has never been formalized so far.

Indicator Matrix and Equivalence Relation

A Boolean matrix s for a partition S is the $N \times m$ matrix having the indicator vector s_t of S_t as its t-th column ($t = 1, ..., m$). The equivalence relation σ for partition S is defined by the condition that $(i, j) \in I \times I$ belongs to σ if and only if both i and j belong to the same class of S. The indicator matrix of the equivalence relation is defined as $\mathbf{S} = (s_{ij})$ of size $N \times N$ where $s_{ij} = 1$ if $(i, j) \in \sigma$, and $s_{ij} = 0$, otherwise. The matrix can be considered as an entity-to-entity similarity matrix assigning similarity 1 to the entities belonging in the same class, and similarity 0 to the entities from different classes.

Linear Subspace and Orthogonal Projector

Yet another simple similarity matrix is suitable to represent partitions: $P_s =$

(p_{ij}) where $p_{ij} = 1/N_t$ if both $i, j \in S_t$ and $p_{ij} = 0$ if i and j belong to different classes of S.

The following statement shows interrelations among the concepts defined.

Statement 5.1. *For a partition S, its distribution, Boolean matrix s, equivalence relation σ, equivalence indicator matrix $\mathbf{S}$, and for similarity matrix P_s, the following relations hold:*

$$\sigma = \bigcup_{t=1}^{m} S_t \times S_t,$$

$$\mathbf{S} = \sum_{t=1}^{m} s_t s_t^T = ss^T,$$

$$P_s = s(s^T s)^{-1} s^T,$$

and matrix $s^T s$ is diagonal, with its diagonal entries equal to N_t, $t = 1, ..., m$, while $|\sigma| = \sum_{t=1}^{m} N_t^2$.

Proof: Obvious. □

To give an example, let us consider partition $S = \{1 - 2 - 3, 4 - 5 - 6, 7 - 8\}$ on the set of eight masterpieces. Then,

$$s = \begin{pmatrix} 1 & 0 & 0 \\ 1 & 0 & 0 \\ 1 & 0 & 0 \\ 0 & 1 & 0 \\ 0 & 1 & 0 \\ 0 & 1 & 0 \\ 0 & 0 & 1 \\ 0 & 0 & 1 \end{pmatrix}$$

$$s^T s = \begin{pmatrix} 3 & 0 & 0 \\ 0 & 3 & 0 \\ 0 & 0 & 2 \end{pmatrix}$$

$$\mathbf{S} = \begin{pmatrix} 1 & 1 & 1 & 0 & 0 & 0 & 0 & 0 \\ 1 & 1 & 1 & 0 & 0 & 0 & 0 & 0 \\ 1 & 1 & 1 & 0 & 0 & 0 & 0 & 0 \\ 0 & 0 & 0 & 1 & 1 & 1 & 0 & 0 \\ 0 & 0 & 0 & 1 & 1 & 1 & 0 & 0 \\ 0 & 0 & 0 & 1 & 1 & 1 & 0 & 0 \\ 0 & 0 & 0 & 0 & 0 & 0 & 1 & 1 \\ 0 & 0 & 0 & 0 & 0 & 0 & 1 & 1 \end{pmatrix}$$

and

$$P_s = \begin{pmatrix} 1/3 & 1/3 & 1/3 & 0 & 0 & 0 & 0 & 0 \\ 1/3 & 1/3 & 1/3 & 0 & 0 & 0 & 0 & 0 \\ 1/3 & 1/3 & 1/3 & 0 & 0 & 0 & 0 & 0 \\ 0 & 0 & 0 & 1/3 & 1/3 & 1/3 & 0 & 0 \\ 0 & 0 & 0 & 1/3 & 1/3 & 1/3 & 0 & 0 \\ 0 & 0 & 0 & 1/3 & 1/3 & 1/3 & 0 & 0 \\ 0 & 0 & 0 & 0 & 0 & 0 & 1/2 & 1/2 \\ 0 & 0 & 0 & 0 & 0 & 0 & 1/2 & 1/2 \end{pmatrix}$$

The equation $P_s = s(s^T s)^{-1} s^T$ shows that the similarity matrix P_s, actually, is the *orthogonal projection* operator onto linear subspace $L(S) = \{x \in R^N : x = sa \text{ for some } a \in R^m\}$ generated by the columns of matrix s. This means that, for any N-dimensional vector $f \in R^N$, the matrix product $P_s f$ is the only minimizer of the L_2-norm difference $\|f - x\|^2 = (f - x, f - x)$ with regard to all $x \in L(S)$. The linear subspace $L(S)$ itself can be considered as yet another representation of S. Indeed, every vector $x \in L(S)$ has its component x_i equal to a_t when $i \in S_t$, thus, it is a recoding of the categories due to the map $t \to a_t$. Therefore, $L(S)$ contains all the transformations of the category codes admissible in the nominal scale. Moreover, the components of a are not required to be mutually different; so, $L(S)$ contains also all many-to-one recodings of the variable, thus allowing merging of some of the categories (classes). Matrix P_s can be considered as a "compact" representation of the infinite space $L(S)$, like the normal vector, c, of a hyperplane $H = \{x : (c, x) = 0\}$.

However, the linear subspace representation has a shortcoming: for any partition S, subspace $L(S)$ contains the vector $\sum_{t=1}^m s_t = u$ having all its components equal to 1; the other vectors with all the components being equal to each other belong to $L(S)$, as well. Thus, all the partition subspaces contain the same unidimensional subspace, the "line" of vectors αu (α is any real), as a common part, which makes a resemblance among them, generated only by the form of representation (via the property that the sum of columns s_t is equal to u, for every partition S). To exclude the irrelevant resemblance, let us consider subspace $L_u(S)$ of $L(S)$, which is orthogonal to the line $\{\alpha u\}$, thus consisting of only vectors $x = sa$ that have the averages equal to zero. Then, a slightly different similarity matrix will substitute for the matrix P_s as the orthogonal projection matrix.

Statement 5.2. *The orthogonal projector on subspace $L_u(S)$ is $P_{us} = P_s - P_u$ where P_u has all its entries equal to $1/N$.*

This means that the similarity matrix P_{us} has its elements p_{ij} equal to $1/N_t - 1/N$ when both $i, j \in S_t$ for some $t = 1, ..., m$, and to $-1/N$ when i and j belong to different classes.

Proof: Indeed, for any $x \in R^N$, $P_u x = \bar{x} u$ where $\bar{x}$ is the average of the components of x, thus providing that $x' = (P_s - P_u)x \in L_u(S)$. The fact that

x' approximates x in $L_u(S)$ follows from the following two properties: (1) $P_s x$ approximates x in $L(S)$, and (2) the line αu is orthogonal to $L_u(S)$. □

The following statement gives characterizations of some of the objects associated with a partition.

Statement 5.3. *(A). A set of integers $\{N_t\}$, $t = 1, ..., m$, for some m, is a partition distribution if and only if $\sum_{t=1}^{m} N_t = N$.*

(B). An $N \times m$ Boolean matrix s is a partition's Boolean matrix if and only if its column-vectors are mutually orthogonal and their sum is equal to the N-dimensional vector u having all its components equal to 1.

(C). A binary relation σ is a partition's equivalence relation if and only if it is reflexive ($(i, i) \in \sigma$ for any $i \in I$), symmetric ($(i, j) \in \sigma$ implies $(j, i) \in \sigma$ for any $i, j \in I$), and transitive ($(i, j) \in \sigma$ and $(j, k) \in \sigma$ implies $(i, k) \in \sigma$ for any $i, j, k \in I$).

(D). A Boolean $m \times m$ matrix $S = (s_{ij})$ is an equivalence indicator matrix if and only if the following conditions hold: (a) $s_{ii} = 1$, (b) $s_{ij} = s_{ji}$, (c) $s_{ij} + s_{jk} \leq 1 + s_{ik}$.

Proof: *(A)* and *(B)* are obvious. For *(C)*, let us consider an only one-way implication: if a binary relation $\sigma \subseteq I \times I$ is reflexive, symmetric and transitive, than it is a partition's equivalence relation. Let us define subsets $S_i = \{j : (i, j) \in \sigma\}$ and prove that they form a partition of I. Obviously, any $i \in I$ belongs to S_i. To prove that any $i \in I$ can belong to only one of these subsets, let us assume that $S_i \cap S_k \neq \emptyset$ for $i \neq k$ and prove that, in this case, $S_i = S_k$. Indeed, in this case, $(i, k) \in \sigma$ since $(i, j) \in \sigma$ and $(j, k) \in \sigma$ for any $j \in S_i \cap S_k$. Thus, any j belonging to S_i must belong to S_k (and vice versa), by transitivity and symmetry properties, which proves the statement. *(D)* is *(C)* expressed in matrix terms. □

5.1.2 Loaded Partitions

A weighted partition is a partition S considered along with real intensity weight coefficients λ_t assigned to each of its classes S_t ($t = 1, ..., m$). Additive representation of a weighted partition is the matrix $S_\lambda = \sum_t \lambda_t s_{it} s_{jt} = (s_{ij}(\lambda))$ where $s_{ij}(\lambda) = \lambda_t$ if both $i, j \in S_t$ for some t and $s_{ij}(\lambda) = 0$ otherwise. When all the intensity weights are equal to the same value λ, the partition will be referred to as a *uniform* one; such a partition has its additive representation equal to λS.

Analogously, a type-cluster partition is defined as a set of partition classes S_t along with corresponding type-vectors c_t of a dimensionality n ($t = 1, ..., m$). Its additive representation is matrix $\sum_{t=1}^{m} s_t c_t^T$ having every row i equal to corresponding class type-vector c_t (when $s_{it} = 1$) (Mirkin 1987, Van Buuren and Heiser

1989).

A pair (S, ω) will be referred to as a *structured partition* if $S = \{S_1, ..., S_m\}$ is a partition on I and ω is a directed graph (binary relation) on the set $\{1, 2, ..., m\}$ of the class indices as the vertex set. The arc $(t, u) \in \omega$ is interpreted as an essential relation from class S_t to class S_u. Structured partitions are called *block models* in social psychology where, usually, a group partition S is associated with several interrelation structures, $\omega_1, ..., \omega_l$, simultaneously (see Arabie, Boorman, and Levitt 1978, Wasserman and Faust 1992).

The structured partition concept can be considered in two ways: 1) as a most general model for the concept of qualitative variable; 2) as a model of interactive parts of a complex system (see p. 113).

A structured partition (S, ω) can be represented by a $N \times N$ Boolean indicator matrix $S_\omega = (s_{ij})$ where $s_{ij} = 1$ for all the pairs (i, j) within the structure and $s_{ij} = 0$ for outer pairs (i, j). More formally, $s_{ij} = 1$ if and only if their respective classes, $S_t \ni i$ and $S_u \ni j$, are in the structure, $(t, u) \in \omega$; otherwise, when $(t, u) \notin \omega$, $s_{ij} = 0$. The corresponding binary relation is defined as $\sigma_\omega = \{(i, j) : i \in S_t \ \& \ j \in S_u \ \& \ (t, u) \in \omega\}$.

Two questions arise. Does a structured partition correspond to any given Boolean matrix (s_{ij})? If yes, how can such a structured partition be determined?

The answers are easy to get based on the concept of structural equivalence. Given an $N \times N$ Boolean matrix (s_{ij}), the entities i and j are said to be *structurally equivalent* if rows i and j coincide, as well as columns i and j. Obviously, the relation of structural equivalence is an equivalence relation, thus defining a partition on I consisting of the classes of structurally equivalent entities. The structure ω on this partition is defined in a natural way: $(t, u) \in \omega$ if and only if there are some $i \in S_t$ and $j \in S_u$ such that $s_{ij} = 1$. This definition is correct since, in this case, $s_{i'j'} = 1$ for all other $i' \in S_t$ and $j' \in S_u$. Indeed, $s_{ij'} = 1$ because the columns j, j' coincide, and then, $s_{i'j'} = s_{ij'} = 1$ because rows i, i' coincide. Thus, every Boolean matrix is the indicator matrix of a structured partition defined by the structural equivalence relation. Actually, the structural equivalence defines the maximum structured partition corresponding to a given Boolean matrix (s_{ij}), which is usually referred to as the *homomorphous* image of the binary relation corresponding to the Boolean matrix.

5.1.3 Diversity

Diversity is an important characteristic of partition distributions. Diversity is minimum when all the N entities belong to the same class, and it is maximum when each of the entities makes a class on its own. To measure the degree of diversity in the intermediate cases, there are, basically, two measures used: *entropy*,

$$H(S) = -\sum_{t=1}^{m} p_t \log p_t \qquad (5.1)$$

and *qualitative variance* or *Gini coefficient*

$$V(S) = 1 - \sum_{t=1}^{m} p_t^2 \qquad (5.2)$$

where $S = \{S_1, ..., S_m\}$, $p_t = N_t/N$, and the logarithm base is usually 2.

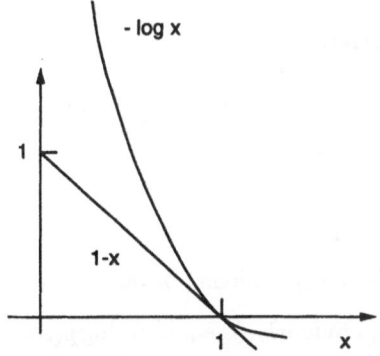

Figure 5.1: Comparing logarithm and linear function: $-\log x$ and $1 - x$.

Obviously, both of these indices have their minimum, 0, when $p_t = 1$ for some t while the other p_u, $u \neq t$, are zeros; both of them have their maximum when all p_t are equal to each other (the so-called uniform distribution). It can be noted, also, that $V(S)$ can be considered a "rough" version of $H(S)$ since $1 - x$ is the linear part of Taylor series decomposition of the function $-\log x$ at $x_0 = 1$ ($0 < x \leq 1$); $H(S)$ is the average value of $-\log p_t$ and $V(S)$ is the average value of $1 - p_t$, due to distribution (S) (the graphs of these two functions are presented in Fig.5.1).

Each of the measures above has some model interpretations that will be presented below.

Models for Entropy

Model 1: Information. In information theory (Brillouin 1962), each class t is considered a code in the data flow transmitted, with its occurrence probability, p_t ($t = 1, ..., m$). The quantity of information in the code is considered equal to $-\log p_t$. This gives $H(S)$ as the average information in the flow.

Model 2: Sample (Clifford and Stephenson 1975). Let S be a pre-fixed partition of I and try to reconstruct it by randomly picking its elements. There are two kinds of sampling considered usually: with return, when the entity picked is put back in I, and without return, when every entity picked is removed from I.

In the first case, the probability that random sampling (with return) produces exactly S is equal to

$$p(S) = p_1^{N_1} p_2^{N_2} ... p_m^{N_m}$$

while in the second case, without return,

$$p(S) = N_1! N_2! ... N_m! / N!.$$

Indeed, the probability of the event that the first N_1 entities picked all belong to the first class is equal to $p_1^{N_1}$ (with return) or, without return,

$$\frac{N_1}{N} \frac{N_1 - 1}{N - 1} ... \frac{1}{N - N_1 + 1}.$$

Continuing these calculations we get the probability expressions above.

The logarithm of the first of the probabilities, obviously, is equal to $\log p(S) = -NH(S)$. To deal with the second probability, we must assume that all the integers N_t are large enough to allow using the well known Stirling formula, $\log n! \approx n(\log n - 1)$, which gives the same result, this time as approximate one.

Model 3: Symmetry (Schreider and Sharov 1982). For a given partition S, let us count the number of one-to-one mappings on I that do not change the partition, which reflects the extent of the symmetry of partition S. Obviously, each of the invariant mappings is a permutation of the entities within classes S_t, which shows that the total number of the mappings is $M(S) = N_1! N_2! ... N_m!$. The proportion of these mappings is $M(S)/N!$ which is exactly the probability of sampling S without return. Entropy $H(S)$ is approximately equal to minus the logarithm of this value divided by N, which means that $H(S)$ measures dissymmetry of the partition.

Models for Qualitative Variance.

Model 1: Error of Proportional Prediction (Somers 1962). For a given S, let an entity $i \in I$ be picked randomly and let there be a recognition device determining which class i belongs to. The recognition rule is referred to as proportional prediction if it identifies classes randomly according to distribution (S): class S_t with probability p_t. Then, since S_t appears with probability p_t, and the probability of error is $1 - p_t$, the average error is equal to $\sum_t p_t(1 - p_t)$, which is exactly $V(S)$.

Model 2: Total Variance. For every class S_t of S, let us define a binary variable, z_t, having value 1 or 0 with respective probability p_t or $1 - p_t$ (binomial distribution). The variance of the variable z_t is equal to $p_t(1 - p_t)$. Then, the summary variance of all the variables, $z_1, z_2, ..., z_m$, considered as mutually independent, is equal to $\sum_t p_t(1 - p_t) = V(S)$.

Model 3: Pair Probability (Rand 1971). For a given S, let us consider probability of the event that, for any pair of the entities $i, j \in I$, i and j belong to different classes of S. Obviously, the event is decomposed into the union of m events, the t-th of which is that $i \in S_t$ while $j \notin S_t$ ($t = 1, ..., m$). The probability of the t-th event is $p_t(1 - p_t)$, which makes the overall probability equal to $V(S)$.

Model 4: Similarity Scatter (Mirkin and Cherny 1970). Let us consider a representation of S through the corresponding equivalence relation, $\sigma \subset I \times I$, or its indicator (similarity) matrix, $S = (s_{ij})$, with $s_{ij} = 1$ when both i and j are in the same class of S and $s_{ij} = 0$ when i and j are in different classes of S. The square (or module) scatter of the matrix S considered as a data matrix is $d(S) = \sum_{i,j} s_{ij}^2 = \sum_{i,j} |s_{ij}| = \sum_{t=1}^m N_t^2$, as follows from Statement 5.1. This implies that $V(S)$ is the complement of $d(S)/N^2$ to unity.

Dual Partitions and the Dissymmetry Principle (Schreider and Sharov 1982).

Let us enumerate classes of an m-class partition S in such a way that $N_m \leq ... \leq N_2 \leq N_1$. Let us pick an entity from each class and put the entities selected as a set of representatives, m-element class T_1. Class T_2 is found the same way in what is left from S on the remaining $N - m$ entities. Repeating this procedure, we exhaust S_m after class T_{N_m} has been collected. The process is continued over the remnant classes S_t. In the end, only S_1 remains nonempty (when $N_1 > N_2$), and the classes T_u selected consist of only one entity. After N_1 repetitions, we get a partition $T = \{T_1, ..., T_{N_1}\}$ of I which is called *dual* to S. Obviously, S is a dual partition to T. The total number of partitions T that are dual to a given S is equal to $M(S) = N_1! N_2! ... N_m!$. All the dual partitions have the same distribution (T) determined by distribution (S).

Regretfully, Yu. Schreider and his coauthors (for references, see Schreider and Sharov 1982) offer no clear interpretation of the duality concept. In the cluster analysis context, the dual classes can be considered anti-clusters when the prime classes are clusters.

In many natural systems, classifications satisfy the so-called Zipf law, $\log N_t = A - B \log t$ or $N_t = at^{-B}$, which means that there is a simple negative (hyperbolic) relationship between the number N_t and its position t ($t = 1, ..., m$). Many socially determined systems like texts (S_t is the set of t-letter words) or settlement systems (classes are sets of settlements (cities, towns or villages) with similar numbers of residents) satisfy the Zipf law. A well-known saying by a celebrated mathematician, S. Banach, may be regarded as a metaphoric expression of the systems nature of the law: "In mathematics, 5% of mathematicians do 95% of mathematics; however, they would not have done even 5% of the mathematics without the other 95% of the mathematicians."

An explanation of this law can be offered in terms of the dual partitions with the claim that, in a complex natural system, the pair, (S, T), of dual partitions emerging must be as dissymmetric as possible. When the symmetry is measured by the product $M(S)M(T)$ or, equivalently, by $H(S) + H(T)$, it can be proved that any dual pair S, T minimizing $M(S)M(T)$ (maximizing $H(S) + H(T)$) satisfies the following: (1) $N_t = a/t$, and (2) $m \log m = bN$, where a and b are positive constants, (3) when N is sufficiently large, there are many singleton classes in S (Schreider and Sharov 1982). Some may claim that all these are properties of "natural" classifications.

5.1.4 Comparison of Partitions

A partition S is *coarser* than a partition T while T is *finer* than S, $T \subseteq S$, if any class of S can be obtained as a union of classes of T; for corresponding equivalence relations this is exactly the case when set-theoretic inclusion holds: $\tau \subseteq \sigma$. Since the set-theoretic intersection of equivalence relations $\sigma \cap \tau$ is an equivalence relation too, it corresponds to a partition denoted as $S \cap T$ which has intersections $S_t \cap T_u$ ($t = 1, ..., m, u = 1, ..., l$) as its classes. This is the coarsest among the partitions that are finer than both S and T. In mathematics, the concept of a partition relates to the case of nonempty classes only, but, in our context, admitting some empty classes seems quite suitable. This allows narrowing the gap between the concepts of partition and nominal scale variable (p. 230). When a partition S is restricted by a subset $I' \subset I$, which means that sets $S'_t = S_t \cap I'$ are considered rather than classes S_t themselves ($t = 1, ..., m$), some of S'_t may be empty; however, we may keep tracking them to consolidate all the information about S with regard to arbitrary I'. In particular, the distribution of $S \cap T$, with this kind of tracking, becomes a synonym for the concept of a contingency table, a most popular notion in the statistics of qualitative data. Such a distribution usually is presented as a table in the format of Table 5.1.

In Table 5.1, N_{tu}, N_{t+}, and N_{+u} denote cardinalities of $S_t \cap T_u$, S_t, and T_u ($t = 1, ..., m, u = 1, ..., l$), respectively. For corresponding proportions (frequencies), usually symbols $p_{tu} = N_{tu}/N$, $p_{t+} = N_{t+}/N$, and $p_{+u} = N_{+u}/N$ are utilized. Obviously, $\sum_{t=1}^{m} N_{tu} = N_{+u}$ and $\sum_{u=1}^{l} N_{tu} = N_{t+}$.

Class	T_1	T_l	Margin
S_1	N_{11}	N_{1l}	N_{1+}
S_m	N_{m1}	N_{ml}	N_{m+}
Margin	N_{+1}	N_{+l}	N

Table 5.1: Distribution of the intersection of two partitions: Contingency table.

In the situation when no particular entity is distinguished from the others, the contingency table contains quite comprehensive information on the interrelation between partitions. The other interrelation concepts used so far can be expressed in terms of the contingency table.

The relation coarser/finer between partitions can be expressed in terms of the other means for presenting partitions as follows.

Statement 5.4. *Partition T is coarser than S if and only if any of the following conditions hold:*

(a) equivalence relation τ includes σ, $\sigma \subseteq \tau$;

(b) every entry of the partition matrix $\mathbf{S}$ is equal to or less than the corresponding entry in $\mathbf{T}$, $\mathbf{S} \leq \mathbf{T}$;

(c) every row of the contingency table (Table 5.1) contains only one nonzero element.

Proof: Obvious. □

When neither S nor T is coarser than the other, some measures of association between partitions are employed. Depending on the underlying representation, there can be distinguished at least four approaches to the problem of measuring association between partitions S and T of I:

(1) The structural approach, based on structural comparison of the corresponding equivalence relations as subsets $\sigma, \tau \subseteq I \times I$;

(2) The contingency modeling approach, based on comparison of the observed contingency table, $(S \cap T)$, with the one based on a mathematical model of association between the partitions;

(3) The geometrical approach, based on calculation of correlation measures

between the similarity matrices associated with the partitions;

(4) The cross-classificational approach, based on combining association measures between classes, S_t and T_u ($t = 1, ..., m; u = 1, ..., l$), of the partitions as subsets of I.

Let us give some of the most important examples within each of the approaches.

Structural Association Approach

The four-fold table for relations $\sigma, \tau \subseteq I \times I$ has form of Table 5.2.

Relation	τ	$\bar{\tau}$	Margin
σ	$\|\sigma \cap \tau\|$	$\|\sigma\| - \|\sigma \cap \tau\|$	$\|\sigma\|$
$\bar{\sigma}$	$\|\tau\| - \|\sigma \cap \tau\|$	$N^2 - \|\sigma\| - \|\tau\| + \|\sigma \cap \tau\|$	$N^2 - \|\sigma\|$
Margin	$\|\tau\|$	$N^2 - \|\tau\|$	N^2

Table 5.2: Four-fold table for relations.

This can be rewritten in terms of the contingency entries, based on Statement 5.1.

Relation	τ	$\bar{\tau}$	Margin
σ	$\sum_{t,u} N_{tu}^2$	$\sum_t N_{t+}^2 - \sum_{t,u} N_{tu}^2$	$\sum_t N_{t+}^2$
$\bar{\sigma}$	$\sum N_{+u}^2 - \sum N_{tu}^2$	$N^2 - \sum N_{t+}^2 - \sum N_{+u}^2 + \sum N_{tu}^2$	$N^2 - \sum_t N_{t+}^2$
Margin	$\sum_u N_{+u}^2$	$N^2 - \sum_u N_{+u}^2$	N^2

Table 5.3: Four-fold table for partitions expressed in terms of their contingency table.

This allows us to formulate two structural coefficients that are complete analogues of the match and mismatch coefficients for subsets.

Equivalence match coefficient:

$$\varepsilon s(\sigma, \tau) = (N^2 - \sum_t N_{t+}^2 - \sum_u N_{+u}^2 + 2\sum_{t,u} N_{tu}^2)/N^2 \qquad (5.3)$$

and

Equivalence mismatch coefficient:

$$e\delta(\sigma, \tau) = (\sum_t N_{t+}^2 + \sum_u N_{+u}^2 - 2\sum_{t,u} N_{tu}^2)/N^2 \qquad (5.4)$$

The latter coefficient represents the relative symmetric-difference distance between σ and τ (Hamming distance between S and T), $d(\sigma, \tau)$, as shown in formula (5.1).

Usually in the literature, only unordered pairs of the entities, $\{i, j\}$, $i \neq j$, are considered. This makes all the squared values x^2 in the four-fold Table 5.3 (numbers of the ordered pairs) to be substituted by values $x(x-1)/2$ (numbers of the unordered pairs), which does not change formulas (5.3) and (5.4) too much: just N^2 must be substituted by $N(N-1)/2$. Modified this way, the formula (5.3) is quite popular as the Rand index (Rand 1971); its complement to unity, formula (5.4) modified was introduced even earlier as the relative symmetric-difference distance (Mirkin and Cherny 1970).

Contingency Modeling Approach

Among the models for contingency tables, the most popular is that of statistical independence. The partitions S and T are called statistically independent, if $p_{tu} = p_{t+}p_{+u}$, for all $t = 1, ..., m$ and $u = 1, ..., l$. This notion is a basic concept in the theory of qualitative data analysis.

Since the distribution observed, (p_{tu}), usually is not statistically independent, the *Pearson chi-squared* coefficient

$$X^2 = \sum_{t=1}^m \sum_{u=1}^l \frac{(p_{tu} - p_{t+}p_{+u})^2}{p_{t+}p_{+u}} \qquad (5.5)$$

is usually used as a measure of deviation of the distribution $(S \cap T)$ from the statistical independence case. It is also referred to frequently as the Pearson goodness-of-fit coefficient.

Obviously, $X^2 = 0$ if and only if the distribution $(S \cap T)$ is statistically independent. If, say, $m < l$, then, X^2 is maximum which is equal to $m - 1$, when every t-th row in the contingency table has one and only one non-zero element, equal to p_{t+}. In this case, obviously, the contingency table allows us unanimously to predict a class of T when a class of S is known since S is coarser than T.

Use of this coefficient is considered substantiated by the following Pearson's theorem. Let the set I be a random independent sample from a statistically independent bivariate distribution with the marginal distributions, (S) and (T), fixed, which means that deviation of X^2 from zero is determined by the sample bias only. Then the probabilistic distribution of NX^2 converges to χ^2 distribution with $(m - 1)(l - 1)$ degrees of freedom (when $N \to \infty$). This allows testing statistical hypotheses of statistical independence of the partitions.

There have been also two normalized versions of X^2 proposed: Kramer's $C^2 = X^2/\min(m - 1, l - 1)$ and Tchouprov's $T^2 = X^2/\sqrt{(m - 1)(l - 1)}$, to make the range of the coefficient between 0 and 1, assuming a wider interpretation of X^2 as just a partition-to-partition association coefficient. It will be shown that such a wider use of X^2 may be quite relevant, due to yet another expression for the coefficient:

$$X^2 = \sum_{t=1}^{m} \sum_{u=1}^{l} \frac{p_{tu}^2}{p_{t+}p_{+u}} - 1, \qquad (5.6)$$

which can be easily proven by squaring the expression in numerator of (5.5) with the subsequent elementary arithmetic transformations.

Geometrical Approach

The central product-moment correlation coefficient between equivalence indicator matrices $\mathbf{S}$ and $\mathbf{T}$ is well-known as the Hubert Γ statistic (see, for example, Hubert 1987, Jain and Dubes 1988).

We limit ourselves here to only covariance and correlation coefficients between projector matrices, P_{us} and P_{ut}, that have a rather simple structure defined in Statement 5.2. and considered as N^2-dimensional vectors. These coefficients were analyzed, in the seventies, in France (see, for example, Saporta 1988).

Statement 5.5. *Covariance coefficient between P_{us} and P_{ut} is their scalar product $(P_{us}, P_{ut}) = X^2$, while the correlation coefficient between them is Tchouprov's coefficient T^2.*

Proof: Indeed, due to Statement 5.2., the matrices P_{us} and P_{ut} are centered and $(P_{us}, P_{ut}) = (P_s, P_t) - (P_s, P_u) - (P_u, P_t) + (P_u, P_u)$.

Since the matrices P_s and P_t both have the (i, j)-th entry nonzero only when $(i, j) \in S_t \cap T_u$ for some t, u (in this case, the entries are equal to $1/N_{t+}$ and $1/N_{+u}$, respectively), their scalar product is equal to $\sum_{t,u} N_{tu}^2/N_{t+}N_{+u}$. Each of the other three scalar products, in the equality above, is equal to 1, as is easy to check. This proves that the covariance coefficient is indeed X^2 due to formula (5.6).

To prove the second part of the statement, it is sufficient to prove that $(P_{us}, P_{us}) = m - 1$ and $(P_{ut}, P_{ut}) = l - 1$, which easily follows from the conclusion above (with $t = u$). $\qquad \square$

A similar association coefficient can be derived for S and T presented in class-indicator form as s and t. In this case, the problem is to measure the resemblance between the linear subspaces, $L_u(S)$ and $L_u(T)$, corresponding to the partitions. In multivariate statistics, the so-called *canonical correlation analysis* is used as a tool for evaluating correlation between two linear subspaces: in each of two subspaces, $\mathbf{X}$ and $\mathbf{Y}$, an orthonormal basis is constructed to be maximally correlated with the other; that is, the first vectors from each basis have their scalar product maximum over all the elements from the subspaces, the scalar product of the second vectors in the bases is maximum with regard to all the elements which are orthogonal to the first ones, etc. It can be shown that the maximum scalar products, actually, are eigenvalues of the matrix $P_\mathbf{X} P_\mathbf{Y}$ where $P_\mathbf{X}$ and $P_\mathbf{Y}$ are orthogonal projectors onto corresponding subspaces. Thus, the total proximity between the subspaces can be characterized by the sum of the eigenvalues, which is well-known to be equal to the sum of the diagonal entries of the matrix, $Tr(P_\mathbf{X} P_\mathbf{Y})$, or, equivalently, to the scalar product of the matrices $P_\mathbf{X}$ and $P_\mathbf{Y}$ themselves being considered as $N \times N$ vectors. In our case, the projectors are matrices P_{us} and P_{ut}; thus, due to the statement above, X^2 is the summary canonical correlation measure between the subspaces associated with indicator matrices s and t representing the partitions, S and T.

Cross-Classificational Approach

In this approach, a measure of overall association between partitions S and T is produced as an average of the values of a subset-to-subset association coefficient, $c(S_t, T_u)$, by all pairs (t, u) $(t = 1, ..., m; \ u = 1, ..., l)$. Sometimes, for each t, only one or few of u-s are taken: just those u_0 providing $c(S_t, T_{u_0})$ to express maximum association between S_t and T_u. From all the variety of the subset-to-subset coefficients averaged, we consider only the three based on overall averaging of the mismatch, δ and the absolute and relative probability change coefficients, Δ and Φ (see Section 4.1.2).

In terms of the contingency table the subset-to-subset coefficients can be expressed as follows:

$$\delta(S_t, T_u) = p_{t+} + p_{+u} - 2p_{tu},$$

$$\Delta(T_u/S_t) = p_{tu}/p_{t+} - p_{+u},$$

$$\Phi(T_u/S_t) = \frac{p_{tu}/p_{t+} - p_{+u}}{p_{+u}}.$$

Then the averaged coefficients are:

$$\delta(S,T) = \sum_{t,u} p_{tu}(p_{t+} + p_{+u} - 2p_{tu}) = \sum_t p_{t+}^2 + \sum_u p_{+u}^2 - 2\sum_{t,u} p_{tu}^2,$$

$$\Delta(T/S) = \sum_{t,u} p_{tu}(p_{tu}/p_{t+} - p_{+u}) = \sum_{t,u} p_{tu}^2/p_{t+} - \sum_u p_{+u}^2,$$

$$\Phi(T/S) = \sum_{t,u} p_{tu}\left(\frac{p_{tu}/p_{t+} - p_{+u}}{p_{+u}}\right) = \sum_{t,u} \frac{p_{tu}^2}{p_{t+}p_{+u}} - 1 = X^2.$$

This can be summarized in the following statement.

Statement 5.6. *The averaged subset-to-subset mismatch, absolute and relative probability change coefficients are the equivalence mismatch, proportional prediction error reduction and Pearson goodness-of-fit coefficients, respectively.*

The only part needing explanation is why $\Delta(T/S)$ is referred to as the reduction of the proportional prediction error. The proportional prediction error for partition T equals $V(T) = 1 - \sum_u p_{+u}^2$, as defined in the preceding subsection. The proportional prediction error within each of the classes S_t, $V(T/S_t)$, can be expressed analogously, $V(T/S_t) = 1 - \sum_u (p_{tu}/p_{t+})^2$. Thus, the average proportional prediction error within classes of S is $V(T/S) = \sum_t p_{t+}V(T/S_t) = 1 - \sum_u p_{tu}^2/p_{t+}$. The difference, $V(T) - V(T/S)$, shows the reduction of the proportional prediction error due to information on the class of S occurred, and is exactly $\Delta(T/S)$.

The coefficient of proportional prediction error reduction, $\Delta(T/S)$, is a core of the category utility function employed in some conceptual clustering algorithms (see p. 151), which is just the sum of $\Delta(T/S)$ by all variable partitions T involved.

Curiously, $\Delta(T/S)$ can be presented as an asymmetric form of the Pearson chi-squared coefficient (5.5):

$$\Delta(T/S) = \sum_{t=1}^m \sum_{u=1}^l {}' \frac{(p_{tu} - p_{t+}p_{+u})^2}{p_{t+}} \tag{5.7}$$

which is proved with elementary arithmetic.

All the coefficients discussed above can be adjusted with a standardizing transformation $b(T/S) = \pm(a(T) - a(T/S))/a(T)$ where $a(T/S)$ is an association measure and $a(T)$ is its particular value. Especially easily such indices are produced when $a(T)$ is a diversity measure and $a(T/S)$ is obtained by averaging $a(T)$ within classes of S, as was done above for the variance $V(S)$. The corresponding standardized coefficient,

$$w(T/S) = (V(T) - V(T/S))/V(T) = \frac{\sum_{t,u} p_{tu}^2/p_{t+} - \sum_u p_{+u}^2}{1 - \sum_u p_{+u}^2}$$

is well known as the Wallis coefficient evaluating relative reduction of the proportional prediction error of classes T_u when S classes become known. Analogously, an entropy based coefficient can be defined based on the conditional entropy $H(T/S) = -\sum_t p_{t+} \sum_u (p_{tu}/p_{t+}) \log[p_{tu}/p_{t+}] = H(S \cap T) - H(S)$. This leads to $h(T/S) = (H(T) - H(T/S))/H(T) = (H(T) + H(S) - H(S \cap T))/H(T)$ as the standardized information-based association measure.

Analogously, some other adjusted measures can be produced. The subject was reviewed by some other authors (see, for example, Goodman and Kruskal 1979, Arabie and Hubert 1985) though this presentation is different.

5.1.5 Discussion

1. A partition can be considered two-foldly: as a form of classification and as a form of nominal scale. In both cases, it concerns the extensional part rather than intensional one which still waits for an adequate formalization.

2. We distinguish between the following formally different, though equivalent, notions for representing the concept of partition: (a) partition S as it is; (b) partition $N \times m$ Boolean indicator matrix s; (c) equivalence relation σ; (d) equivalence Boolean similarity matrix $S = ss^T$, (e) linear m-dimensional and $(m-1)$-dimensional subspaces, $L(S)$ and $L_u(S)$, of the N-dimensional vector space; (f) orthogonal projection onto $L(S)$ or $L_u(S)$ $N \times N$ with similarity matrix P_s or P_{us}.

3. The diversity of a partition is its important characteristic. There is a common opinion that rationally designed partitions (as, for example, made for coding of data or separating technical devices) tend to be uniformly distributed while the classes in naturally emerged systems (for example, the biological genera partitions or settlement types), in contrast, are distributed nonuniformly, Zipf-wisely. We consider two basic characteristics of diversity, the entropy and qualitative variance, and present several interpretations for each of them. A principle of maximum dissymmetry of the dual partitions in complex "natural" systems, expressed in the entropy terms, supports the idea of the relevance of Zipf distribution.

4. Comparing partitions, especially in cross-classificational form, is a subject having almost a 100-year history of research. We distinguish between the

following basic approaches to that: (a) structural, (b) contingency modeling, (c) geometrical, and (d) cross-classificational ones. Some of the coefficients can be produced within all of the approaches, which should be considered as a substantiation of their universal applicability: equivalence match/mismatch (Rand coefficient, symmetric-difference distance), Pearson chi-squared (goodness-of-fit, average relative change of probability, between-subspace covariance), and reduction of the proportional prediction error (Wallis coefficient, average absolute change of probability).

5.2   Admissibility in Agglomerative Clustering

5.2.1   Space and Structure Conserving Properties

Let us consider the agglomerative clustering procedure in the following form. Let $D = (d_{ij})$ be a dissimilarity entity-to-entity matrix. Initially, each of the entities (cases) is considered as a single cluster (singleton). The main steps of the algorithm are as follows.

Step 1. Find the minimum value d_{uv} in D and merge clusters u and v.

Step 2. Transform D merging both the rows and columns u and v into a new row (and column) $u \cup v$ with its dissimilarities defined as

$$d_{t,u\cup v} = F(d_{tu}, d_{tv}, d_{uv}) \tag{5.8}$$

where F is a pre-fixed numeric function.

If the number of the clusters obtained is larger than 2, go to Step 1, else End.

The Lance and Williams family of agglomerative clustering algorithms $LW(\alpha, \beta, \gamma)$ is defined with F in (5.8) of the following "quasi-linear" form:

$$d_{t,u\cup v} = \alpha(r_t, r_u)d_{tu} + \alpha(r_t, r_v)d_{tv} + \beta(r_t, r_u, r_v)d_{uv} + \gamma(r_t)|d_{tu} - d_{tv}| \tag{5.9}$$

where r_t, r_u, and r_v are the ratios of the cardinalities of t, u, and v, respectively, to the cardinality of $u \cup v$. This format of the coefficients considered in Chen and Van Ness 1995 covers many clustering algorithms; for example, Ward's method in Table 3.2 satisfies formula (5.9) with $\alpha(x, y) = (x + y)/(1 + x)$, $\beta(x, y, z) = -z/(x + y + z)$ and $\gamma = 0$ for the Euclidean distance squared. An agglomerative clustering algorithm will be referred to as a LW-algorithm if it can be described by formula (5.9).

Lance and Williams 1967 made a suggestion to reduce the potential number of LW-algorithms in such a way that only those of them should be considered that satisfy some supplementary admissibility conditions. They suggested concepts of space-conserving and space-distorting algorithms, formally analyzed later by Dubien and Warde 1979. The following exposition is based on a recent work by Chen and Van Ness 1995 who analyzed the following notions, among the others.

Let us denote as m_{tuv} and M_{tuv} the minimum and maximum of the two dissimilarities, d_{tu} and d_{tv}. An agglomerative algorithm will be referred to as *space-conserving* if $m_{tuv} \leq d_{t,u \cup v} \leq M_{tuv}$; *space-contracting* if $d_{t,u \cup v} \leq m_{tuv}$; and *space-dilating* if $d_{t,u \cup v} \geq M_{tuv}$, for any agglomeration step involving clusters u and v to be merged. The properties mean that a space-conserving algorithm somehow "averages" the distances to the clusters merged while space-contracting algorithm puts the merged cluster nearer to the others and a space-dilating one moves it farther from the other clusters. Curiously, the nearest neighbor (single linkage) algorithm is space-contracting while the farthest neighbor (complete linkage) algorithm is space-dilating.

The property of space-conserving is related to yet another useful property of the clustering algorithms. Let us refer to a partition as a *clump structure* if all the within-cluster dissimilarities are smaller than all the between-cluster dissimilarities. Obviously, for any $m \leq N$, it can be no more than one m-cluster clump structure; and if S and T are the clump structures, then either $S \subseteq T$ or $T \subseteq S$. An agglomerative algorithm will be referred to as *clump structure admissible* if it always finds the clump structures if they exist.

Statement 5.7. *If an agglomerative clustering algorithm is space-conserving, then it is clump structure admissible.*

Proof: Let a partition $S = \{S_1, ..., S_m\}$ be a clump structure; thus the maximum within-cluster distance, M_w, is smaller than the minimum between-cluster distance, m_b. Obviously, the agglomerative procedure starts combining two entities within a cluster S_t, $t = 1, ..., m$. The distances between the new cluster and the other clusters in S_t will be less than or equal to M_w while the distances between the new cluster and any of the clusters out S_t will be greater than or equal to m_b since the property of space-conserving holds. When the agglomeration step is reiterated, the algorithm never combines any two clusters which belong to different clusters S_t and S_u until all the clusters $S_1, ..., S_m$ are combined; this completes the proof. $\square$

When LW-algorithms are considered, these two properties are equivalent (Chen and Van Ness 1995). To make the original proof of the statement less technical, we introduce a weaker admissibility concept related to situations when distinct entities are considered different even if they are in the same location. More explicitly, let us consider the case when, in nonnegative indifference matrix d, d_{ij} can be zero when $i \neq j$. However, $d_{ik} = d_{jk}$ is presumed for all $k \in I$ when $d_{ij} = 0$ (evenness). Having this in mind, an agglomerative clustering algorithm will be called *proper* if $d_{t,u \cup v} = d_{tu} = d_{tv}$ whenever $d_{uv} = 0$. The proper LW-algorithms can be characterized by the following.

Statement 5.8. *An LW-algorithm is proper if and only if* $\alpha(x, y) + \alpha(1 - x, y) = 1$ *for any rational* x, $0 < x < 1$, *and* $y > 0$.

Proof: Let us consider two entities, i_1 and i_2, such that $d_{i_1 i_2} = 0$ and, thus, $d_{ji_1} = d_{ji_2}$ for any $j \in I$. Then, by formula (5.9), $d_{j,i_1 \cup i_2} = \alpha(x, y) d_{ji_1} + \alpha(1 - x, y) d_{ji_2}$ where $x = y = 1/2$. The properness of the algorithm, obviously, means that $\alpha(x, y) + \alpha(1 - x, y) = 1$. Applying the same operation repeatedly, we'll have $\alpha(x, y) + \alpha(1 - x, y) = 1$ for any rational x, $0 < x < 1$, and $y > 0$. □

Statement 5.9. *An LW-algorithm is proper if it is clump structure admissible.*

Proof: Let us take three entities, i_1, i_2, and j, such that $d_{i_1 i_2} = \delta$ and $d_{ji_1} = d_{ji_2} = 1 + \delta$ where $\delta > 0$. Then, by (5.9), $d_{j,i_1 \cup i_2} = (\alpha(x, y) + \alpha(1 - x, y))(1 + \delta) + \beta \delta$ where $x = y = 1/2$. Let $\delta \to 0$, then $d_{j,i_1 \cup i_2} \to \alpha(x, y) + \alpha(1 - x, y)$. Let us take yet another entity k with $d_{kj} = 1 + \epsilon$ and $d_{ki_1} = d_{ki_2} \geq 2 + \epsilon$ where $\epsilon > \delta$. Since subsets $\{i_1, i_2, j\}$ and $\{k\}$ are clumps, $d_{j,i_1 \cup i_2} < d_{kj}$, which implies that $\alpha(x, y) + \alpha(1 - x, y) \leq 1$ which is proved by letting $\epsilon \to 0$. To prove that $\alpha(x, y) + \alpha(1 - x, y) \geq 1$, let us suppose $d_{kj} = 1 - \epsilon$ and $d_{ki_1} = d_{ki_2} \geq 1 + \delta$. Then, the subsets $\{i_1, i_2\}$ and $\{j, k\}$ are clumps, which implies the opposite inequality. Repeating this conclusion for many entities having zero distances from i_1 and i_2, we get the same equality, $\alpha(x, y) + \alpha(1 - x, y) = 1$ for any rational x, $0 < x < 1$, and $y > 0$. □

Now, we are ready to prove the main result.

Statement 5.10. *For an* $LW(\alpha, \beta, \gamma)$*-algorithm, the following properties are equivalent:*

1. *The algorithm is space-conserving;*

2. *The algorithm is clump structure admissible;*

3. *For any rational* x, $0 < x < 1$, *and* $y > 0$, *the following relations hold:*

   (a) $\alpha(x, y) + \alpha(1 - x, y) = 1$;

   (b) $\beta(x, 1 - x, z) = 0$;

   (c) $|\gamma(y)| \leq \alpha(x, y)$.

Proof: Let us prove that $(2) \to (3)$. The equality (a) follows from Statements 5.8. and 5.9.. To prove that $\beta(x, 1 - x, z) = 0$, let us consider a configuration of four points, i_1, i_2, j, k, such that $d_{ji_1} = d_{ji_2} = 1$, $d_{i_1 i_2} = \delta$ where δ is a constant satisfying $\delta > \epsilon$. Let $x = a/b$ and $y = c/d$ where a, b, c and d are integers. Let us put ad points in i_1, $(b-a)d$ points in i_2, cb points in j, and cb points in k. Being clump structure admissible, the algorithm combines initially all the points located within i_1, i_2, j and k without changing the nonzero distances by Statement 5.9. This implies that the derivation is valid for any rational x and y.

When $d_{jk} = 1 + \epsilon$, it will be two clumps, $\{i_1, i_2, j\}$ and $\{k\}$, leading to $d_{j, i_1 \cup i_2} < d_{kj}$ and, thus, $1 + \delta\beta < 1 + \epsilon$ which implies $\beta \leq 0$. Taking $d_{jk} = 1 - \epsilon$, we similarly have $\beta \geq 0$ which proves (b).

To show $\gamma(y) \leq \alpha(x, y)$, let us consider the same numbers of points put in i_1, i_2, j, and k having $d_{i_1 i_2} < d_{ji_1} < d_{ji_2}$ and $d_{jk} = d_{ji_2} + \epsilon$. Then, there are two clumps, $\{i_1, i_2, j\}$ and $\{k\}$, leading to $d_{j, i_1 \cup i_2} < d_{kj}$ and, thus, $\alpha(x, y)d_{ji_1} + (1 - \alpha(x, y))d_{i_2 j} + \gamma(d_{ji_2} - d_{ji_1}) < d_{ji_2} + \epsilon$, which leads to $\gamma(y) \leq \alpha(x, y)$ when $\epsilon \to 0$. Now, let us take $d_{jk} = d_{ji_1} - \epsilon$, all the other inter-distance relations unchanged. There are two clumps, $\{i_1, i_2\}$ and $\{j, k\}$, leading to $d_{j, i_1 \cup i_2} > d_{kj}$. Letting $\epsilon \to 0$, it gives $\gamma(y) \geq -\alpha(x, y)$ which completes the proof that $(2) \to (3)$.

Let us prove now that $(3) \to (1)$. Let $d_{tu} \leq d_{tv}$; then, (3) implies $d_{t, u \cup v} = \alpha(r_u, r_t)d_{tu} + \alpha(r_v, r_t)d_{tv} + \gamma(r_t)(d_{tv} - d_{tu}) = d_{tu} + (\alpha(r_v, r_t) + \gamma(r_t))(d_{tv} - d_{tu})$. The algorithm is space-conserving since $0 \leq \alpha(r_v, r_t) + \gamma(r_t) \leq 1$.

The proof is over because $(1) \to (2)$ is proved in Statement 5.7. $\qquad\qquad \Box$

Chen and Van Ness 1994 proved that space-dilating WL-algorithms are characterized by the inequality:

$$\alpha(x, y) + \alpha(1 - x, y) - 1 \leq \max[0, -\beta(x, 1 - x, y), \alpha(1 - x, y) - \gamma(y)]$$

while the space contracting algorithms satisfy the converse inequality for $\alpha(x, y) = \alpha(1 - x, y) - 1 = -\gamma(y)$.

5.2.2   Monotone Admissibility

In agglomerative clustering, the cluster index function is defined for every cluster u in the recursive way: $h(\{i\}) = 0$ for all $i \in I$ and $h(u \cup v) = d_{uv}$. It is the index function which is employed in drawing the tree representation for the results of agglomerative clustering: for every cluster S, its index on the picture is proportional to $h(S)$. This is why it is considered important that the index function would be monotone: $h(S) > h(T)$ whenever $T \subset S$.

It appears, LW-algorithms can lead to non-monotone index functions. Let us call an agglomerative clustering algorithm *monotone* if the corresponding index function is monotone. Using Milligan 1979 result, Batageli 1981 found the following characterization of monotone LW-algorithms.

Statement 5.11. *An $LW(\alpha, \beta, \gamma)$-algorithm is monotone if and only if*

$$(a) \; \alpha(x, y) + \alpha(1 - x, y) \geq 0,$$

$$(b) \; \alpha(x, y) + \alpha(1 - x, y) + \beta(x, 1 - x, z) \geq 1, \; and$$

$$(c) \; \gamma(y) \geq -\min[\alpha(x, y), \alpha(1 - x, y)],$$

for any x, y, $0 < x < 1$, $y > 0$.

Proof: The monotonicity of the algorithm means that, for u and v merged, $d_{t,u\cup v} \geq d_{uv}$ for any t in (5.9) because $d_{t,u\cup v} < d_{uv}$ for some t would mean that, at the next step, t and $u \cup v$ must be merged since d_{uv} is the minimum of all the between-cluster distances by definition. Let us suppose that (a) to (c) hold and prove that $d_{t,u\cup v} \geq d_{uv}$ for any t. Without any loss of generality, let $d_{tu} \geq d_{tv}$. Then, (5.9) becomes:

$$d_{t,u\cup v} = (\alpha(x, y) + \gamma(y))d_{tu} + (\alpha(1 - x, y) - \gamma(y))d_{tv} + \beta(x, 1 - x, y)d_{uv}.$$

Since $\alpha(x, y) + \gamma \geq 0$ by (c), $d_{t,u\cup v} \geq (\alpha(x, y) + \gamma(y) + \alpha(1 - x, y) - \gamma(y))d_{tv} + \beta(x, 1 - x, y)d_{uv}$. Since $d_{uv} \leq d_{tv}$, $d_{t,u\cup v} \geq (\alpha(x, y) + \alpha(1 - x, y) + \beta(x, 1 - x, y))d_{uv}$, by (a), which implies $d_{t,u\cup v} \geq d_{uv}$, by (b).

Conversely, suppose that the algorithm is monotone, that is, $d_{t,u\cup v} \geq d_{uv}$ for all t. This means that

$$(\alpha(x, y) + \gamma(y))d_{tu} + (\alpha(1 - x, y) - \gamma(y))d_{tv} \geq (1 - \beta(x, 1 - x, y))d_{uv}.$$

Suppose that (c) does not hold, which means that $\alpha(x, y) + \gamma(y) < 0$ (in the assumption that $d_{tu} \geq d_{tv}$). In this case, let us take d_{tu} satisfying inequality

$$d_{tu} > \max[d_{tv}, ((1 - \beta(x, 1 - x, y)d_{uv} - (\alpha(1 - x, y) - \gamma(y))d_{tv})/(\alpha(x, y) + \gamma(y))];$$

thus leading to non-monotonicity. The contradiction implies that $\alpha(x, y) + \gamma(y) \geq 0$ and, by monotonicity, $(\alpha(x, y) + \alpha(1 - x, y))d_{tv} \geq (1 - \beta)d_{uv}$. Again, suppose $\alpha(x, y) + \alpha(1 - x, y) < 0$. Then, taking, for some t, $d_{ut} = d_{vt} > \max[(1 - \beta)/(\alpha(x, y) + \alpha(1 - x, y)), 1]d_{uv}$, we have a counterexample to the latter inequality; therefore, (a) holds. This implies $d_{t,u\cup v} \geq (\alpha(x, y) + \alpha(1 - x, y) + \beta(x, 1 - x, y))d_{uv}$. If (b) does not hold, that is, $\alpha(x, y) + \alpha(1 - x, y) + \beta(x, 1 - x, y) < 1$, then $d_{t,u\cup v} < d_{uv}$ which contradicts the monotonicity. Thus, all conditions, (a), (b), and (c), hold, which proves the statement. □

Combining this result with Statement 5.10., we have

Corollary 5.1. *Any space-conserving (clump structure admissible) LW-algorithm is monotone.*

5.2.3   Optimality Criterion for Flexible LW-Algorithms

To substantiate the recalculation formula (5.8), a general criterion of optimality for partitions can be sought such that, at each step, the agglomeration is made to minimize the criterion increment, as it is done in Ward's method. To formulate the problem explicitly, let us restrict ourselves to the criteria having the format of summary functions $F(S) = \sum_{t=1}^{m} f(S_t)$ where f is a set function and $S = \{S_1, ..., S_m\}$ is a partition. The between-cluster distance $d_{S_1 S_2}$ will be called *criterion-generated* if there exists a set function $f(S_t)$ such that $d_{S_1, S_2} = f(S_1 \cup S_2) - f(S_1) - f(S_2)$ for any pair of nonoverlapping clusters. Obviously, this expression represents an increment of the corresponding criterion, $F(S)$, when clusters S_1 and S_2 in a partition are merged, the other clusters unchanged.

The problem is to describe the criterion-generated LW-algorithms and corresponding criteria. We do not know any general answer to the problem. In this section, a more modest problem will be analyzed related to the class of the so-called flexible LW-algorithms.

Let us refer to an LW-algorithm as a *flexible* one if the recalculations are driven by the following version of formula (5.9):

$$d_{t,u\cup v} = \alpha d_{tu} + \alpha d_{tv} + \beta d_{uv} \qquad (5.10)$$

In the formula, $\gamma = 0$ and α and β are arbitrary constants, which is a quite strong restriction (usually non-required): in practical computations, the coefficients may heavily depend on the clusters as we have seen in Section 3.2.3. Another modification concerns the situations when recalculations are made: formula (5.10) is supposed to be applicable for every pair of clusters u and v merged, not for the optimal pair only.

It appears, the set of criteria generating flexible LW-algorithms is quite narrow; actually, it can be considered a unique criterion corresponding to a unique LW-algorithm when the number of the entities is greater than 4.

Statement 5.12. *A flexible LW-algorithm is criterion generated if and only if $\alpha = 1$ and $\beta = 0$, and the criterion is equal to set function*

$$f(S) = (1/2) \sum_{i,j \in S} f_{ij} - (|S| - 2) \sum_{i \in S} f_i$$

where f_{ij} and f_i $(i \in I$ and $N \geq 4)$ are some reals.

Proof: Let us consider a flexible criterion-generated LW-algorithm satisfying equality (5.10) with all the distances expressed through a set function $f(S)$, which leads the function to satisfy the following equation:

$$f(t \cup u \cup v) = \alpha[f(t \cup u) + f(t \cup v)] + (1+\beta)f(u \cup v) - (\alpha+\beta)[f(u)+f(v)] - (2\alpha-1)f(t).$$

Having (5.10) applied to the distance $d(u, t \cup v)$ (with t and u exchanged), we obtain yet another equation:

$$f(t \cup u \cup v) = \alpha[f(t \cup u) + f(u \cup v)] + (1+\beta)f(t \cup v) - (\alpha+\beta)[f(t)+f(v)] - (2\alpha-1)f(u).$$

Subtracting one equation from the other, we get: $0 = (\alpha - \beta - 1)(f(t \cup v) - f(u \cup v))$ which implies that $\alpha = \beta + 1$ and

$$f(t \cup u \cup v) = \alpha(f(t \cup u) + f(t \cup v) + f(u \cup v)) - (2\alpha-1)(f(t) + f(u) + f(v)).$$

Applying the latter formula to $f(t \cup u \cup v \cup w)$ twice (the first time $v \cup w$ is considered as a single cluster), we get: $f(t \cup u \cup v \cup w) = \alpha f(t \cup u) + \alpha^2(f(t \cup v) + f(t \cup w)) + (2\alpha^2 - 2\alpha + 1)f(v \cup w) - 2\alpha(2\alpha-1)(f(v) + f(w)) - (\alpha+1)(2\alpha-1)(f(t)+f(u))$ which can be true only when $\alpha = 1$, by the symmetry-based considerations. This leads to

$$f(t \cup u \cup v) = f(t \cup u) + f(t \cup v) + f(u \cup v) - (f(t) + f(u) + f(v))$$

which implies, by induction, the formula of the criterion set function in the statement.   □

Corollary 5.2. *The only flexible criterion-generated WL-algorithm above involves the aggregate dissimilarity* $d(t, u) = \sum_{i \in t} \sum_{j \in u} d_{ij}$ *where* $d_{ij} = f_{ij} - f_i - f_j$, $i, j \in I$.

Corollary 5.3. *When the entity weight is constant,* $f_i = f$ *for all* $i \in I$, *the formula for the only criterion generating the flexible WL-algorithms becomes*

$$f(S) = (1/2) \sum_{i,j \in S} f_{ij} - (|S| - 2)|S|f = (1/2) \sum_{i,j \in S} (f_{ij} - \pi) + |S|f$$

where $\pi = 2f$ *(here, the diagonal dissimilarities* d_{ii} *and* f_{ii} *are excluded from the data). Applied to a partition* $S = \{S_1, ..., S_m\}$, *this leads us to criterion*

$$F(S) = (1/2) \sum_{t=1}^{m} \sum_{i,j \in S} (f_{ij} - \pi) + Nf \qquad (5.11)$$

to be maximized by S.

A similar summary threshold linkage-partition criterion will appear in the next section, though in a different context.

5.2.4 Discussion

1. Agglomerative clustering may be considered in two ways: as a set of particular techniques for partitioning or as a method for revealing a hierarchical structure in the data. In this chapter, we discuss the agglomeration clustering as a partitioning technique; the other aspect will be considered in Chapter 7.

2. The major line of discussion goes in the framework suggested by Lance and Williams 1967 along with both some general properties of the agglomeration steps, like space-conserving and clump cluster admissibility.

3. Three subjects are developed in some detail:

    (a) The study of interrelation among a few general properties of the agglomeration algorithms, in general, and WL-algorithms, in particular (based mostly on the recent work of Chen and Van Ness 1994, 1995).

    (b) The characterization of the so-called monotone admissibility for LW-algorithms (Milligan 1979, Batageli 1981) which makes the cluster hierarchy found have a proper graphical representation. The property is considered usually as a necessary condition for an agglomerative algorithm to be good one; however, in some cases, when d_{uv} has no meaning of "diameter", the requirement becomes irrelevant, which will be illustrated in the next Section 5.3.

    (c) Finding a partitioning criterion associated with Lance-Williams formula in such a manner that the formula provides for the optimal increment of the criterion while merging. It appears that among flexible LW-algorithms, there is, actually, only one such criterion and only one Lance-Williams formula pattern, corresponding to each other.

4. The direction of research outlined deserves to be continued and extended at least in the following directions:

    (a) Formulating general properties of the agglomerative algorithms and investigation of relationship among the properties, aiming, at least partly, at eventually getting an axiomatic description of some practically important algorithms or classes of the algorithms;

    (b) Extending the research into different classes of clustering algorithms, as the moving-center or seriation;

    (c) Producing general computing formulas other than that of Lance and Wiiliams.

5.3   Uniform Partitioning

5.3.1   Data-Based Validity Criteria

Milligan 1981 considered a representative list of thirty data-based validity criteria. Each of the criteria was examined in a simulation study as a goodness-of-fit measure between the input data and partitions found with clustering algorithms. Since the input data were generated to represent a clear nonoverlapping (although some noisy) cluster structure, performances of the criteria could be fairly considered as the testing scores of their recovery characteristics. It turned out, the best six criteria involve only two kinds of validity measures of the cluster structures in their relation to the entity-to-entity distances: metric and non-metric correlations. Let $D = (d_{ij})$, $i, j \in I$, be a distance matrix and $S = \{S_1, ..., S_m\}$ a partition of I. The metric measures evaluate correlation between D, and the equivalence indicator function $1 - S = (1 - s_{ij})$ where $s_{ij} = 1$ if both i and j belong to the same class of S and $s_{ij} = 0$ if not; transformation $1 - s_{ij}$ is used to transform the similarity measure S into a dissimilarity measure without any change of the absolute correlation value. The non-metric measures are based on rank correlation between these matrices. The matrix correlations themselves are among the six best criteria. Their formulas can be expressed as follows:

$$r(d_{ij}, 1 - s_{ij}) = (d_b - d_w)(n_b n_w)^{1/2}/N\sigma_d \qquad (5.12)$$

and

$$rr(d_{ij}, 1 - s_{ij}) = (s_+ - s_-)/\{N(N-1) - n_S)N(N-1)\}^{1/2} \qquad (5.13)$$

where d_b, n_b and d_w, n_w are the averages (d) and numbers (n) of the distances between and within the clusters, respectively, while σ_d is the standard deviation of the distances; s_+ represents the number of times when two entities clustered together have a smaller distance than two points which are in the different clusters, s_- counts for the reverse outcomes, and n_S is the number of quadruples of the entities (i, j, k, l) consisting of two pairs, (i, j), (k, l), both within or both between the clusters.

In the present author's opinion, the emergence of the matrix correlation measures as the best cluster validity indices can be interpreted in framework of the linear approximation clustering. The following concerns the metric product-moment correlation only; the rank correlation index should be considered just as a good approximation of the metric one since the order structure of the distances has been found holding much of the metric structure (Shepard 1966).

To make situation more clear, let us recall the approximational meaning of the product-moment correlation coefficient (p. 65). It determines the slope and the residual variance of the linear regression equation $y = ax + b + e$ connecting two variables, y and x, with the residuals e least-squares minimized, as follows: $a = \rho \sigma_y/\sigma_x$, $\delta^2 = (1 - \rho^2)\sigma_y^2$. Here, δ^2 is the residual variance defined as the average of the residuals squared. This shows that the least-squares linear regression criterion is equivalent to the criterion of maximizing the determination coefficient ρ^2.

Let us put this in the clustering context. The distance matrix (d_{ij}) stands for y while the sought equivalence indicator matrix $(1 - s_{ij})$ serves as x, and ρ^2 shows the relative decreasing of the variance of the distances after they have been approximated by a linear function of the indicator matrix (s_{ij}). The value of ρ here must be positive (when the clusters are in accordance with the distances), thus reflecting quality of the partition in the problem of linear approximation of the distances by the equivalence indicator matrix. This allows considering Milligan's 1981 results as an empirical confirmation of validity of the following approximation clustering model.

5.3.2 Model for Uniform-Threshold Partitioning

Let the data matrix be a similarity matrix $A = (a_{ij})$, $i, j \in I$. Partition $S = \{S_1, ..., S_m\}$ to be found is associated with its equivalence indicator matrix $S = (s_{ij})$. Let us define, for every $S \subseteq I$, $\nu(S) = |S|$ or $\nu(S) = |S| - 1$ depending on the diagonal elements a_{ii} of A are given or not, respectively. The linear transformation $\lambda S + \mu$ translates the equivalence indicator values 1 and 0 into $\lambda + \mu$ and μ, respectively.

When A can be considered as noisy information on $\lambda S + \mu$, the following model seems appropriate:

$$a_{ij} = \lambda s_{ij} + \mu + e_{ij} \tag{5.14}$$

where e_{ij} are the residuals to be minimized. The value of μ relates to a shift of the origin of the similarity measurement scale. Sometimes, no shift is considered necessary thus leaving $\mu = 0$; in this latter case, the matrix A is supposed to be centered (thus having zero as its grand mean).

When both λ and μ are to be adjusted using the least-squares criterion, the linear regression analogy underscored above remains fair. In this case, the least-squares criterion (with both λ and μ optimally adjusted) equals $L^2 = (1 - r(a_{ij}, s_{ij})^2)s_A^2$ where $r(a_{ij}, s_{ij})$ is defined as (5.12) multiplied by -1 (since s_{ij}, not $1 - s_{ij}$, is involved here). The optimal λ equals the numerator of the correlation coefficient, $\lambda = a_w - a_b$, while the optimal $\mu = a_b$, which shows the meaning of the optimal equivalence indicator matrix values: a_w within S and a_b between the clusters. With these λ and μ substituted, the equivalent maximized determination coefficient becomes proportional to

$$(a_w - a_b)^2 n_w n_b \tag{5.15}$$

The maximizing criterion (5.15) is equivalent to maximizing a somewhat simpler criterion

$$a_w^2 n_w + a_b^2 n_b \tag{5.16}$$

which is proved by squaring the parenthesis in (5.15) and using equalities, $n_b = N\nu(I) - n_w$ and $n_w = N\nu(I) - n_b$, when appropriate.

This form of the criterion shows that maximizing both values a_w and a_b is welcome which seems somehow confusing: intuition tells us that a reasonable partition must have a_w large and a_b small. The coefficients, n_w and n_b, do not eliminate the problem although they somewhat diminish it: small n_w and large n_b help keep the criterion value up along with properly rated a_w and a_b.

When μ is restricted to be zero, the partition indicator matrix has $\lambda + \mu = a_w$ and $\mu = 0$ as its within and between entries, respectively. The least-squares criterion, in this case, reflects only the within partition pairs; it is equivalent to the problem of maximizing the following expression for a partition's contribution to the square data scatter:

$$a_w^2 n_w \tag{5.17}$$

which resembles the additive single cluster criterion in (4.28), Section 4.5.1. This criterion does not involve the between pairs and, in this aspect, seems better than (5.16) thus substantiating the model with $\mu = 0$.

When both $\lambda > 0$ and μ are pre-specified, the least-squares approximation becomes equivalent to the problem of maximizing yet another criterion,

$$SU(\pi, S) = \sum_{t=1}^{m} \sum_{i,j \in S_t} (a_{ij} - \pi) \tag{5.18}$$

where the threshold $\pi = \lambda/2 + \mu$ is, actually, the average of within and between average similarities. Since the threshold value is the same for every cluster S_t, criterion (5.18) can be referred to as the *uniform threshold* criterion. We can see that criterion (5.18) falls into formula (5.11) derived in Corollary 5.3. as the only criterion leading to flexible LW-algorithms.

Analyzing this simpler criterion can give an insight into the nature of the other two criteria, (5.15) and (5.17), differing from (5.18) by particular choices of the coefficients, λ and μ. The properties of the uniform threshold criterion in the case when all the possible partitions are considered feasible, can be summarized as follows (see Kupershtoh, Mirkin, Trofimov 1976).

1. Value π is a "soft" threshold determining clusters as connected with the "crisp" threshold graph $G_\pi = \{(i,j) : a_{ij} > \pi\}$: the objects i,j having $a_{ij} > \pi$ tend to be put in the same cluster while the objects with $a_{ij} < \pi$ tend to be kept apart, though this is only a tendency, not a rule, which is disclosed more clearly in the items 2) and 3) to follow.

2. If S is optimal, then the total proximity, $a(t,t) = \sum_{i,j \in S_t}(a_{ij} - \pi)$, within every cluster S_t is nonnegative while every total between-proximity, $a(t,u) = \sum_{i \in S_t} \sum_{j \in S_u}(a_{ij} - \pi)$, is nonpositive ($t \neq u$). Admitting the opposite, $a(t,u) > 0$ for some $t \neq u$, we'll have a contradiction to the optimality assumption since after merging S_t and S_u in $S_t \cup S_u$, the maximum criterion value will be increased by $2a(t,u) > 0$ which is impossible if S is optimal.

3. The optimal partition satisfies the following "compactness" properties with regard to the threshold π:

   (a) $a_b \leq \pi \leq a_w$;

   (b) $a_{tt} \leq \pi \leq a_{tu}$ $t \neq u$

   where a_{tt} and a_{tu} are the average similarities in corresponding blocks ($S_t \times S_t$ and $S_t \times S_u$, respectively).

4. When π increases, the number of clusters in an optimal partition, in general, grows, though there are examples when the number of optimal clusters can decrease when π is increased. In the extreme case when π is so large that all $a_{ij} < \pi$, the optimal partition consists of the singletons (in the opposite case, when $a_{ij} \geq \pi$ for all $i,j \in I$, the only optimal "partition" consists of the unique class I). Yet it is another characteristic of S which is co-monotone to π: the qualitative variance $V(S) = \sum_t p_t(1 - p_t)$ of the optimal partition increases when π grows. The proof closely follows the proof of Statement 4.6. in Section 4.2.3.

5.3.3 Local Search Algorithms

The following three techniques will be presented, initially, for the case when the threshold π in the criterion (5.18) is constant.

Agglomerating/Strewing

Based on the properties above, let us define two neighborhoods: $N_1(S) = S(t,u) : t \neq u$} where $S(t,u)$ differs from S only in that classes S_t and S_u of S are merged in one class $S_t \cup S_u$ in $S(t,u)$, and $N_2(S) = \{S^t : t = 1, ..., m\}$ where S^t differs from S only in that class S_t of S has been strewed in separate singletons in S^t. Then, let $N(S) = N_1(S) \cup N_2(S)$. With this neighborhood, an iteration of the local search algorithm, starting with a partition S as its input, checks all the values $a(t,u) = \sum_{i \in S_t} \sum_{j \in S_u}(a_{ij} - \pi)$ and takes the maximum of $a(t,u)$ with regard to $t \neq u$ and of $-a(t,t)$ by all $t = 1, ..., m$. If the maximum holds for $t \neq u$, the agglomerated partition $S(t,u)$ is taken as S for the next iteration. If $-a(t,t)$ is maximum, the partition S^t is taken as the input for the next iteration. The computation stops when the maximum value is not positive anymore. In this case, the current partition S is the result; it satisfies the "compactness" properties above.

If the starting partition is the trivial partition consisting of N singletons, the values $a(t, t)$ are always non-negative and, thus, the operation of strewing never applies. In this case, the algorithm can be considered a hierarchical agglomerative algorithm, except for the stopping rule which is applied when the criterion (5.18) begins decreasing (with all $a(t, u) \leq 0$). Thus, the criterion does not allow continued merging when all the clusters become both "cohesive" and "isolated" (up to the threshold π).

There is no need to calculate the matrix $a(t, u)$ after every agglomeration step; a simplest version of the Lance-Williams formula is valid:

$$a(t, u \cup v) = a(t, u) + a(t, v)$$

as we have seen in Section 5.2.3.

Exchange

Let us consider yet another neighborhood, $N_3(S)$ which consists of the partitions $S(i, t)$ obtained from S by moving entity $i \notin S_t$ in S_t. With this neighborhood, the local search algorithm for criterion (5.18) can be described in terms of the average linkage function, p. 178. For every $i \in I$ and S_t not containing i, the values $al(i, S_t)$ are calculated and their maximum (with regard to t), $al(i, S_{t(i)})$ is determined. Then, if it is larger than π then i must be moved into $S_{t(i)}$, and the partition $S(i, t(i))$ is the result of the iteration; if not, the partition S is final. The reader is invited to prove that this is really a local search algorithm for criterion (5.18).

Obviously, the resulting partition, in the exchange process, satisfies the following "compactness" condition: for every $i \in I$ its average linkage to its cluster is larger than to any other cluster.

The following two modifications of the algorithm seem straightforward: 1) all i having $al(i, S_{t(i)})$ larger than π are moved in $S_{t(i)}$ simultaneously; 2) at each step, the values $al(i, S_t)$ are calculated for one i only; after that moving i into $S_{t(i)}$, if qualified, is performed, and a next i is considered for the next iteration.

In the present author's experiments, the best results have been found with all the three neighborhoods united; that is, with $N(S) = N_1(S) \cup N_2(S) \cup N_3(S)$.

Seriation Techniques

A seriation algorithm with the summary threshold function as the criterion can be utilized for sequentially obtaining the uniform-threshold clusters. After separating a cluster, seriation is made again, this time for the reduced set of entities (with all the already clustered entities being removed). This is repeated until no unpartitioned objects remain.

All the techniques considered can be employed for criteria (5.15) and (5.17) as well. These criteria differ from (5.18) in that they involve optimal values of the threshold $\pi = \lambda/2 + \mu$ rather than a constant. Thus, at every step, after partition S is updated, the threshold must be recalculated as $\pi = (a_w + a_b)/2$, for (5.15), or $\pi = a_w/2$, for (5.17).

Let us apply the agglomeration algorithm for uniform partitioning the Functions and Confusion data sets. The Functions data set is a similarity matrix between 9 elementary algebraic functions; the grand mean, 2.72, is subtracted from all the entries. In Table 5.4, the uniform partitions corresponding to different threshold values (two of them optimal) are presented.

We can see that both $\pi = 1$ and $\pi = 1.326$ lead to four class partitioning, though the partitions are different.

Threshold	m	Partition	Residual Variance
-1	2	1-2-5-6-7-8-9, 3-4	0.482
0	3	1-2, 3-4, 5-6-7-8-9	0.436
1	4	1-2, 3-4, 5-6-7-8, 9	0.448
2	5	1-2, 3-4, 5-6, 7-8, 9	0.511
3	5	1-2, 3-4, 5-6, 7-8, 9	0.511
4	7	1-2, 3-4, 5, 6, 7, 8, 9	0.705
0.507	3	1-2, 3-4, 5-6-7-8-9	0.436
1.326	4	1-2, 3-4, 5-6-9, 7-8	0.557

Table 5.4: Uniform partitions of 9 elementary functions (the Functions data set) with the residual variance estimated due to complete model (5.14); the last two rows correspond to optimal threshold values based on either complete or zero constant form of the model.

In Table 5.5, the results of uniform partitioning for the centered matrix Confusion (between 10 symbolic integer digits) are presented. The matrix has been preliminarily symmetrized and its diagonal entries excluded.

Curiously, the iterative adjustment of the threshold π in the complete model starting with $\pi = 0$ leads to the stationary $\pi = 21.0375$ corresponding to the relative residual variance 0.476 while the optimal threshold is $\pi = 41.927$. This is an example of a locally, not globally, optimal solution.

Threshold	m	Partition	Residual Variance
-20	2	1-4-7, 2-3-5-6-8-9-0	0.754
0	4	1-4-7, 2, 3-5-9, 6-8-0	0.476
30	6	1-7, 2, 3-9, 4, 5-6, 8-10	0.439
50	6	1-7, 2, 3-9, 4, 5-6, 8-10	0.439
60	7	1-7, 2, 3-9, 4, 5, 6, 8-10	0.468
90	8	1-7, 2, 3-9, 4, 5, 6, 8, 10	0.593
41.927	6	1-7, 2, 3-9, 4, 5-6, 8-10	0.439
46.460	6	1-7, 2, 3-9, 4, 5-6, 8-10	0.489

Table 5.5: Uniform partitions of 10 segmented digits (the Confusion data set) with the residual variance estimated due to the complete model (5.14); the last two rows correspond to optimal threshold values based on either complete or zero constant form of the model.

5.3.4   Index-Driven Consensus Partitions

There exists yet another model leading to the same uniform threshold clustering criterion: index-driven consensus partition. The model can be formulated as follows.

Let $x_1, .., x_n$ be nominal descriptors of the entities $i \in I$ and $x_k(i)$ be a symbol assigned to i by k-th descriptor ($k = 1, ..., n$). A partition S^k on I corresponds to each descriptor x_k, $k = 1, ..., n$; its classes consist of the entities i having the same category $x_k(i)$. The problem is to find such a partition S of I which, in some sense, could be considered as a "compromise" or "consensus" partition equally reflecting all the given descriptor partitions S^k ($k = 1, ..., n$). The problem of consensus partition can be treated in either an axiomatic or index-driven way (see p. 123). Here, only the latter approach is considered.

Let $\mu(S, T)$ be a dissimilarity index between partitions on I; then, partition S will be referred to as a consensus with regard to $\{S^1, ..., S^n\}$ if it minimizes criterion $\mu(S) = \sum_{k=1}^{n} \mu(S, S^k)$.

It turns out, for some indices μ, the consensus problem is equivalent to a uniform-threshold partitioning problem. Let us consider two such indices: (1) equivalence mismatch coefficient $ed(S, T)$, and (2) average probability change $\Delta(S/S^k)$. The latter index being nonsymmetrical can be interpreted as an index of predictive ability of descriptor S^k toward the consensus partition; it is a correlation index and should be maximized rather than minimized.

Let us recall that the entries of the equivalence indicator matrix $\mathbf{S}^k = (s_{ij}^k)$ are defined as follows: $s_{ij}^k = 1$ if i, j belong to the same class of S^k and $= 0$ if not.

The entries p_{ij}^k of the orthogonal projector $\mathbf{P}^k=(p_{ij}^k)$ are defined as 0 if i, j are in different classes of S^k, and $1/|S_t^k|$ if both i, j belong to class S_t^k. Then, let $A = (a_{ij})$ and $B = (b_{ij})$ be similarity matrices defined as $A = \sum_k \mathbf{S}^k$ and $B = \sum_k \mathbf{P}^k$: a_{ij} equals the number of the descriptors coinciding for i and j while b_{ij} represents the sum of the weights of these descriptors, the weights being inversely proportional to the descriptors' frequencies.

Statement 5.13. *The equivalence mismatch consensus partition is a maximizer of criterion (5.18) $SU(\pi, S)$ for the similarity matrix A and $\pi = n/2$, while the absolute probability change consensus partition maximizes criterion (5.18) for similarity matrix B and $\pi = n/N$ which is the average similarity.*

Proof: Equivalence mismatch consensus partition S minimizes $ed(S) = \sum_{k=1}^{n} \sum_{i,j \in I} (s_{ij}^k - s_{ij})^2$. Since squaring does not change Boolean values, $(s_{ij}^k - s_{ij})^2 = s_{ij}^k + s_{ij} - 2s_{ij}^k s_{ij}$. This and equality $a_{ij} = \sum_k s_{ij}^k$ imply that $ed(S) = \sum_{i,j \in I}(a_{ij} - 2a_{ij}s_{ij} + ns_{ij}) = \sum_{i,j \in I} a_{ij} - 2SU(n/2, S)$, which proves the first part of the statement. The second part can be proved analogously because $\Delta(S/S^k) = (\mathbf{S}, \mathbf{P}^k - uu^T/N)$ where (A, B) is the scalar product of matrices A and B considered as $N \times N$-dimensional vectors, and u is the N-dimensional vector having all its components equal to unity. $\square$

To illustrate the meaning of the criteria derived, let us consider an example of the entities $i \in I$, each painted by one of n colors (see Table 5.6 where nine entities are presented as painted with four colors).

Entity	Color			
	Red	Green	Blue	Yellow
1	+	-	-	-
2	+	-	-	-
3	+	-	-	-
4	-	+	-	-
5	-	+	-	-
6	-	-	+	-
7	-	-	+	-
8	-	-	+	-
9	-	-	-	+

Table 5.6: Indicator matrix of the color descriptor for 9 entities.

Let us consider n binary variables corresponding to each color (columns of Table 5.6) and ask ourselves: what is the consensus partition for these binary descriptors? Due to the statement above, for ed and Δ, the answer can be done in terms of the corresponding similarity matrices. The general formulas for the similarities can be suggested as follows. If the entities i and j are of the same color, they belong to the same class in each of the bi-class partitions S^k, thus yielding $a_{ij} = n$; if they are of different colors, they belong to different classes only for the corresponding two descriptors, thus providing $a_{ij} = n - 2$. Comparing these values with the threshold $\pi = n/2$, we can see that all $a_{ij} - n/2$ are non-negative if $n \geq 4$ (as in the Table 5.6) thus leading to the universal cluster I as the only equivalence mismatch consensus partition. Still, when there are two or three colors, the only positive values $a_{ij} - n/2$ are just for i, j having the same color, which corresponds to the correct consensus partition, in this particular case.

Similarity matrix A, for the data in Table 5.6, is as follows:

$$
A = \begin{pmatrix}
1 & 2 & 2 & 2 & & & & & \\
2 & 2 & 2 & 2 & & & & & \\
3 & 2 & 2 & 2 & & & & & \\
4 & & & & 2 & 2 & & & \\
5 & & & & 2 & 2 & & & \\
6 & & & & & & 2 & 2 & 2 \\
7 & & & & & & 2 & 2 & 2 \\
8 & & & & & & 2 & 2 & 2 \\
9 & & & & & & & & 2
\end{pmatrix}
$$

Matrix B for index Δ can be represented analogously. The consensus partition here is also unsatisfactory when $n \geq 4$.

This result shows that the index-based consensus partition concept can be misleading, even when such nice indices as $e\delta$ and Δ have been employed. Yet the original matrix-approximation approach presented by criteria (5.15) to (5.18) works well in this situation. The grand mean a of A is, obviously, between $n - 2$ and n, which leads to the centered similarities being positive only within the color classes; thus, the color descriptor is the only optimal solution for criterion (5.18) with $\pi = a$.

5.3.5   Discussion

1. Though a major portion of the uniform partitioning method was described long ago (see Kupershtoh, Mirkin, and Trofimov 1976, and Mirkin 1985, for a complete version), it is still unknown in the international cluster analysis community. Its connection with well-known Milligan's 1981 experiments has never been discussed. The Milligan's 1981 results can be interpreted as a strong empirical evidence supporting meaningfulness of the partitioning approximation model with all the classes weighted with the same intensity weight: the winning criteria are based on the product-moment correlation

between the data and partition matrices, which is the maximizing criterion of the model.

2. The basic notion in the uniform partitioning method, "soft" threshold π is subtracted from the entity-to-entity similarities rather than traditionally eliminating all the lesser values. This leads to a certain flexibility: threshold π determines that it is the average (not maximum or minimum) within and between similarities must comply with it, which underscores the threshold's soft action.

3. The threshold determines the number of clusters in the optimal partition; changing the threshold changes the optimal partition (its qualitative variance is proved to follow threshold changes). Since the average "diameter" of all the clusters is determined by the same threshold value, it may lead to inconvenient results when the "real" data structure contains clusters of really different sizes.

4. Agglomerative clustering is an especially simple procedure when applied to the problem of uniform partitioning: this is done with just simple summation of the rows and columns involved. Moreover, there is a natural stopping rule: when all the summary between-entries become negative and all the summary within-entries are positive. Non-monotonicity of the method may be considered an example when the monotonicity requirement, in the context of LW-algorithms, looks neither mandatory nor natural.

5. Some index-driven consensus partition problems can be reformulated as those of uniform partitioning. The symmetric-difference distance consensus partitioning, which has received recognition (see, for example, Barthélemy, Leclerc, and Monjardet 1986), is of this kind. However, as the example considered has shown, its flexibility might need to be improved somehow.

5.4 Additive Clustering

5.4.1 The Model

Real-world cluster patterns may show a great difference in cluster "diameters". This kind of structure may lead to ill-structured clusters when the uniform threshold criterion is employed because the cluster sizes, in average, are to be bound by the same value π.

Let us consider some more realistic model with the clusters having distinct intensity weights and, thus, thresholds:

$$a_{ij} = \sum_{t=1}^{m} \lambda_t s_{it} s_{jt} + \mu + e_{ij} \tag{5.19}$$

where $s_t = (s_{it})$ is the indicator vector of the sought cluster S_t, $t = 1, ..., m$. The model was introduced by Shepard and Arabie 1979. Simultaneously, in 1976 – 1980, the author and his collaborators in Russia developed what they called qualitative factor analysis methods as based on the same model (actually, the model was even more general, see Mirkin 1987b for detail and references).

The model can be considered when no "intercept" μ is present in the equations (or, equivalently, when $\mu = 0$); in this case, matrix A should be preliminarily centered, that is, the grand mean a subtracted from all the similarities.

The model (5.19) can be employed in both of the cases: when clusters S_t are assumed to be non-overlapping and when cluster overlaps are admitted. The latter case has been covered in Section 4.6. In this section, only the case of partitioning (no clusters overlapping) will be considered. In this case, matrices $s_t s_t^T$ presented as items of the matrix model (5.19) are mutually orthogonal, which allows for the following Pythagorean decomposition (provided that λ_t are optimal):

$$\sum_{i,j \in i} (a_{ij} - \mu)^2 = \sum_{t=1}^{m} \lambda_t^2 \sum_{i,j \in I} s_{it} s_{jt} + \sum_{i,j \in I} e_{ij}^2.$$

where $\lambda_t^2 = a_t - \mu$, a_t is the average similarity within cluster S_t and μ is zero or optimal ($\mu = a_b$) depending on the assumption of the model.

The sums $\sum_{i,j \in I} s_{it} s_{jt}$ can be expressed as $|S_t| \nu(S_t)$ where $\nu(S_t) = |S_t|$ if the diagonal similarities are given or $= |S_t| - 1$, if not. This leads to the following form of the decomposition:

$$\sum_{i,j \in i} (a_{ij} - \mu)^2 = \sum_{t=1}^{m} \lambda_t^2 |S_t| \nu(S_t) + \sum_{i,j \in I} e_{ij}^2. \tag{5.20}$$

Equation (5.20) shows the cluster contribution, $\lambda_t^2 |S_t| \nu(S_t)$, to the square scatter of the similarities (around μ). Moreover, it shows the actual criterion, $g(S)$, to be maximized with regard to the sought partition (when the least-squares criterion is employed):

$$g(S) = \sum_{t=1}^{m} \lambda_t^2 |S_t| \nu(S_t) \tag{5.21}$$

5.4.2 Agglomerative Algorithm

The additive clustering model usually is considered as a model with overlapping clusters; apparently no particular additive clustering algorithm for the nonoverlapping case has been published. Let us show how the standard agglomeration techniques can be utilized to partition I by maximizing criterion (5.21).

The algorithm starts with the trivial partition consisting of N singletons and it merges, at each step, two classes, S_u and S_v, that maximize the increment of criterion (5.21):

$$L(u,v) = g(S(u,v)) - g(S),$$

where partition $S(u,v)$ is obtained from S by merging its classes S_u and S_v together. The algorithm stops when the increment becomes negative.

The agglomeration procedure, in this case, cannot be presented with the Lance-Williams formula since different pairs u, v may lead to different values of μ involved in the computations. Yet it does not mean that the main computation advantage of the Lance-Williams formula, iterative recalculation of the intercluster similarities, cannot be kept, though in a somewhat modified version. A computationally safe agglomeration procedure is presented below.

Additive Nonoverlapping Agglomerative Clustering Algorithm

Step 1 (Initial Setting).

Set $m = N$ and consider interclass similarity matrix $(a(t,u))$ where $a(t,u) = a_{tu}$, vector (N_t) of the cluster cardinalities (all $N_t = 1$ in the beginning), and vector (a_t) of the cluster averages (equal to corresponding a_{tt} or to 0 if the diagonal similarities are not given), all of dimension m. Calculate $n_b = N\nu(I) - \sum_{t=1}^{m} N_t\nu(N_t)$ (number of between-cluster pairs of entities) and $A_b = \sum_{i,j\in I} a_{ij} - \sum_{t=1}^{m} a_t N_t\nu(N_t)$ (the sum of between similarities). Then let $\mu = A_b/n_b$, $\lambda_t = a_t - \mu$, and $g = \sum_{t=1}^{m} \lambda_t^2 N_t\nu(N_t)$. As usual, $\nu(N_t)$ denotes N_t if the diagonal similarities are given and $N_t - 1$ if not.

Step 2 (Agglomeration).

Phase 1. (Finding a pair of clusters to merge).

Find maximum of values $l(u,v)$, $u, v = 1, ..., m$, defined as

$$l(u,v) = \sum_{t\neq u,v} (a_t - \mu(u,v))^2 N_t\nu(N_t) + (a_{uv} - \mu(u,v))^2 N_{uv}\nu(N_{uv})$$

where $N_{uv} = N_u + N_v$, $a_{uv} = (a_u N_u\nu(N_u) + a_v N_v\nu(N_v) + 2a(u,v))/N_{uv}\nu(N_{uv})$, and $\mu(u,v) = \mu - 2(n_b a(u,v) - A_b N_u N_v)/(n_b^2 - 2n_b N_u N_v)$. If (u^*, v^*) is a maximizing cluster pair, check the inequality $l(u^*, v^*) > g$. If yes, go to the next phase. If no, end calculation; the partition considered is the result.

Phase 2. (Merging the clusters).

Merge clusters $S_{u\bullet}$ and $S_{v\ast}$ into the aggregate cluster $S_{u\bullet v\bullet} = S_{u\bullet} \cup S_{v\bullet}$. Calculate its characteristics $N_{u\bullet v\bullet} = N_{u\bullet} + N_{v\ast}$ and $a_{u\bullet v\bullet} = (a_{u\bullet} N_{u\bullet} \nu(N_{u\bullet}) + a_{v\bullet} N_{v\bullet} \nu(N_{v\ast}) + 2a(u^*, v^*))/N_{u\bullet v\bullet} \nu(N_{u\bullet v\bullet})$. Then, recalculate general values: $A_b \leftarrow A_b - 2a(u^*, v^*)$, $n_b \leftarrow n_b - 2N_{u\bullet} N_{v\bullet}$, $\mu \leftarrow \mu - 2(n_b a(u^*, v^*) - A_b N_{u\bullet} N_{v\bullet})/(n_b^2 - 2n_b N_{u\bullet} N_{v\bullet})$, and $g = l(u^*, v^*)$.

Finally, recalculate interclass similarity matrix $(a(u, t))$ just adding up the row v^* to the row u^* component-wise, then analogously adding up the column v^* to the column u^*, deleting then both row and column v^* and considering the row and column u^* updated as related to cluster $S_{u\bullet v\bullet}$.

Phase 3. (Loop.) If $m > 2$, substitute m by $m - 1$ and go to Phase 1. If $m = 2$, end.

Correctness of the algorithm follows from the fact that, actually, $l(u, v)$ equals $g(S(u, v))$ since $\mu(u, v)$ is the average between-cluster similarity in partition $S(u, v)$. The latter is easy to prove using the following numerical identity:

$$\frac{a - c}{b - d} = \frac{a}{b} - \frac{bc - ad}{b(b - d)}.$$

Obviously, in the agglomeration process, the value μ decreases at each step, thus ensuring that the scatter (5.20) increases.

5.4.3   Sequential Fitting Algorithm

Another approach to fitting the model (5.18) is sequential finding of clusters one-by-one, each step using residual similarities computed due to the model, as a special case of the SEFIT procedure (see Section 4.6 for detail). An advantage of the approach is that it can be applied equally to both of the situations, overlapping and nonoverlapping clusterings.

Let us consider only the case when $\mu = 0$ in the model (5.18). The problem of finding a satisfactory μ with the sequential fitting approach still has no satisfactory solution. The present author takes it just equal to the average similarity thus admitting that the universal cluster I is to be extracted at the initial step.

When the clusters must be nonoverlapping, the sequential fitting procedure works as follows: at each step a cluster is found minimizing criterion

$$L^2 = \sum_{i,j \in I'} (a_{ij} - \lambda s_i s_j)^2$$

by unknown real (positive) λ and Boolean s_i, $i \in I'$. Here, I' is a current entity set obtained from I by removing all the entities clustered in the previous steps. The found cluster S_t is removed from I' making $I' \leftarrow I' - S_t$. This provides a natural end of the clustering process when no unclustered entities remained, $I' = \emptyset$; the number of clusters thus defined afterward.

When all clusters S_t and their intensities λ_t are found, a kind of adjusting procedure for the clusters could be suggested based on the idea of iteratively fixing all the elements of the model (5.18) except for only one or two, and following up optimizing the criterion by the relaxed elements.

Applying the sequential fitting method for partitioning ten styled digits by the Confusion data (the diagonal entries removed, the matrix symmetrized, and the grand mean subtracted), we find three nonsingleton clusters presented in Table 5.7.

Cluster	Entities	Intensity	Contribution, %
1	1-7	131.04	25.4
2	3-9	98.04	14.2
3	6-8-0	54.71	13.3

Table 5.7: Non-singleton clusters for 10 segmented digits (Confusion data).

Entities 2, 4, and 5 form singletons having no close connections with the clusters, though they are quite close to some of the other entities. For instance, 5 is near 9 ($a_{59} = 98.04$) while it is not too distant from 3 ($a_{53} = -2.96$), which makes $al(5, 3-9) = (98.04 + 2.96)/2 = 50.5$ and shows that it would be quite appropriate to join 5 to cluster 3-9. This is not the case of entity 2: though 3-9 (and 3-5-9) is its closest cluster, still the average similarity between them is negative, which means that 2 must be a singleton. Entity 4 is quite close to cluster $1 - 7$ with $al(4, 1-7) = 22.04$ and can be added to it. This produces the set of clusters presented in Table 5.8. Their contribution to the square scatter of data is somewhat less (42% rather than 52% done by the original clusters), however the clusters themselves are somewhat larger.

5.4.4 Discussion

The model presented is intermediate between the uniform partitioning and genuine additive cluster models: it has a format of the latter while only nonoverlapping clusters are admitted, as in the former model. This makes the cluster sizes flexible

Cluster	Entities	Intensity	Contribution, %
1	1-4-7	58.37	15.1
2	3-5-9	55.21	13.5
3	6-8-0	54.71	13.3

Table 5.8: Corrected non-singleton clusters for 10 segmented digits (Confusion data).

since the thresholds can be different for different clusters, and, still, the procedures remain quite simple and stable. The sequential fitting approach can be considered as a quite close mate to the seriation techniques in Section 4.2: it just requires the seriation process to start all over again, after each of the clusters separated.

5.5   Structured Partition and Block Model

5.5.1   Uniform Structured Partition

Let $A = (a_{ij})$ be an association matrix and (S, ω) a structured partition on I represented by the Boolean matrix $\mathbf{S}_\omega = (s_{ij})$ where $s_{ij} = 1$ if and only if $(t, u) \in \omega$ for $i \in S_t$ and $j \in S_u$. Then, the linear model of the proximities approximated by $\mathbf{S}_\omega$, is as follows:

$$a_{ij} = \lambda s_{ij} + e_{ij} \tag{5.22}$$

There is no constant term μ here, firstly, for the sake of simplicity, and, secondly, because the criticisms concerning criterion (5.16) on p.256 seem relevant here, too. Instead, we suggest the matrix A to have its grand mean subtracted from all the entries preliminarily.

This model assumes an approximation of the real pattern of association to the structural equivalence. It suggests uniting in the same class those entities that identically interact with the others. Let us point out that this does not mean that the structurally equivalent entities must interact among themselves; such a within-non-interacting class may correspond to a subsystem purported to serve the others or to be served by the others; in industries, it could be energy or maintenance facilities.

When λ is positive, the least squares fitting problem for model (5.22) can be equivalently represented as the problem of maximizing

$$SU(\pi, S, \omega) = \sum_{(u,t) \in \omega} \sum_{i \in S_t} \sum_{j \in S_u} (a_{ij} - \pi) \qquad (5.23)$$

by (S, ω) for $\pi = \lambda/2$.

When there is no constraints on ω and S is fixed, the optimal ω (for given S) can be easily identified depending on the summary proximity values

$$a(\pi, t, u) = \sum_{i \in S_t} \sum_{j \in S_u} (a_{ij} - \pi).$$

Statement 5.14. *The structure ω maximizing (5.23) for given S is*

$$\omega(S) = \{(t, u) : a(\pi, t, u) > 0\}.$$

Proof: Obviously, $SU(\pi, S, \omega) = \sum_{(t,u) \in \omega} a(\pi, t, u)$. Thus, only positive $a(\pi, t, u)$ must be included in ω to make (5.23) maximum, thus leaving the negative ones out of the structure.   □

Actually, $\omega(S)$ may be considered as a threshold graph on the set of the clusters defined by "similarities" $a(\pi, t, u)$ with the threshold equal to zero. The following corollary is an extended form of that.

Corollary 5.4. *When there are no constraints imposed on (S, ω), the optimal structured partition consists of the singleton clusters connected by the structure of the π-threshold graph $G_\pi = \{(i, j) : a_{ij} > \pi\}$.*

Proof: The structure of the threshold graph $\omega = G_\pi$ involves all the positive entries $a_{ij} - \pi$ and no negative ones, which provides the global maximum to the value of (5.23).   □

Thus, no aggregating is needed to optimize the criterion, which is not an uncommon phenomenon in clustering. The square-error (WGSS) criterion for partitioning also has better values for the smaller clusters leading to the singletons as the best partition.

Yet another conclusion from the statement is as follows.

Corollary 5.5. *With no constraints on ω, maximizing criterion (5.23) is equivalent to maximizing criterion*

$$AS(\pi, S) = \sum_{t,u=1}^{m} |a(\pi, t, u)|.$$

Proof: Since, for a given π, $\sum_{t,u=1}^{m} a(\pi, t, u) = const$, maximizing $a(\omega(S)) = \sum_{(t,u)\in\omega(S)} a(\pi, t, u)$ keeps its complement to the constant, $\sum_{(t,u)\notin\omega(S)} a(\pi, t, u)$, minimal, which corresponds to the maximum of $\sum_{(t,u)\notin\omega(S)} |a(\pi, t, u)|$ because $a(\pi, t, u) \leq 0$ for every $(t, u) \notin \omega(S)$. $\qquad\square$

This shows that criterion (5.23) can be considered as depending on partition S only, defining the structure afterward as $\omega = \omega(S)$. Therefore, optimizing criterion (5.23) can be done with any local search partitioning algorithm. Let us consider the traditional agglomeration algorithm, at each step merging those two clusters S_t and S_u which give the minimal decrease of criterion $AS(\pi, S)$. It appears, that difference $\Delta(t, u) = AS(\pi, S) - AS(\pi, S(u, t))$ where $S(u, t)$ is partition obtained from S by merging its classes S_t and S_u, can be expressed as follows:

$$\Delta(t, u) = \sum_{v \neq t, u}[sgn_{tuv} \min(|a(\pi, t, v)|, |a(\pi, u, v)|)+$$

$$sgn_{vtu} \min(|a(\pi, v, t)|, |a(\pi, v, u)|)] + 2\min(a_+, |a_-|), \text{ where}$$

$sgn_{tuv} = |sgn \, a(\pi, t, v) - sgn \, a(\pi, u, v)|$, $sgn_{uvt} = |sgn \, a(\pi, v, t) - sgn \, a(\pi, v, u)|$, and a_+ (or a_-) is the sum of all positive (or negative) values in the quadruple $a(\pi, t, t), a(\pi, t, u), a(\pi, u, t), a(\pi, u, u)$ related to the associations in the fissioned class.

The proof of the equality is based on simple arithmetic considerations and can be found in Kupershtoh and Trofimov 1975 (see also Mirkin and Rodin 1984, p. 116-117).

It remains now to consider the case when λ and threshold $\pi = \lambda/2$ are not fixed but must be optimally adjusted as the average association within the structure: $\lambda = a_w = \sum_{i,j\in I} a_{ij}s_{ij}/\sum_{i,j} s_{ij}$.

To do that, we need to consider the problem of finding the optimal λ when partition S is fixed. That can be done with the following iterative process.

Finding Optimal Threshold for S Fixed
1. Let π, initially, be zero.
2. Find $\omega(S)$ with the fixed threshold, then calculate the within average value a_w and take $\pi = a_w/2$.
3. If newly defined π does not coincide with π at the preceding iteration, go to step 2; else end.

The process converges since the least-squares criterion is decreased at every step and the number of structures is finite. Moreover, the stationary point corresponds to the global optimum since the criterion is a convex function of λ.

This algorithm can be utilized beyond clustering, for instance, when a suitable threshold graph is sought.

The agglomeration procedure with the optimal λ can be defined as follows.

Agglomeration with Optimal λ
1. Find an optimal threshold value for the trivial partition consisting of N singleton clusters.
2. With the threshold π and partition S fixed, find the best pair of clusters, t and u, to merge (based on values of $\Delta(t, u)$ above). For the agglomerate partition, $S(t, u)$, find the optimal threshold π and, with this π, repeat search for the best pair of clusters of S, until π does not vary.
3. Merge clusters S_t and S_u, take corresponding optimal threshold and go to step 2 until the number of clusters becomes two.

Let us consider the Confusion data (between 10 segmented integer digits) from Table 1.16, p. 46, with the diagonal entries eliminated. The matrix A centered by subtracting its grand mean, 33.4556 (no diagonal entry is considered), is as follows:

−	−26.5	−26.5	−11.5	−29.5	−18.5	26.5	−33.5	−29.5	−29.5
−19.5	−	13.5	−29.5	2.5	13.5	−19.5	−4.5	−26.5	−15.5
−4.5	−4.5	−	−26.5	−15.5	−33.5	6.5	−4.5	118.5	−18.5
115.5	−11.5	−29.5	−	−29.5	−22.5	−3.5	−26.5	7.5	−33.5
−19.5	−7.5	9.5	−19.5	−	45.5	−26.5	−26.5	92.5	−19.5
−8.5	−19.5	−26.5	−22.5	63.5	−	−29.5	121.5	−22.5	9.5
235.5	−29.5	−12.5	−12.5	−26.5	−33.5	−	−33.5	−29.5	−26.5
−22.5	−5.5	−5.5	−15.5	−15.5	36.5	−22.5	−	33.5	138.5
−8.5	−4.5	77.5	12.5	48.5	−22.5	−12.5	48.5	−	9.5
−15.5	−29.5	−26.5	−22.5	−26.5	−15.5	−8.5	37.5	−12.5	−
1	**2**	**3**	**4**	**5**	**6**	**7**	**8**	**9**	**0**

The structure of 24 positive entries (threshold graph G_0) is presented in Fig.5.2 (a). The graph looks overcomplicated; it is not easy to realize what are the main flows of confusion, by this picture. Let us find an optimal threshold π. To do that, let us find the average proximity in the graph presented in Fig.5.2 (a). It is 53.64, which gives $\pi = 53.64/2 = 26.82$. This threshold cuts out some of the arrows in that graph since there are only 14 entries, in A, that are larger than this π. The smaller set of arrows gives a larger average proximity, 84, leading to new $\pi = 42$. There are only 11 entries larger than 42 in A, which leads to a new, recalculated average, 97.13. With the new $\pi = 48.56$, there are 8 remaining entries to be larger than π, which leads to a larger average again. This new average, 115.72, is final since exactly the same 8 entries remain larger than the new value $\pi = 57.86$. Thus, we have got an optimal threshold graph corresponding to threshold $\lambda/2$ for optimal $\lambda = 115.72$ (see Fig.5.2 (b)).

It is not difficult to estimate the contribution of the structure found to the proximity a_{ij} variance: it equals λ squared multiplied by the number of positive entries (ones in matrix (s_{ij})), 8, related to the square scatter of the proximities: 59.6%.

Interpretation of this graph does not seem difficult: the main line of confusion is shown quite unambiguously (say, digits 4 and 7 go for 1, etc.) However, for illustrative

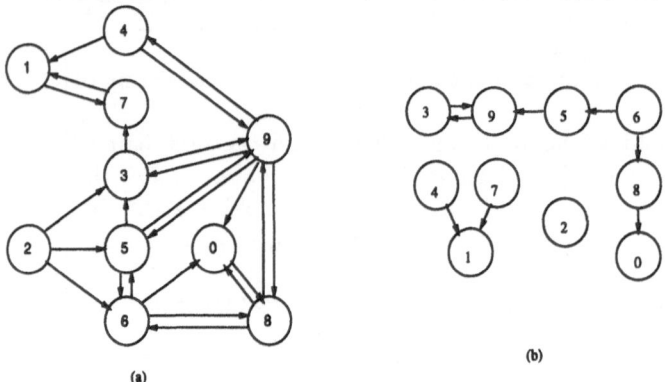

Figure 5.2: Finding an optimal threshold graph for the Confusion data: from (a) to (b).

purposes, let us try further aggregate the graph. Such an aggregate graph is presented in Fig.5.3 (a): the non-singleton classes, 4-7, 5-8, and 6-9, unite unconnected entities. The structure comprises 18 entries in A, some of them being negative, such as, for instance, $a_{05} = -26.5$. This structure is far from optimal, for $\pi = 0$. The optimal structure, $\omega(S)$ with $\pi = 0$, must include more connections as shown in Fig.5.3 (b). The average of all the 25 within structure entries is equal to 46.3, which makes $\pi = 23.15$ to cut out of the structure the weakest connections, such as from 2 to 3, with $a_{23} = 13.5 < \pi$. Removing them, we obtain the structured partition presented in Fig.5.3 (c), which almost coincides with the threshold-graph-derived structure (a). This is the final structured partition since its intensity weight, 68.44, does not suggest further cutting out any of its arcs.

Some other examples of structured partitioning (in molecular genetics) can be found in Mirkin and Rodin 1984.

5.5.2   Block Modeling

Let us recall that a *block model* is a partition S along with several structures, $\omega_1, ..., \omega_n$, which represent an aggregate structure of a system with several kinds of interrelation between its subsystems. In the literature, only some heuristic algorithms for finding block models have been published so far (see, for example, Arabie, Boorman, and Levitt 1978, Wasserman and Faust 1992). To put block modeling in the approximation framework, we suggest a particular model based on

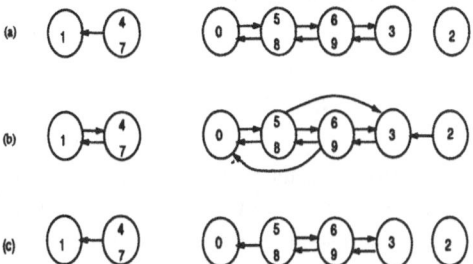

Figure 5.3: Structured partitions for the Confusion data.

different weights assigned to partition S and structures ω_k.

Let the data be represented by a set of association matrices $A_k = (a_{ij,k})$ where different k, $k = 1, ..., n$, relate to different aspects of the relationship between the entities or to different periods of time or to different locations, etc.

The resulting block model will be represented by a partition S on I along with a set of structures ω_k corresponding to respective matrices A_k, $k = 1, ..., n$.

Let us consider the following model assuming that the associations are just "noisy" structural interrelations:

$$a_{ij,k} = \lambda s_{ij} + \mu_k r_{ij,k} + e_{ij} \qquad (5.24)$$

where matrix (s_{ij}) is the equivalence indicator matrix of partition S (no structural information included), and $(r_{ij,k})$ is the structure indicator matrix: $r_{ij,k} = 1$ if and only if $i \in S_t$ and $j \in S_u$ for $t \neq u$ and $(t, u) \in \omega_k$; the intensity weights λ and μ_k may be fixed a priori or found with least-squares adjusting.

Optimal values of λ and μ_k are equal to the averages of corresponding associations: within the partition and within each of the structures ω_k, respectively: $\lambda = (1/n) \sum_{i,j,k} a_{ij,k} s_{ij} / \sum_{i,j} s_{ij}$, and $\mu_k = \sum_{i,j} a_{ij,k} r_{ij,k} / \sum_{i,j} r_{ij,k}$.

With these optimal values put in, the following decomposition holds:

$$\sum_{i,j,k} a_{ij,k}^2 = n\lambda^2 \sum_{i,j} s_{ij} + \sum_{k=1}^{n} \mu_k^2 \sum_{i,j} r_{ij,k} + \sum_{i,j,k} e_{ij,k}^2 \qquad (5.25)$$

demonstrating that the least-squares fitting criterion requires maximizing the contribution to the square data scatter of the block model sought:

$$c(S, \omega_k) = n\lambda^2 \sum_{i,j} s_{ij} + \sum_{k=1}^{n} \mu_k^2 \sum_{i,j} r_{ij,k} \qquad (5.26)$$

Again, to analyze the meaning of the criterion, let us assume that the intensity weights, λ and μ_k, $k = 1, ..., n$, are constant. It is not difficult to see that minimizing the least-squares criterion (with the constant weights) is equivalent to maximizing

$$CS(\lambda, \mu_k, S, \omega_k) = \lambda SU(A, \lambda n/2, S) + \sum_k \mu_k AS(A_k, \mu_k/2, S, \omega_k),$$

or

$$CS(\lambda, \mu_k, S, \omega_k) = \lambda \sum_{i,j \in I} (a_{ij} - n\lambda/2)s_{ij} + \sum_k \mu_k \sum_{i,j} (a_{ij,k} - \mu_k/2)r_{ij,k} \qquad (5.27)$$

where $A = (a_{ij}) = \sum_{k=1}^{n} A_k$. Criterion (5.27) is a linear combination of the uniform-threshold partition criterion and structured partitioning criteria taken with different thresholds.

For a fixed partition S, an optimal block model is defined based on the aggregate threshold graph as in Statement 5.14. above (all the intensity weights are assumed positive): $E(S) = \{(t, t) : \sum_{i,j \in S_t} (a_{ij} - n\lambda/2) > 0\}$ (the loop ties) and $\omega_k(S) = \{(t, u) : \sum_{i \in S_t} \sum_{j \in S_u} (a_{ij,k} - \mu_k/2) > 0\}$. This means that the criterion, actually, can be optimized by S only since all the structural elements can be defined afterward with the formulas above.

This allows application of the same agglomeration method as in the case of structured partitioning. Change $\Delta(t, u)$ of the criterion (5.27) when S is substituted by $S(t, u)$, with S_t and S_u merged into the united cluster $S_{tu} = S_t \cup S_u$, can be expressed as:

$$\Delta(t, u) = \lambda B(t, u) - \sum_{k=1}^{n} \mu_k B_k(t, u)$$

where

$$B(t, u) = (a(\pi, t, t) + a(\pi, t, u) + a(\pi, u, t) + a(\pi, u, u))^+ - a(\pi, t, t)^+ - a(\pi, u, u)^+$$

where $x^+ = \max(0, x)$ and $\pi = n\lambda/2$, and

$$B_k(t, u) = \sum_{v \neq t, u} [sgn_{ktuv} M(k, t, v) + sgn_{kvtu} M(k, v, t) + a_k(\pi_k, t, u)^+ + a_k(\pi_k, u, t)^+$$

where

$$sgn_{ktuv} = |sgn\ a_k(\pi_k, t, v) - sgn\ a_k(\pi_k, u, v)|,$$

$$sgn_{kuvt} = |sgn\ a_k(\pi_k, v, t) - sgn\ a_k(\pi_k, v, u)|,$$

$$M(k, t, v) = \min(|a_k(\pi_k, t, v)|, |a_k(\pi_k, u, v)|),$$

$$M(k, v, t) = \min(|a_k(\pi_k, v, t)|, |a_k(\pi_k, v, u)|),$$

$\pi_k = \mu_k/2$, and the index k in $a_k(\pi, u, v)$ means that it is computed by $A_k = (a_{ij,k})$.

To prove the formulas, let us note, initially, that all the items in criterion (5.27) having no relation to either S_t or S_u are mutually eliminated in $\Delta(t, u)$. Moreover, if $sgn\ a_k(\pi_k, t, v) = sgn\ a_k(\pi_k, u, v)$ then $(t, v) \in \omega_k(S)$ and $(u, v) \in \omega_k(S)$ implies $(tu, v) \in \omega_k[S(t, u)]$, thus, the corresponding items are mutually excluded also. If $sgn\ a_k(\pi_k, t, v) \neq sgn\ a_k(\pi_k, u, v)$, then (tu, v) belongs or does not belong to $\omega_k[S(t, u)]$ depending on which value's module is larger. If, for instance, $a_k(\pi_k, t, v) > 0$ and $a_k(\pi_k, t, v) > |a_k(\pi_k, u, v)|$ then $(tu, v) \in \omega_k[S(t, u)]$ and only the other, negative quantity, $a_k(\pi_k, u, v)$ is added to $\Delta(t, u)$. If the opposite holds, $a_k(\pi_k, t, v) < |a_k(\pi_k, u, v)|$, then $(tu, v) \notin \omega_k[S(t, u)]$ and $-a_k(\pi_k, t, v)$ is added in $\Delta(t, u)$. All this is represented in $B_k(t, u)$ by the item

$$-\mu_k/2|sgn\ a_k(\pi_k, t, v) - sgn\ a_k(\pi_k, u, v)|\min(|a_k(\pi_k, t, v)|, |a_k(\pi_k, u, v)|).$$

The items for arcs (v, tu) are treated analogously. $\qquad\square$

Since all $B(t, u)$, $B_k(t, u)$ are positive, merging the classes increases the criterion if and only if $\lambda B(t, u) > \sum_{k=1}^{n} \mu_k B_k(t, u)$.

5.5.3 Interpreting Block Modeling as Organization Design

Although the block modeling problem considered has been developed entirely in the framework of approximation clustering, it can be interpreted also as a particular problem of organization design. Such an interpretation seems useful both for better understanding the meaning of the problem and for applying the model in organization design.

Let an industrial system consist of the elementary working units $i \in I$. Its organization structure involves a partition of the units in m nonoverlapping divisions and a line-staff control system. Line control structure is based on direct hierarchical subordination, while staff units perform numerous assisting activities of which only coordinating of interaction between the divisions will be considered (for terminology, see, for instance, Hutchinson 1967).

Let us form an index for measuring the intensity of control efforts in the organization. Line control activity is twofold since it provides "interior" control within divisions and assists in connecting divisions with their counterparts in the organization. Assuming that the major control effort within division S_t is being done towards pair-wise interactions, it can be evaluated as $c_1|S_t|^2$ where c_1 is a scale coefficient. The external control effort is assumed to depend on the intensity of interactions between the division and its within-enterprise partners. The more interactions, the higher effort (since the probability of any kind of conflicts, fall-outs, break-downs, etc. grows). In an industrial system, the intensity of the "external" interactions can be evaluated based on estimates of pair-wise technological interactions, a_{ij}, between the elementary units $i, j \in I$. For instance, a_{ij} may be just the number of manufactured articles received by j from i in a production cycle. In these terms, the external control effort can be evaluated as $d_1 \sum_{i \in S_t} \sum_{j \notin S_t} a_{ij}$ where d_1 is a coefficient. Thus, the whole effort of the line control in S_t is $c_1|S_t|^2 + d_1 \sum_{i \in S_t} \sum_{j \notin S_t} a_{ij}$.

Then, let us assume that the staff coordinating activity is organized in two ways: some interactions are controlled "individually", through particular staff officers, while the others are maintained through an "administrative" system based on some standard rules and procedures. Let us measure the tension of the efforts to coordinate interaction between divisions S_t and S_u by $c_2|S_t||S_u|$ if it is made within the "administrative" control subsystem or by $d_2 \sum_{i \in S_t} \sum_{j \in S_u} a_{ij}$ if it is made within the "individual" control. The difference arises since the effort of the "individual" control depends on the volume of interactions while the "administrative" effort is determined by the number of interactions.

Let ω denote the set of interacting pairs of divisions controlled in the "administrative" fashion. Then the total control effort is equal to:

$$E = c_1 \sum_t |S_t|^2 + d_1 \sum_{i \in S_t} \sum_{j \notin S_t} a_{ij} + c_2 \sum_{(t,u) \in \omega} |S_t||S_u| + d_2 \sum_{(t,u) \notin \omega} \sum_{i \in S_t} \sum_{j \in S_u} a_{ij}$$

where the coefficients reflect the "relative costs" of distinct control techniques.

With elementary arithmetic, E can be rewritten as

$$E = (d_1 + d_2) \sum_{i,j \in I} a_{ij} - EE$$

where

$$EE = (d_1 + d_2) \sum_t \sum_{i,j \in S_t} \left(a_{ij} - \frac{c_1}{d_1 + d_2} \right) + d_2 \sum_{(t,u) \in \omega} \sum_{i \in S_t} \sum_{j \in S_u} \left(a_{ij} - \frac{c_2}{d_2} \right) \quad (5.28)$$

Equation (5.28) implies that the problem of designing the organization structure which minimizes the total effort E is equivalent to the problem of maximizing EE, which is very similar to the approximation block modeling criterion (5.27) (for $n = 1$).

The thresholds here are related to cost ratios $c_1/(d_1 + d_2)$ and c_2/d_2, respectively. This gives a bilateral relationship between the empirical association data and organization effort: the thresholds can be defined by the costs and, conversely, if the thresholds are given, cost ratios can be estimated. This latter dependence can be utilized in organization design decisions concerning, for instance, maintenance or energy or transportation facilities: should they be united in a specialized maintenance/energy/transportation division or, in contrast, assigned each to a particular production division?

Yet the two models — for organization design and for block modeling — are different. For instance, the former involves four cost parameters while there are only two intensity weights in the latter. To reduce the four-parameter diversity to make the organization design criterion EE be of the block modeling format, it is necessary and sufficient that the following equalities be held: $c_1 = (d_1 + d_2)^2$ and $c_2 = d_2^2/2$.

Obviously, the present model does not take into account many real organizational phenomena (other control goals exist, real-world associations are neither constant nor homogeneous, etc.), however, some of them are involved quite clearly (line-staff control subsystems or "administrative" and "individual" ways of coordinating control), which suggests that there is potential in the model for further elaboration.

5.5.4 Discussion

1. The concept of a structured partition is a model for representing interrelated subsystems in a complex system. It can be used also for further formalizing various types of qualitative variables depending on the type of structure of relationship between the categories: for example, a question from a questionnaire with several ordered categories, such as from "very likely" to "very unlikely", may have a category "don't know" which is completely out of the order.

2. The approximation approach leads to a problem which admits a threshold graph as the best solution when no restrictions are imposed on the partition or the between-class relation structure. With any m-class structured partition admitted, the problem becomes equivalent to an unstructured partitioning problem, though its criterion involves all the within and between class summary entries; the soft threshold plays as important a role here as it does in the uniform partitioning problem.

3. In the block model constructing aspect, the approximation approach produces a partitioning criterion which is an additive mixture of both the structured and uniform partitioning criteria. This latter criterion admits an interpreta-

tion in terms of organization design, which allows us to look at its features from a substantive perspective.

5.6   Aggregation of Mobility Tables

5.6.1   Approximation Model

Let $P(I) = (p_{ij})$, $i, j \in I$, denote an intergenerational occupation mobility table such as Mobility 5 and 17 data in Tables 1.21, p. 55, and 1.7, p. 33. Set I is a set of occupations and p_{ij} is proportion of the cases when, in a family, the (first) son's occupation has been $j \in I$ while $i \in I$ has been his father's occupation. The marginal values, $p_{i+} = \sum_{j \in I} p_{ij}$ and $p_{+j} = \sum_{i \in I} p_{ij}$, are proportions of fathers' i and sons' j occupations, respectively ($i, j \in I$). The problem of aggregation of such a table may arise from practical reasons (for instance, when set I appears too large) or because of theoretical considerations (for instance, to analyze the social class structure supposedly reflected in intergenerational moves, see Breiger 1981).

For a partition $S = \{S_1, ..., S_m\}$ on I, let us consider its $N \times m$ indicator matrix $s = (s_{it})$, $i \in I, t = 1, ...m$, with its entries

$$s_{it} = \begin{cases} 1, & if \quad i \in S_t \\ 0, & if \quad i \notin S_t \end{cases}$$

Let us consider a pair of columns, $s_t = (s_{it})$, $s_u = (s_{iu})$, and $N \times N$ matrix $s_t s_u^T$ with the entries $s_{ij,tu} = s_{it} s_{ju}$. Obviously, the matrix has all its entries equal to zero except for the entries (i, j) in box $S_t \times S_u$, each equal to unity. Multiplying this matrix by any real μ_{tu}, we obtain a matrix of the same structure where μ_{tu}, not 1, stands for nonzero elements. This gives meaning to the following representation of the RCP values $q_{ij} = p_{ij}/p_{i+}p_{+j} - 1$ through partition S and a set of intensity weights μ_{tu}, $t, u = 1, ..., m$:

$$q_{ij} = \sum_{t=1}^{m} \sum_{u=1}^{m} \mu_{tu} s_{it} s_{ju} + r_{ij} \tag{5.29}$$

where r_{ij} are residuals; that is, any r_{ij} is defined as the difference between q_{ij} and the double sum in (5.29).

Since the mobility data are aggregable, let us approximate the RCP values q_{ij} with a better adjustment of both partition S and the intensity values $\mu_{tu}(t, u = 1, ..., m)$ by minimizing the following weighted least-squares criterion:

$$E^2 = \sum_{i,j} p_i p_j (q_{ij} - \sum_{t,u=1}^{m} \mu_{tu} s_{it} s_{ju})^2 \tag{5.30}$$

with respect to unknown s_{it}, s_{ju}, μ_{tu} for given $P(I)$.

Let us consider the aggregate $m \times m$ table $P(S)$ corresponding to a partition S, with its entries defined as $p_{tu} = \sum_{i \in S_t} \sum_{j \in S_u} p_{ij}$ $(t, u = 1, ..., m)$. The initial data table is just $P(I)$, corresponding to the trivial partition of I with singletons as the classes. Pearson goodness-of-fit criterion 5.5, p. 241, will be denoted by $X^2(S)$ or $X^2(I)$ when it is calculated for $P(S)$ or $P(I)$, respectively.

The model (5.29)–(5.30) connects the original and aggregate tables as follows from the two statements below.

Statement 5.15. *For any given partition $S = \{S_1, ..., S_m\}$ of I, the optimal values of $\mu_{tu}, t, u = 1, ..., m$, are the aggregate RCP values:*

$$\mu_{tu} = q_{tu} = \frac{p_{tu} - p_{t+}p_{+u}}{p_{t+}p_{+u}}$$

calculated for $P(S)$.

Proof: Obviously seen by setting zero the derivatives of E^2 (5.30) with respect to μ_{tu} for each pair (t, u). □

Statement 5.16. *For any S and the optimal $\mu_{tu} = q_{tu}$ $(t, u = 1, ..., m)$, the following decomposition holds:*

$$X^2(I) = X^2(S) + E^2 \tag{5.31}$$

Proof: Square the parenthesis in (5.30), with q_{tu} substituted for μ_{tu}, and take into account that the Boolean indicator vectors s_t are mutually orthogonal and that X^2 can be expressed through RCP values by formula (2.8), p. 85. □

A conclusion from Statement 5.16. is that minimizing criterion (5.30) is equivalent to maximizing the goodness-of-fit statistic $X^2(S)$ with respect to partitions S and corresponding aggregate matrices $P(S)$. Moreover, formula (5.31) can be considered as a decomposition of the scatter of the initial data (measured by $X^2(I)$) into "explained" and "unexplained" parts, which means that the value $X^2(S)$ represents the contribution of the partition S into the original value of the contingency coefficient. This gives yet another interpretation to the coefficient, to be added to those discussed in Section 5.1.4. In the framework of the approximation model

(5.29), the Pearson chi-squared coefficient has nothing to do with statistical independence or description of one partition through another one: it is the ratio $X^2(S)/X^2(I)$ which should be used for estimating similarity between the aggregate and original mobility patterns rather than the statistical independence indices involving χ^2-distribution. .

Decomposition (5.31) may lead also to various computational strategies for fitting the approximation model (5.29) with criterion (5.30).

For example, the agglomerative algorithm based on a step-by-step merging of pairs of classes is applied to the rows and columns simultaneously, as follows.

Agglomeration Chi-Squared Algorithm
Each iteration starts with a partition $S = \{S_1, ..., S_m\}$ and corresponding aggregate table $P(S)$ Due to the criterion in (5.31), those two classes are merged that make the decrement of $X^2(S)$ minimum.
The result of the agglomeration process can be represented by a dendrogram along with the index values E^2, assigned to its internal nodes, each associated with a particular partition S and corresponding table $P(S)$. All leaves (singletons) have zero index value while the root's index value is $X^2(I)$.

The dendrogram in Fig.5.4 represents the result of the agglomeration chi-squared algorithm applied to the Mobility 17 data in Table 1.21, p. 55. The tree differentiates the three major divisions well: Nonmanual (1 to 7), Manual (8 to 15) and Farm (16-17). Five classes produced by the algorithm, basically, coincide with the Featherman-Hauser aggregation presented in the Mobility 5 data (Table 1.7, p. 1.7). However, there is a difference in the partitioning of the manual workers (occupations 8 to 15): the algorithm separates manufacturing worker class, 8-13-14, rather than maintaining Featherman-Hauser's Upper-Lower division (the separation of the manufacturers has been suggested in Breiger 1981). Our 5-class partition is somewhat better than that of Featherman-Hauser by the value of X^2 accounted for: it takes 74.8% of the original X^2 value while the latter partition accounts for 72.5%. There is also the 8-class Breiger's 1981 partition present in Fig.5.4 as compared to that found with the agglomerative chi-squared algorithm; again, there are not many differences, and the algorithm's results can be substantiated.

5.6.2   Modeling Aggregate Mobility

The approximation aggregation model presented can be considered yet another example of a clustering model for which a substantively motivated meaning can be provided.

Let us assume that the intergenerational mobility process runs according to the aggregate matrix $P(S)$ in such a way that the individual transition frequencies p_{ij}

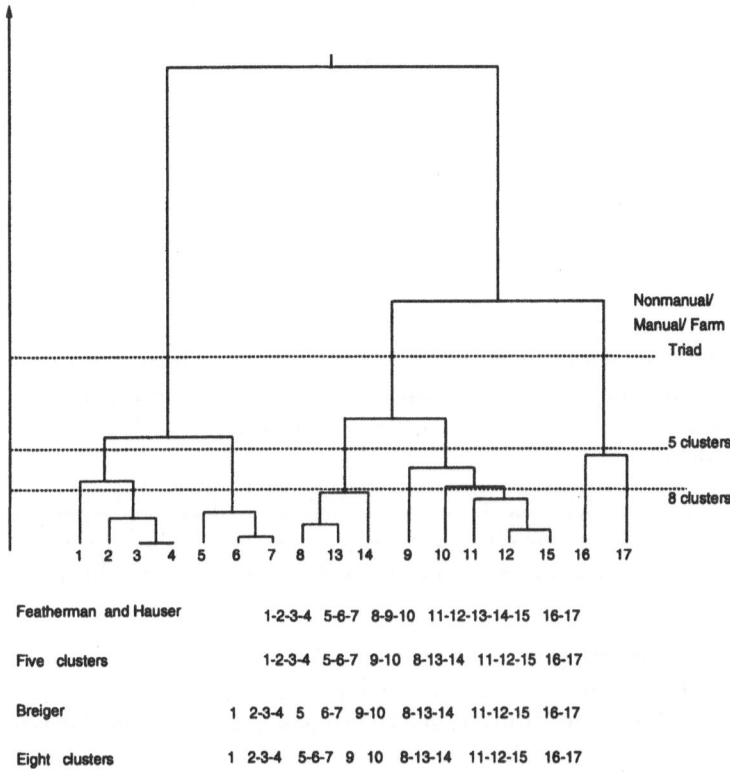

Figure 5.4: Chi-squared agglomeration tree for Mobility 17 data.

are determined by the theoretical aggregate frequencies p_{tu} and conditional probabilities, $p_{i/t} = p_{i+}/p_{t+}$ and $p^{j/u} = p_{+j}/p_{+u}$, of picking up the father's occupation, i, or the occupation of son, j, from their classes S_t or S_u, respectively. This means that the parent's and son's occupation distributions within the S classes are assumed proportional to the proportions observed. The following hypothetical value of the frequency entry for $i \in S_t$ and $j \in S_u$ follows from these assumptions:

$$F_{ij,tu} = p_{i/t} p_{tu} p^{j/u}, \tag{5.32}$$

Although this model has never been formulated explicitly, it, actually, underlies most developments concerning the mobility aggregation problem (Breiger 1981, Goodman 1981).

If the observed values p_{ij} are not equal to the model values (5.32), any of the traditional χ^2 statistics, likelihood-ratio L and goodness-of-fit X^2, can be used to estimate the deviation. The likelihood-ratio statistic is

$$L(p_{ij}, F_{ij,tu}) = \sum_{i,j} p_{ij} \log \frac{p_{ij}}{F_{ij,tu}} \qquad (5.33)$$

and the goodness-of-fit is

$$X^2(p_{ij}, F_{ij,tu}) = \sum_{i,j \in I} \frac{(p_{ij} - F_{ij,tu})^2}{F_{ij,tu}} \qquad (5.34)$$

Although the formulas look quite similar, there is a great difference between them: the first can be decomposed into the values of L for a traditional statistical hypothesis of null association (independence), the second cannot. Let us discuss the subject in more detail.

The independence (null association) between the father's and son's occupations is expressed with the traditional equality $p_{ij} = p_{i+}p_{+j}$; in the mobility studies it is frequently referred to as *perfect mobility* meaning that the people's behavior is perfectly random (Hout 1986).

Under the hypothesis of perfect mobility, the expected transition data must be $F_{ij} = p_{i+}p_{+j}$, for the individual occupations, and $F_{tu} = p_{t+}p_{+u}$, for the aggregate ones. Using these symbols and substituting (5.32) into (5.33), one can easily derive that

$$L(p_{ij}, F_{ij,tu}) = L(p_{ij}, F_{ij}) - L(p_{tu}, F_{tu}) \qquad (5.35)$$

The equality (5.35) means that $L(p_{ij}, F_{ij,tu})$, as a measure of discrepancy between the observed data and the values in (5.32) expected under the model, can be calculated through values of L as a measure of the sample bias of the observed and aggregated data from the perfect mobility hypothesis. The number of degrees of freedom in $L(p_{ij}, F_{ij,tu})$ is equal to the difference $(N-1)^2 - (m-1)^2$ between the degrees of freedom of the terms in the right side of (5.35). This allows using standard sampling bias χ^2-based reasoning to deal with somewhat more complicated model (5.32).

In contrast to L in (5.35), the goodness-of-fit value (5.34) cannot be decomposed into the difference of goodness-of-fit values calculated under the perfect mobility hypothesis for the original and aggregate matrices, $P(I)$ and $P(S)$. Thus, the model in (5.32) cannot be employed to substantiate the correspondence-analysis-based method of aggregation suggested above. However, another aggregation model can be formulated which admits X^2, not L, decomposed.

The alternative model is based upon the assumption that the observed mobility is a result of two distinct processes, perfect mobility and imperfect mobility, so that perfect mobility runs in terms of the original occupations while imperfect mobility is governed by the "imperfectness" of the aggregate process. More explicitly, let the value $p_{tu} - p_{t+}p_{+u}$ in the theoretical aggregate matrix $P(S)$ reflect the difference between the "real" and "perfect", at the aggregate level, mobility. In terms of the original occupations, this gives $p_{i/t}(p_{tu} - p_{t+}p_{+u})p^{j/u}$ as the imperfect part of the overall mobility. Combining this with the perfect mobility part we have the alternative model as follows:

$$p_{ij} = p_{i+}p_{+j} + p_{i/t}(p_{tu} - p_{t+}p_{+u})p^{j/u} \tag{5.36}$$

Although the right part in (5.36) is equal to that in (5.32), this formulation allows us further to specify the model. Let us assume that all the difference between the left part (empirical) and right part (theoretical) in (5.36) is due to the term related to perfect mobility, thus, admitting that there are no errors in the imperfect mobility term. This allows us to rewrite the model in the following format:

$$e_{ij} = p_{ij} - p_{i/t}(p_{tu} - p_{t+}p_{+u})p^{j/u} \tag{5.37}$$

where e_{ij} is that part of the observed mobility from i to j which counts for moves that are subject to the hypothesis of perfect mobility. This implies that: (1) in a "natural" process, the values e_{ij} must be nonnegative, and (2) the χ^2 statistic for e_{ij} must admit values $F_{ij} = p_{i+}p_{+j}$ as expected in the corresponding entries (i, j), $i, j \in I$. Specifically, the goodness-of-fit statistic, in this case, is equal to

$$X^2 = \sum_{i,j \in I} \frac{(e_{ij} - p_{i+}p_{+j})^2}{p_{i+}p_{+j}} \tag{5.38}$$

with the $(N-1)^2 - (m-1)^2$ degrees of freedom. This value is indeed equal to the difference between the goodness-of-fit values for matrices $P(I)$ and $P(S)$, under the perfect mobility hypothesis. Indeed, X^2 (5.38) is the minimized criterion E^2 in decomposition (5.31) where the other two items are exactly the goodness-of-fit values discussed.

The likelihood-ratio statistic for the alternative model (5.37) cannot be decomposed correspondingly.

For Featherman-Hauser 5-class aggregation, there are three negative values of e_{ij} (5.37) present. For the 5-class aggregation produced by the algorithm (Fig.5.4), there is only one negative e_{ij}, which is just quite small, some -0.00001. This can be considered yet another argument in favor of the aggregation found.

5.6.3 Discussion

1. The aggregation model presented can be considered as an alternative to the concept of a structured partition: the final aggregate table $P(S)$ itself repre-

sents both the partition and all within- and between-cluster associations in terms quite expressive of the relative change of probability (RCP) values.

2. The model admits a natural aggregation criterion, in terms of the least decreasing of the value X^2 which expresses here the (weighted) data scatter, not a probabilistic-model-based bias from the perfect mobility.

3. It appears, the approximation model can be further elaborated in such a way that it becomes connected with aggregate modeling of mobility as a process. The model developed admits the goodness-of-fit criterion, in contrast to the traditional maximum-likelihood based considerations. This has quite a practical appeal: the goodness-of-fit criterion, as a data scatter index, can be applied to data containing zero entries (which is a frequent situation, especially, with detailed original categories) while the maximum-likelihood criterion cannot since it is based on the entry logarithms.

4. The aggregation method developed can be applied to any other aggregable interaction data, not only mobility.

Chapter 6

Partition: Rectangular Data Table

FEATURES

- Bilinear clustering for mixed – quantitative, nominal and binary – variables is proved to be a theory-motivated extension of K-Means method.

- Decomposition of the data scatter into "explained" and "residual" parts is provided (for each of the two norms: sum of squares and moduli).

- Contribution weights are derived to attack machine learning problems (conceptual description, selecting and transforming the variables, and knowledge discovery).

- The explained data scatter parts related to nominal variables appear to coincide with the chi-squared Pearson coefficient and some other popular indices, as well.

- Approximation (bi)-partitioning for contingency tables substantiates and extends some popular clustering techniques.

6.1   Bilinear Clustering for Mixed Data

6.1.1   Bilinear Clustering Model

In Section 2.1.3, it was shown how to transform a column-conditional data table containing mixed variables (quantitative, nominal and binary) into a quantitative rectangular matrix; the rows of the matrix still correspond to the entities while its columns are expanded to include all the nominal categories. More strictly, the set V of columns of the transformed matrix is obtained from the original set K by removing all the columns corresponding to nominal variables k and adding columns for all their categories $v \in k$ instead. With this transformation done, still the information about which category v belongs to which variable k is kept. Thus, the data is represented as a data matrix $Y = (y_{iv}), i \in I, v \in V$, where rows $y_i = (y_{iv}), v \in V$, correspond to the entities $i \in I$ and the entries y_{iv} are quantitative values associated with corresponding variables/categories $v \in V$.

Let the entities be partitioned into groups (clusters) presented by an additive type cluster structure which is a set of m clusters, any cluster t, $t = 1, ..., m$, being defined with two objects: 1) its membership function $z_t = (z_{it}), i \in I$, where z_{it} is 0 or 1 characterizing, thus, cluster set $S_t = \{i \in I : z_{it} = 1\}$, 2) its standard point, or centroid vector, $c_t = (c_{tv}), v \in V$, to be combined in an $N \times |V|$ cluster-type matrix with elements $\sum_{t=1}^{m} c_{tv} z_{it}$.

The cluster-type matrix is compared with the given $N \times |V|$ matrix Y via equations

$$y_{iv} = \sum_{t=1}^{m} c_{tv} z_{it} + e_{iv} \tag{6.1}$$

where residual values e_{iv} show the difference between the data and the type clusters. When the clusters are not given a priori, they can be found in such a way that the residuals are made as small as possible, thus minimizing a criterion of form

$$\Phi(\{|e_{iv}|\}) \tag{6.2}$$

where Φ is an increasing monotone function of its arguments. The equations in (6.1) along with criterion in (6.2) to be minimized by unknown parameters, c_{tv}, z_{it}, e_{iv}, for y_{iv} given, will be referred to as *bilinear clustering model*.

Though the model is quite similar to that of the principal component analysis (the only difference is that z_t are Boolean, not arbitrary, vectors), it has a meaning on its own, just as a clustering model: every data row $y_i = (y_{iv})$, $v \in V$, is the sum of the standard points c_t of the clusters t containing i, up to the residuals. When the clusters are required to be nonoverlapping, that means that the membership functions must be mutually orthogonal or, in the other words, $S_t \cap S_{t'} = \emptyset$ for $t \neq t'$. In the nonoverlapping case, the type-cluster matrix $\sum_{t=1}^{m} c_{tv} z_{it}$ has a very simple structure: its rows are the vectors $c_t = (c_{tv})$ only, and every i-th row equals c_t for that specific cluster t which contains the entity $i \in I$.

Two specific Minkovski forms of criterion in (6.2) for minimizing the residuals are mostly considered here:

$$L_2 = \sum_{i \in I} \sum_{v \in V} e_{iv}^2,$$

$$L_1 = \sum_{i \in I} \sum_{v \in V} |e_{iv}|,$$

The strategies based on minimizing these two norms have rather long histories in statistics: K.F. Gauss (1777-1855) was the most influential proponent of the least-squares criterion while P.S. Laplace (1749-1827)) is usually credited for promoting least-moduli.

With the non-overlapping restriction, the Minkovski criteria become especially simple:

$$L_p = \sum_{i \in I} \sum_{v \in V} |y_{iv} - \sum_t c_{tv} z_{it}|^p = \sum_{v \in V} \sum_{t=1}^{m} \sum_{i \in S_t} |y_{iv} - c_{tv}|^p \qquad (6.3)$$

which shows that L_p, actually, is $L_p = \sum_{t=1}^{m} \sum_{i \in S_t} d^p(y_i, c_t)$ where d^p is the p-th power of the Minkovski distance $l_p(x - y)$ associated with Minkovski p-norm.

Due to formula (6.3), when the membership functions are given, the optimal c_{tv} is determined only by the values y_{iv} within S_t. For the case when $p = 1$ or $p = 2$, it is quite simple. The least-squares optimal c_{tv} is the average of y_{iv} in S_t, $c_{tv} = \sum_{i \in S_t} y_{iv}/|S_t|$, while the least-moduli optimal c_{tv} is a median of $y_{iv}, i \in S_t$; that is, the mid-term in the ordered series of the variable values. The optimal cluster vector c_t, $t = 1, ..., m$, will be referred to as the average or the median (vector), respectively.

The criterion (6.2) value, when its argument is the data matrix $Y = (y_{iv})$ itself, $\Phi(\{|y_{iv}|\})$, may be considered as a measure of the scatter of the data. Indeed, due to the model in (6.1) and (6.2), $\Phi(\{|y_{iv}|\}) = \Phi(\{|e_{iv}|\})$ when no clusters are presented (that is, when $y_{iv} = e_{iv}$). This definition allows for using the data scatter concept in the traditional meaning as admitting decomposition in two major parts: that "explained via the model" and an "unexplained" one. When $\Phi(|y_{iv}|)$ is taken as the scatter of the data and value $\Phi(|e_{iv}|)$ in (6.2) is considered a measure of the "unexplained" scatter, their difference, $\bar{\Phi} = \Phi(|y_{iv}|) - \Phi(|e_{iv}|)$ will be nonnegative for any appropriate minimizer of (6.2) since $\Phi(|e_{iv}|) = \Phi(|y_{iv}|)$ when all $c_{tv} = 0$ which is not an optimal solution. Value $\bar{\Phi}$ can be interpreted as the "explained" part of the data scatter $\Phi(|y_{iv}|)$, which gives the sought decomposition of the data scatter in the two parts, $\Phi(|y_{iv}|) = \bar{\Phi} + \Phi(|e_{iv}|)$. Such a decomposition looks especially appropriate when the "explained" part, $\bar{\Phi}$, can be further decomposed through the elements of the cluster structure found, as will take place with criteria L_1 and L_2.

In this setting, it is the data scatter which is decomposed into explained and unexplained parts due to the bilinear model; moreover, the unexplained part is nothing but the minimized criterion of the model. This explains why the data scatter has been chosen as the base of the data standardization procedures in Section 2.1.2.

Let us see how the solutions to the bilinear clustering model (6.1)–(6.2) depend on the standardization parameters.

The dependence certainly exists when the scale parameters are involved. Indeed, change of the scale of a variable is equivalent to introducing corresponding weight coefficient for this variable in the criterion (6.3).

> The principle of equal contribution, **P1**, makes all the variables have the same contribution to the scatter of the data, which makes meaningful comparison of the variables by their contributions to the explained (or unexplained) part of it. Such comparison may reveal the most contributing, thus salient, variables and categories.

As to the influence of the shift parameters, the situation here somehow differs. On one hand, we may introduce a constant item in the model equation (6.1) to eliminate any dependence of the solution on the origin of the data space. Using this option for a non-overlapping cluster structure, we can see that the optimal value of such a constant item μ must be zero. Indeed, the optimal value of μ added to the right parts of equations (6.1) is determined via the same equations based on the derivative of criterion L_p (6.3) that determine the optimal values c_{tv}; the only difference is that now item $c_{tv} + \mu$ instead of c_{tv} is involved, for every cluster t. This shows that $\mu = 0$ when c_{tv} are optimal; no shift of the origin is needed. This looks rather obvious when put in the cluster analysis terms: when the standard points $c_t = (c_{tv})$ are defined, assigning the entities to the clusters depends on their distances to the standard points only (as expressed in criterion (6.3)).

On the other hand, when the bilinear model is set forth in a sequential way with the "factor" axes z_t identified one-by-one, not simultaneously, as it is done in Section 2.3, the solution heavily depends on the origin of the variable/category space. Indeed, the axes z_t are drawn through the origin in the directions of maximal variance, which can change drastically when the origin is changed.

> The principle of minimizing the data scatter, **P2**, concerns the sequential fitting methods when the solution elements are obtained one-by-one.

6.1.2 Least-Squares Criterion: Contributions

With the least-squares criterion, the following decomposition holds.

Statement 6.1. *When values c_{tv} are optimal for a partition $S = \{S_t\}$ of I, the following decomposition of the data scatter holds:*

$$\sum_{i\in I}\sum_{v\in V} y_{iv}^2 = \sum_{t=1}^{m}\sum_{v\in V} c_{tv}^2 |S_t| + \sum_{i\in I}\sum_{v\in V} e_{iv}^2, \qquad (6.4)$$

Proof: This is proved by squaring the parenthesis and putting the optimal values $c_{tv} = \sum_{i\in S_t} y_{iv}/|S_t|$ in criterion L_2 in form (6.3). $\qquad \square$

The decomposition is quite well-known in cluster analysis (see, for example, Jain and Dubes 1988), though quite under-employed. Usually, it is interpreted in terms of analysis of variance: divided by N, equation (6.4) shows how the overall variance of the data (the sum of the single variable variances) is decomposed into within-group variance, $\sum_v \sum_t p_t \sum_{i\in S_t}(y_{iv} - c_{tv})^2/N_t$, and inter-group variance, $\sum_{v\in V}\sum_{t=1}^{m} p_t c_{tv}^2$, where p_t and N_t are frequencies and cardinalities in distribution (S).

In cluster analysis, interpretation of (6.4) in terms of the contributions to data scatter seems more helpful. We can see that the contribution of a pair variable-cluster (v, t) to the explained part of the data scatter is expressed with a rather simple formula, $c_{tv}^2|S_t|$: it is proportional to the cluster cardinality and to the squared distance from the grand mean of the variable to its mean (standard value) within the cluster.

Corollary 6.1. *Relative contributions of the elements of the cluster structure, due to criterion L_2, are as follows:*

a) Variable v to the data scatter: $w(v) = \sum_{t=1}^{m} c_{tv}^2|S_t|/\sum_{i,v} y_{iv}^2$;

b) Cluster t to the data scatter: $w(t) = \sum_{v\in V} c_{tv}^2|S_t|/\sum_{i,v} y_{iv}^2$;

c) Variable v to cluster t: $w(v/t) = c_{tv}^2/\sum_{v\in V} c_{tv}^2$;

d) Entity i to cluster t: $w(i/t) = \sum_{v\in V} y_{iv} c_{tv}/\sum_{v\in V} c_{tv}^2 = (y_i, c_t)/(c_t, c_t)$.

Proof: All the formulas here are just obvious implications from the decomposition in (6.4), except for $w(i/t)$ which is found with $c_{tv} = \sum_{i\in S_t} y_{iv}/|S_t|$ put into $c_{tv}^2|S_t|$:

$$c_{tv}^2|S_t| = (\sum_{i\in S_t} y_{iv}/|S_t|)c_{tv}|S_t| = \sum_{i\in S_t} y_{iv} c_{tv},$$

which proves the statement.   □

Curiously, $w(i/t)$ is the only contribution which can be negative: this occurs when the scalar product of vectors y_i and c_t is negative, that is, the angle between y_i and c_t (from zero) is obtuse which might be interpreted that the i-th entity is foreign to its cluster.

Another corollary concerns the contributions of the nominal variables and their categories to the scatter part "explained" via cluster partition S.

Let us consider, initially, what the standard values c_{tv} are for the qualitative categories. To do that, let us recall that we denote frequency (proportion of ones) of the category v in all the set I by p_v while using p_{tv} for the proportion of the entities simultaneously having category v and belonging to cluster S_t. Then, it is not difficult to see that, for any category v standardized by formula $y_{iv} = (x_{iv} - a_v)/b_v$, its mean within cluster S_t is equal to

$$c_{tv} = (p_{vt} - p_t a_v)/(p_t b_v).$$

This follows from the fact that the average of the binary variable x_{iv} by $i \in S_t$ equals p_{tv}/p_t. The contribution of a category-cluster pair (v, t) to the explained part of the data scatter is equal to

$$s(v, t) = c_{tv}^2 |S_t| = N(p_{vt} - p_t a_v)^2/(p_t b_v^2), \qquad (6.5)$$

which can be considered a measure of association between category v and cluster t. In particular,

$$s(v, t) = N(p_{vt} - p_t p_v)^2/(p_t p_v),$$

when $b_v = \sqrt{p_v}$, or,

$$s(v, t) = N(p_{vt} - p_t p_v)^2/p_t$$

when $b_v = 1$, etc.

Since every nominal variable k is considered as the set of its categories v, the joint contribution of k and the set of the clusters S_t to the scatter of the data is equal to $F(k, S) = \sum_t \sum_{v \in k} s(v, t)$ which is

$$F(k, S) = N \sum_{t=1}^{m} \sum_{v \in k} \frac{(p_{vt} - p_t a_v)^2}{p_t b_v^2} \qquad (6.6)$$

by (6.5). Substituting the appropriate values of $a_v = p_v$ and b_v, we arrive at the following.

Corollary 6.2. *For criterion L_2, the contribution of a nominal variable $k \in K$ to that part of the square scatter of the square standardized data which is explained by* the (sought or found or expert-given) cluster partition $S = \{S_1, ..., S_m\}$, is equal to

$$M(S/k) = \frac{N}{\#k - 1} \sum_{v \in k} \sum_{t=1}^{m} \frac{(p_{vt} - p_v p_t)^2}{p_v p_t} \tag{6.7}$$

when $b_v = \sqrt{p_v(\#k - 1)}$ (the first standardizing option), or

$$\Delta(S/k) = N \sum_{v \in k} \sum_{t=1}^{m} \frac{(p_{vt} - p_v p_t)^2}{p_t} \tag{6.8}$$

when $b_v = 1$ (no normalizing), or

$$W(R/k) = N \sum_{v \in k} \sum_{t=1}^{m} \frac{(p_{vt} - p_v p_t)^2 / p_t}{1 - \sum_{v \in k} p_v^2} \tag{6.9}$$

when $b_v = \sqrt{1 - \sum_{v \in k} p_v^2}$ (second standardizing option).

All three of the coefficients relate to well known indices of contingency between the nominal variables: $M(S/k)$ is a normalized version of the Pearson chi-squared coefficient, $\Delta(R/k)$ is proportional to the coefficient of reduction of the error of proportional prediction, and $W(R/k)$ is nothing but the Wallis coefficient (see Section 4.1.2 where all three are discussed in various settings). Here, each of the coefficients turns out to have yet another meaning of a measure of the contribution of cluster partition S to the square scatter of the indicator matrix of the other variable, k. Amazingly, it is the method of data standardization which determines which of the coefficients is produced as the contribution-to-scatter.

Note that the normalization of the Pearson chi-squared coefficient here involves only the number of categories in k, not the number m of classes in S, which adds yet another, asymmetrical, normalization to the two well-known normalized forms of the Pearson coefficient: Cramer's and Tchouprov's, both taking into account both of the numbers, $\#k - 1$ and $m - 1$ (see Section 4.1.2).

These formulas should be recommended also as the contribution-based measures of association between partition S and a multiple choice variable k which is just a set of (overlapping) categories v, since no requirement of nonoverlapping vs has been utilized in the analysis above.

Taking different values of the shift and scale coefficients, a_v, b_v, other contingency measures can be derived as special cases of the formula (6.6).

Let us finish the discussion with analysis of contribution of a quantitative variable into the explained part of the data scatter. There have not been as many indices of association between partitions and quantitative variables developed as for partition-to-partition case. A most important measure is the so-called correlation ratio (squared) coefficient defined as

$$\eta^2(S,k) = \frac{\sigma_k^2 - \sum_{t=1}^m p_t \sigma_{tk}^2}{\sigma_k^2} \qquad (6.10)$$

where $p_t = N_t/N$ is the proportion of entities in S_t, and σ_k^2 or σ_{tk}^2 is the variance of the variable k in all the set I or within cluster S_t, respectively.

The larger correlation ratio, the better association between S and k; it equals 1 when the variable is constant within each of the classes S, and it is 0 when there is no reduction of the overall variance within the clusters.

It turns out, when the variable k is standardized,

$$\eta^2(k,S) = \sum_t c_{tk}^2 p_t$$

which means that the contribution of pair (k, S) to the explained part of the data scatter in decomposition (6.4) is exactly $N\eta^2(k,S)$.

> The part of the square data scatter explained by a partition, S, appears to be equal to the sum of the correlation measures between the partition and the variables. The correlation measure is $N\eta^2(k,S)$ when k is a quantitative variable, or $M(S/k)$, $\Delta(S/K)$, or $W(S/K)$ when k is nominal, depending on the standardization option accepted.

From the theoretical point of view, the result links different lines in clustering and statistics and may be considered a support for the standardizing options suggested in Section 2.1.3 for treating mixed variables.

From a practical point of view, the result gives an equivalent reformulation of the least-squares criterion and can be used in conceptual clustering: finding a partition S in terms of the variables $k \in K$ can be done with the sum of the correlation coefficients to be maximized; the conceptual clustering results will be consistent with those found by traditional square-error clustering.

6.1.3 Least-Squares Clustering: Equivalent Criteria

After having the contributions analyzed, we can formulate several equivalent forms of the least-squares criterion as applied to the problem of partitioning:

A. *Distant Centers:*

$\max_S \sum_{v \in V} \sum_{t=1}^m c_{tv}^2 |S_t|$

where c_{tv} is the average of the category/variable v in S_t.

B. *Semi-Averaged Within Similarities:*

$\max_S \sum_{t=1}^{m} \sum_{i,j \in S_t} a_{ij}/|S_t|$

where $a_{ij} = (y_i, y_j) = \sum_v y_{iv} y_{jv}$.

C. *Consensus Partition:*

$\max_S \sum_{k \in K} \mu(S, k)$

where $\mu(S, k) = N\eta^2(S, k)$ (6.10) when k is a quantitative variable and $\mu(S, k)$ is a contingency coefficient $F(k, S)$ in (6.6); in particular, it can be the modified Pearson $M(S/k)$ or Wallis $W(S/k)$ coefficient depending on the standardizing option selected.

D. *Within Variance Weighted:*

$\min_S \sum_{t=1}^{m} p_t \sigma_t^2$

where p_t is proportion of the entities in S_t and $\sigma_t = \sum_v \sum_{i \in V} (y_{iv} - c_{tv})^2/|S_t|$ is the total variance in S_t.

E. *Semi-Averaged Within Distances Squared:*

$\min_S \sum_{t=1}^{m} \sum_{i,j \in S_t} d^2(y_i, y_j)/|S_t|$

where $d^2(y_i, y_j)$ is the Euclidean distance (squared) between the row-points corresponding to entities $i, j \in I$.

F. *Distance-to-Center Squared:*

$\min_S \sum_{t=1} \sum_{i \in S_t} d^2(c_t, y_i)$

where $d^2(c_t, y_i)$ is the Euclidean distance (squared) between the standard point c_t and an entity i's row-vector.

G. *(Within Group) Error Squared:*

$\min_{S,c} \sum_{t=1}^{m} \sum_{v \in V} \sum_{i \in S_t} (c_{tv} - y_{iv})^2$.

H. *Bilinear Residuals Squared:*

$\min_{c,z} \sum_{i \in i} \sum_{v \in V} e_{iv}^2$

where c, z, e are the variables defined according to the bilinear equations (6.1).

Each of these criteria expresses a clustering goal, each time involving a different clustering concept:

(a) finding types (centroids) as far from the grand mean as possible (item A);

(b) minimizing within-cluster variances (item D);

(c) maximizing within cluster similarities (item B) or minimizing within cluster dissimilarities (item E);

(d) minimizing the difference between the cluster structure and the data (items G, H);

(e) maximizing total correlation/contingency between the sought partition and the variables given (item C).

All the goals above become equivalent when they are explicated with the criteria considered.

Statement 6.2. *The criteria A. through H. are equivalent to each other.*

Proof: No proof is needed since all the nontrivial equivalences have been already proven above.   □

Equivalence of the items D to G has been known for quite a long time; what may be considered non-standard formulations are A, B, C, and H.

6.1.4   Least-Moduli Decomposition

The least-moduli criterion, however mysterious it seems to be, can be accompanied by a similar additive decomposition of the data scatter into the explained and unexplained parts.

Let S_t be an entity subset, and $S_{tv} = \{i \in S_t : |y_{iv}| < |c_{tv}| \ \& \ sgn \ y_{iv} = sgn \ c_{tv}\}$ where, as usual, $sgn \ x$ is 1 if $x > 0$, 0 if $x = 0$, and -1 if $x < 0$. This means that $S_{tv} = \{i \in S_t : 0 \le y_{iv} \le c_{tv}\}$ if c_{tv} is positive or $S_{tv} = \{i \in S_t : c_{tv} \le y_{iv} \le 0\}$ if c_{tv} is negative. Having a value c_{tv} fixed, the set S_t is partitioned into three subsets by the variable/category v depending on relations between y_{iv}, $i \in S_t$, and c_{tv}. For $c_{tv} > 0$, let us denote the cardinalities of the subsets where y_{iv} is larger than, equal to or less than c_{tv} by n_{tv1}, n_{tv2} and n_{tv3}, respectively. Then, let $n_{tv} = n_{tv1} + n_{tv2} - n_{tv3}$. For $c_{tv} < 0$, the symbols n_{tv1} and n_{tv3} are interchanged along with corresponding change of n_{tv}. If c_{tv} is the median of values y_v in S_t and all the values y_{iv}, $i \in S_t$, are different, then $n_{tv1} = n_{tv3}$ and $n_{tv2} = 0$ or $= 1$ depending on the cardinality of S_t (even or odd, respectively).

Statement 6.3. *When values c_{tv} are L_1-optimal for a partition $S = \{S_t\}$ of I, the following decomposition of the module data scatter holds:*

$$\sum_{i \in I} \sum_{v \in V} |y_{iv}| = \sum_{v \in V} \sum_{t=1}^{m} (2 \sum_{i \in S_{tv}} |y_{iv}| - n_{tv}|c_{tv}|) + \sum_{i \in I} \sum_{v \in V} |e_{iv}|. \qquad (6.11)$$

Proof: Since

$$|a - b| = |a| + |b| - |sgn \ a + sgn \ b| \min(|a|, |b|)$$

for any real a and b, L_1 in (6.3) can be expressed as

$$\sum_v \sum_t \sum_{i \in S_t} |y_{iv} - c_{tv}| = \sum_v \sum_t \sum_{i \in S_t} (|y_{iv}| + |c_{tv}| - |sgn\ y_{iv} + sgn\ c_{tv}| \min(|y_{iv}|, |c_{tv}|)).$$

Let us consider the right part of this. The first term equals $\sum_v \sum_t \sum_{i \in S_t} |y_{iv}| = \sum_{i,v} |y_{iv}|$ and the second is $\sum_v \sum_t |S_t||c_{tv}|$. To analyze the third term, let us assume $c_{tv} > 0$. Then, the expression $|sgn\ y_{iv} + sgn\ c_{tv}| \min(|y_{iv}|, |c_{tv}|)$ is equal to $2c_{tv}$ when $c_{tv} \le y_{iv}$ (which counts for $n_{tv1} + n_{tv2}$ elements $i \in S_t$), $2y_{iv}$ when $0 \le y_{iv} \le c_{tv}$, and 0 when $y_{iv} < 0$ since in that latter case the signs of c_{tv} and y_{iv} are different. This exactly corresponds to the equality in (6.11). The case of negative c_{tv} is considered analogously. □

Let us denote the contribution of a variable-cluster pair (v, t) to the module scatter by $s(t, v) = 2 \sum_{i \in R_{tv}} |y_{iv}| - n_{tv} |c_{tv}|$. Based on this, various relative contribution measures can be defined as in Section 6.1.2: (a) variable to scatter, $w(v) = \sum_t s(t, v)/\sum_{i,v} |y_{iv}|$; (b) cluster to scatter, $w(t) = \sum_v s(t, v)/\sum_{i,v} |y_{iv}|$; (c) variable to cluster, $w(v/t) = s(t, v)/\sum_v s(t, v)$; (d) entity to cluster, $w(i/t) = |sgn\ y_{iv} + sgn\ c_{tv}| \min(|y_{iv}|, |c_{tv}|) - |c_{tv}|$. The latter expression follows from the proof of the statement above.

Let us consider the case when v is a category.

Statement 6.4. *For any category v standardized (with arbitrary a_v and b_v), its median in cluster S_t is equal to*

$$c_{tv} = \begin{cases} -a_v/b_v & if & p_{tv} < 0.5p_t \\ (1 - 2a_v)/2b_v & if & p_{tv} = 0.5p_t \\ (1 - a_v)/b_v & if & p_{tv} > 0.5p_t \end{cases}$$

The contribution of the category-cluster pair, (v, S_t), to the module data scatter is equal to

$$s(t, v) = N|2p_{tv} - p_t||c_{tv}|.$$

Proof: Let us see what the central value is of the initial binary variable, $x_{iv}, i \in S_t$, for category v. Obviously, the median is 0, 1/2 or 1 depending on the proportion of ones, p_{tv}/p_t, among $x_{iv}, i \in S_t$: whether it is less than, equal to or greater than 1/2, respectively. With those values transformed as $y_v = (x_v - a_v)/b_v$, the formulas for c_{tv} are proven.

To derive the formula for $s(t, v)$, let us see that $S_{tv} = \emptyset$ since the values c_{tv} and y_{iv} must have different signs if they are not equal to each other (y_{iv} may have one of two values only since v is a category). Thus, $n_{tv1} + n_{tv2} = Np_{tv}$ and

$n_{tv3} = N(p_t - p_{tv})$ when $c_{tv} = (1 - a_v)/b_v > 0$ where a_v, b_v are the values used in the module standardization rule. Analogously, $n_{tv1} + n_{tv2} = N(p_t - p_{tv})$ and $n_{tv3} = Np_{tv}$ when $c_{tv} = -a_v/b_v < 0$.   □

Now we can see what the value is of the part of a nominal variable $k \in K$ explained by the cluster partition S.

Corollary 6.3. *The contribution of a nominal variable $k \in K$ to the absolute scatter of the module standardized data, as explained by the clustering partition $S = \{S_1, ..., S_m\}$, is equal to*

$$A(S/k) = \frac{N}{\#k} \left(\sum_{(v,t) \in A_+} \frac{2p_{vt} - p_t}{p_v} + \sum_{(v,t) \in A_-} \frac{p_t - 2p_{vt}}{1 - p_v} + \sum_{v \in A_=} |2p_{vt} - p_t| \right) \quad (6.12)$$

where $A_+ = \{(v,t) : p_v < 0.5 \text{ and } p_{vt}/p_t > 0.5\}$, $A_- = \{(v,t) : p_v > 0.5 \text{ and } p_{vt}/p_t < 0.5\}$, and $A_= = \{v : p_v = 0.5\}$.

The coefficient $A(S/k)$ takes into account the situations when the patterns of occurrences of the categories $v \in k$ in the clusters t differ from those in the entire set I. Such a difference appears when v is frequent in S_t ($p(v/t) > 0.5$) and rare in I ($p_v < 0.5$), or, conversely, v is rare in S_t and frequent in I. If, for instance, the number of categories $v \in k$ is five or more, in a common situation, every p_v will be less than 0.5. Then, $A(S/k)$ is high if each cluster collects most of a corresponding category v, and $A(S/k)$ is zero if the categories v are distributed more or less uniformly among the clusters.

> $A(R/k)$ has a relevant operational meaning and can be considered as an interesting, though nontraditional, contingency coefficient.

6.1.5 Discussion

Bilinear clustering model for column-conditional data has been employed by the author for a decade (see Mirkin 1987a, 1990, Mirkin and Yeremin 1991). Some may say that the model is just a mathematically complicated cover for a simple square-error clustering strategy; that the equivalent reformulations of the criterion are mostly well-known; that the square scatter decomposition is well-known, also; and that though the strategy includes generic forms of K-Means and agglomerative clustering, it involves also two major drawbacks of the square-error clustering: (1) only spheroidal clusters can be revealed while, in the real-world data, quite elongated and odd shapes may be present sometimes, and (2) the results depend much on the weights/scales of the variables. There is truth in that, however there is truth beyond, too.

Let us start with the drawbacks mentioned.

First, a spheroidal (or cubic, with L_1 or L_∞ criterion) shape seems quite a good shape for typology making! Type clusters must surround corresponding prototypes. If it is not so, look at the variables. The fact that the types are dispersed throughout the variable space and not coherent means that the variables are not appropriate for explaining the types and must be changed or transformed somehow. As to the other shapes, they should not be revealed heuristically: a similar way of modeling must be developed for every particular kind of shape. For example, if a cluster S_t must reflect a linear relation between some variables, thus satisfying equation $(c_t, y_i) = 0$ for some particular c_t and for $y_i \in S_t$, this can be modeled with equation $\sum_t \sum_{i \in I}(c_t, y_i)s_{it} = e$ where e is the minimized error (with regard to centroids c_t and indicators s_t unknown).

Second, the fact that the solution depends on the data standardization should imply not giving up but developing an adequate data standardization system. The bilinear model serves as a vehicle for developing such a standardization rule. The model allows shifting attention from the error minimizing criterion to the data scatter as the quantity which is to be explained with the model and, thus, is to be clarified itself, which is the purpose of the standardization principles employed.

On the other hand, the model has led us to several new clustering options.

With the bilinear model, extending the standardizing rules to the mixed data case has been quite natural. Moreover, analysis of the nominal variable contribution to the data scatter has shown the contribution's meaning in terms of the contingency coefficients, which was previously not obvious at all. Even less obvious was the dependence of the coefficients themselves on the normalizing option. Who could imagine that normalizing categories by $\sqrt{p_v}$ versus non-normalizing changes the coefficient from the reduction of proportional prediction error to the Pearson chi-squared? In textbooks, it was written that the two coefficients were quite different; the former one bore much more operational meaning than the latter (see, for example, Reynolds 1977, Goodman and Kruskal 1979, Agresti 1984).

The fact that the bilinear clustering model much resembles that of principal component analysis has led us to the method of principal cluster analysis having a nice geometrical interpretation in terms of the reference-point concept as a particular parameter in clustering. Usually, the variable space properties are considered as not depending on its origin. It is not so in principal clustering which is greatly affected by the origin location (as the principal component analysis is) and allows thus to model an important classification making phenomenon connected with difference in classes depending on the viewing point (more on that in the next section). Principal clustering is also connected with some other linear learning models (like perceptron) as we saw in Statement 4.16., p. 196.

Shifting to the error-criterion perspective from that of the distance chosen has made possible a detailed analysis of the least-moduli clustering which is connected with non-standard measures of entity-to-entity similarity and cross-classification contingency.

Last, but not least, the bilinear model allows treating some nonstandard classification structures such as overlapping or fuzzy clusters. Fuzzy clustering via the bilinear model has produced a non-traditional kind of cluster modeling ideal type concept: the standard points of the bilinear fuzzy clusters appear to be extreme, not "average", points of the entity cloud in the variable space.

6.2   K-Means and Bilinear Clustering

6.2.1   Principal Clustering and K-Means Extended

Principal cluster analysis method as applied to the model in (6.1), is, actually, the sequential fitting procedure with single clusters sought at each step. Let us consider it in both of the two versions: 1) hard (crisp) clusters defined by Boolean indicator function, and 2) fuzzy clusters defined by fuzzy indicator functions having the interval between 0 and 1 as their ranges.

Hard/Fuzzy Principal Cluster Partitioning
1. Set t=1 and define data matrix Y_t as the initial data matrix standardized.
2. For $Y = Y_t$ find a principal cluster as described in Section 4.4.1, in the case of hard clustering, and in Section 4.4.2, in the case of fuzzy clustering. Define z_t, c_t as the cluster solution found (membership function and the standard point [centroid], respectively); compute its contribution w_t to the data scatter. In the hard clustering case, the clusters are presumed to be nonoverlapping, thus, the search should be made only among those entities which are unassigned to the preceding clusters. In the fuzzy clustering case, the sequential fitting procedure is quite straightforward when no restriction on the membership functions is imposed. However, when a fuzzy partition is sought, that means that the membership functions must satisfy the "unity" condition, $\sum_t z_{it} = 1$, for every $i \in I$. Thus, at any step t, for any $i \in I$, its cumulative membership $\alpha_{it} = \sum_{u<t} z_{iu}$ must be taken into account: criterion $L(c, z)$ is minimized by c and z satisfying inequality $z_i \leq 1 - \alpha_{it}$ rather than just $z_i \leq 1$.

---

3. Stop-Condition. If there must be nonoverlapping hard clusters, check whether there are yet unclustered entities remaining. (In the other case, check the stopping rule described in Section 2.3.4): if yes, go to 4; else go to 5.

4. Compute the residual data $y_{iv}^{(t+1)} = y_{iv}^t - c_{tv}z_{it}$, increase t by 1, and go to 2. (For hard partitioning, this operation is unnecessary because the entities, once clustered, never appear again.)

5. End: in the case of hard clustering, present the solution found along with the contribution weights associated as its interpreting aids. In the case of fuzzy clustering, find the final fuzzy cluster $t + 1$ by setting $z_{i,t+1} = 1 - \sum_{u<t} z_{iu}$ and, subsequently, calculating $c_{t+1,v}$ by formula (4.24), p. 203. This final fuzzy cluster is a "ground" cluster since it relates, mostly, to the area around the grand mean of the entity set, in contrast to the extreme "ideal types" corresponding to the preceding fuzzy clusters.

To give a more practical image to the algorithm, let us rephrase it (for the case of hard clustering) in terms of the K-Means method. Let us recall that this method starts with an m class partition of I or with m somehow selected tentative standard points or "seeds", c_t. Then the algorithm repeatedly performs the following two-step iteration: (1) update the partition based on the standard points: when all c_t are given, make each S_t the set of y_i that are nearest (by Euclidean distance) to c_t, $t = 1, ..., m$; (2) update the standard points: when all S_t are given, compute c_t as the mean of the within-cluster vectors. The algorithm stops when the updating procedure does not change the clustering.

The principal cluster analysis can be considered as a technique that exploits many of the same mechanisms as the moving-center method, but which mitigates the need for prior knowledge, and separates clusters from the set of instances one by one. First, an initial cluster $S_1 \subset I$ is extracted with its standard point c_1; the complementary set represents the main "body" of instances, which serves as the source for separating additional clusters one by one. This is reflected in that fact that the main body's standard point is fixed at 0, given the square scatter standardization, and it is not changed during the entire clustering computation. The principal clustering procedure, in these terms can be reformulated as a kind of "separate-and-conquer" strategy considered by Pagallo and Haussler 1990.

Algorithm SCC (Separate-and-Conquer Clustering)
Step 0. $t \leftarrow 1$.
Step 1 (Selection of an extreme point). Pick a point, y_{i*}, maximizing Euclidean distance $d(0, y_i)$, $i \in I$, from the origin of the variable space. Take $c_t = y_{i*}$ as the initial center (seed) of the t-th cluster.

Step 2 (Separating the cluster).   Find a cluster (S_t, c_t) with the Reference-point-based moving center method in Section 4.3.2 (with $\alpha = 1$ and zero as the reference point).
Step 3 (Excluding the entities). Set $I = I - S_t$. If $I = \emptyset$, end; else set t=t+1 and go to Step 1.

To give intuition to the algorithm, let us consider a situation when there is only one variable, uniformly distributed across its range. Then, having the zero point in the midrange, SCC separates initially one fourth of the range at one extreme, then one fourth at the other extreme, with one half of the range left to be cut at extremes again. A traditional divisive version of the method, with both of the standard points updated, will produce, initially, a split just in the midrange, splitting then each of the clusters by half, etc. (see Section 7.6.2).

This example reflects a general property that the size of an SCC-designed cluster depends on its distance from the origin (which is just the reference point) as stated in (5.35): the nearer to that point, the less the diameter of the cluster! Thus, SCC could be modified to allow the user to specify the reference-point origin based on the user's knowledge of the variable space: the better the knowledge, the smaller the classes. It can be useful, for instance, for a robot-planning system: the robot must learn and classify the nearest part of the world in more detail than more distant objects. Independent use of a reference-point-based approach in a different substantive study has been made by M. Damashek 1995.

Positioning the reference point as the grand mean causes SCC to separate a subset of instances corresponding to an extreme combination of the variable values. In this respect, SCC models a typology-making process based on the assumption that the extreme combinations of the variable values correspond to some "theoretical" types while points around the grand mean are just a noise.

Let us see how principal partitioning works with the Points data (comprising six 2-dimensional points presented in Table 2.1, p. 62, and Fig.2.1).

Initially, let us consider the data as it is (with no standardizing). In this case, the reference point is the origin, zero, and, obviously, 6 is the most distant (from zero) point to start with. It can be seen easily that 5 is the only point which is closer to 6 than to zero, which makes the first cluster to contain, currently, 5 and 6, with point (3, 2.5) being their gravity center. All the other points are farther from this gravity center than from zero, which ends the process of forming S_1: we have $S_1 = 5 - 6$. Analogously, starting with 1, we get all the other points to belong in the second cluster, $S_2 = 1 - 2 - 3 - 4$. Thus, putting the reference point at zero and having no standardization done, we find principal cluster partition $S = \{1 - 2 - 3 - 4, 5 - 6\}$, which seems obvious by the picture presented in Fig.2.1 (a).

However, the situation becomes quite different after the data have been square scatter standardized (see Fig.2.1 (b)). Though point 6 still is the farthest from zero, point 5 does not join to the first cluster anymore because its distance from zero, 1.36, is less than its distance to 6, 1.46. Thus, point 6 alone forms a cluster. Similarly, all the other points (except for 1 and 2 joined in the same cluster) form singleton clusters on their own, because their nearest neighbors are closer to the origin than to them. On the first glance, such a result, five clusters obtained in the set of six points, looks disappointing. However, the standardized data present a picture (Fig.2.1 (b)) which does not show as unambiguous a two-cluster pattern as Fig.2.1 (a) does: some people may see three clusters formed by pairs: 5, 6; 2, 4; and 1, 3; others may say that any point is so distant from the others that it should be considered as a singleton cluster, etc. Lack of any specific cluster pattern can be seen especially clearly if the observer gazes at the coordinate axes intersection point in Fig.2.1 (b).

However, if the number of the clusters is specified as 2, the extended K-Means algorithm produces exactly the two clusters that have been found for the raw data unstandardized.

What remains important about principal cluster analysis, is that, in all the considered modifications (for hard overlapping or nonoverlapping clusters and for fuzzy clusters, as well), the data scatter is additively decomposed into the sum of the cluster contributions w_t and the unexplained part of it. The decomposition looks as follows:

$$\sum_{i \in I} \sum_{v \in V} y_{iv}^2 = \sum_{t=1}^{m} \sum_{v \in V} c_{tv}^2 \sum_{i \in I} z_{it}^2 + \sum_{i \in I} \sum_{v \in V} e_{iv}^2,$$

for both, fuzzy and hard clustering, when the least-squares criterion is applied. In the case when the least-moduli criterion is used, the square scatter must be substituted by the absolute scatter while the cluster contribution w_t will be expressed by the function $g_1(S_t) = 2 \sum_{i \in S_t} (b_t(i, c_t) - |c_t|)$ from p. 201.

The principal cluster analysis option is nothing but a sequence of repeated computations with the reference-point-based moving center method, and can be used also to extend the K-Means method for a wider class of situations when the user can fix a few (not all) tentative centers even if she/he does not know the total number of the clusters or the total number is larger than the number of tentative centers the user is able to specify.

Let $l \geq 0$ be the number of the centers specified by the user, and $m \geq l$ or $m = *$ (unknown), the number of the clusters. Then, the extended K-Means method can be formulated as follows.

Extended K-Means Method

Step 0 (Analysis of the prior information). Standardize the data set. Take the central point of the data set as the reference point a. Set $A = I$. If $0 < l$ and m is not fixed ($m = *$) or $m > l$, then go to Step 1, if $l = 0$ then go to Step 2, if $l = m$ then go to Step 3.

Step 1 (Reducing the data set when $l > 0$). For any of prior tentative centroids, $c_t, t = 1, ...l$, find sets $A_t = \{i : d(y_i, c_t) \leq d(y_i, a)\}, t = 1, ...l$. Exclude the found sets; that is, let $A = I - \cup_{t=1}^{l} A_t$.

Step 2 (Principal clusters for initial setting). Repeatedly find a cluster with the reference-point-based algorithm (with $\alpha = 1$) for set A of the entities until $m - l$ ($l \geq 0$) clusters is found (or until set A is exhausted when $m = *$). If the value $m - l$ (when m is pre-fixed) is not reached until set A is exhausted, then increase α in proportion to the ratio of $m - l$ to the number of clusters found, and repeat the step with this new α. If $m - l$ has not been reached yet, α is again increased, until $m - l$ clusters along with their centroids are found. These $m - l$ centroids along with the l prior ones are considered initial setting.

Step 3 (Parallel K-Means). Perform parallel K-Means algorithm.

Let us apply the extended K-Means algorithm to the Masterpiece data standardized, with prefixed number of clusters, 3. The data is a 8×7 matrix:

$$
Y = \begin{pmatrix}
-0.775 & -0.816 & -0.444 & -1.291 & 0.722 & -0.354 & -0.433 \\
-1.247 & -0.898 & -1.154 & -1.291 & -0.433 & 1.061 & -0.433 \\
-1.404 & -0.891 & -1.154 & -1.291 & -0.433 & 1.061 & -0.433 \\
0.041 & 1.428 & -0.444 & 0.775 & -0.433 & -0.354 & 0.722 \\
0.151 & 1.732 & 0.976 & 0.775 & -0.433 & -0.354 & 0.722 \\
1.470 & 0.413 & -0.444 & 0.775 & -0.433 & -0.354 & 0.722 \\
0.622 & -0.652 & 0.976 & 0.775 & 0.722 & -0.354 & -0.433 \\
1.141 & -0.317 & 1.686 & 0.775 & 0.722 & -0.354 & -0.433
\end{pmatrix}.
$$

Upon calculation of all the distances from the row-vectors to zero, the farthest point appears to be 3. Point 2 is closer to 3 than to zero, which leads to cluster $S_1 = \{2, 3\}$ separated. Among the remaining six elements, 8 is the most distant from zero, which leads to separation of another cluster $S_2 = \{7, 8\}$. The three elements 4, 5, 6 are joined to the farthest entity, 5, forming another, third cluster. Iterations of parallel K-Means lead to joining 1 to S_1 without any other change. (The principal partitioning algorithm with no m specified, SCC, produces 4 clusters, separating also entity 1 [the novel in verses, EugOnegin by A. Pushkin]).

The clusters found correctly identify the authors. Each of the clusters is represented by three rows in Table 6.1: the first is the cluster centroid in terms of the raw data matrix X as it is presented in Table 2.3, 74; the second, centroid in the standardized form of matrix Y; the third presents the squares of the standardized values multiplied by the cluster cardinality, $c_{tv}^2 |S_t|$, which are the contributions of the variable-cluster pairs, due

Cluster	LenS	LenD	NumC	InMon	Dire	Beha	Tho	Sum
	12.67	12.27	1.33	0	0.33	0.67	0	
Pushkin	-1.14	-0.87	-0.62	-1.29	-0.05	0.59	-0.43	
	3.91	2.26	1.16	5.00	0.01	1.04	0.56	13.94
	25.55	44.10	4.5	1	1	0	0	
Tolstoy	0.88	-0.48	1.33	0.775	0.72	-0.35	-0.43	
	1.45	0.47	3.54	1.20	1.04	0.25	0.37	8.44
	23.47	183.13	2.67	1	0	0	1	
Dostoevski	0.55	1.19	0.03	0.775	-0.43	-0.35	0.72	
	0.92	4.25	0.00	1.80	0.56	0.38	1.56	9.48
Sum	6.29	6.99	4.70	8.00	1.61	1.67	2.50	31.86

Table 6.1: Cluster structure of the Masterpiece data; in any cluster, the averages of the variables in real and standardized scales are shown in the first and second rows; the third row contains the contributions (weighted averages squared).

to Corollary 6.1., p. 289.

Summing up all the contributions in a row, we get the cluster contribution to the square data scatter. The square data scatter itself is equal to 40 (the number of entities by the number of variables) due to the standardizing option applied. All the values along with the relative contributions of the variables (per cent) to the clusters are presented in Table 6.2.

Cluster	LenS	LenD	NumC	InMon	Dire	Beha	Tho	Total
Pushkin	28.06	16.22	8.30	35.86	0.04	7.47	4.02	34.86
Tolstoy	17.24	5.57	42.01	14.25	12.35	2.96	4.43	21.09
Dosto	9.71	44.86	0.03	19.01	5.92	3.95	16.48	23.71
Total	15.72	17.46	11.76	20.00	4.02	4.16	6.25	79.66

Table 6.2: Contribution weights of the variables to clusters along with the contributions of the clusters and the variables to the square data scatter (all per cent).

Table 6.2 shows that the three clusters count for almost 80% of the data scatter. Although the algorithm used tends to design sequential clusters with their contributions decreased, the third cluster has greater contribution than the second due to the local, "greedy" nature of the algorithm. Among the variables, InMon is an obvious leader contributing all its 20% initial weight to the cluster structure. This occurs because the variable is constant in each of the clusters. Another qualitative variable's, Presentat, contribution is only $4.02 + 4.16 + 6.25 = 14.43\%$ of the data scatter, because it has different values for Pushkin's novels. On the other hand, this variable differentiates between Tolstoy and Dostoevski very clearly, and category Thought is characteristic for Dostoevski. Why that does not give higher scores? Because, in this example, we don't consider the categories as independently meaningful elements: it is all three, not each, of them get the weight of a variable, $N = 8$. Thus, for a particular category to get a higher score, it should be standardized differently. For instance, if we do not put factor $1/(\#k - 1)$ in b_v, the weight of category Thought becomes 5, and it becomes 8 if we consider the category as a particular Boolean variable.

The total contribution of the three categories, $1.609 + 1.666 + 2.498 = 5.773$, must be exactly half the Pearson chi-squared coefficient between Presentat and the cluster partition, due to the theory presented in Section 6.1.2. To test this, let us consider the corresponding contingency table:

$$P = \begin{pmatrix} 1 & 2 & 0 \\ 2 & 0 & 0 \\ 0 & 0 & 3 \end{pmatrix}$$

Due to a simple calculation formula (5.6), p. 242, Pearson's coefficient is the sum of the squared contingency table elements divided by the marginals (minus one) multiplied by $N = 8$:

$$X(S/k) = 8[1/(3 \times 3) + 4/(2 \times 3) + 4/(2 \times 3) + 9/(3 \times 3) - 1] = 11.556,$$

thus, $M(S/k) = 5.778$ which matches the contribution value (up to the computation errors). The relative value of $M(S/k)$ (without factor N) is equal to 0.722.

Let us take a look at the value of coefficient $A(S/k)$ emerged in the least-moduli context (see Corollary 6.3.), with S being the author clusters, and k the unique nominal variable Presentat. Since all the categories of Presentat have their frequencies $p_v < 0.5$, only set A_+ is involved in the calculation and, obviously, $A = \{$(Pushkin, Behavior), (Tolstoy, Direct), (Dostoevski, Thought)$\}$. This leads to the relative value of $A(R/k)/N = [(4/8 - 3/8)/(2/8) + (4/8 - 2/8)/(3/8) + (6/8 - 3/8)/(3/8)]/3 = [1/2 + 2/3 + 1]/3 = 0.72$.

When SCC was applied to the Disorders data, the algorithm produced four clusters coinciding with the four mental disorder classes in Table 1.11, p. 39 (in the same order), except that entity 21 was clustered with the fourth class; the same phenomenon has been reported in Mezzich and Solomon (1990), p. 69 - 73, as occurred for complete linkage, ISODATA and K-Means clustering.

For the Iris data set, the algorithm finds sequentially 6 clusters, some of which are parts of the predefined ones while the others are mixed, which is not a wonder since the Iris predefined classes are quite overlapped in the variable space. The Confusion matrix for the 6 SCC discovered clusters and 3 predefined classes is presented in Table 6.3.

| Predefined classes | SCC discovered clusters | | | | | | |
of the Iris data	1	2	3	4	5	6	Total
1		49	1				50
2			12	17	2	19	50
3	26		2	7	15		50
Total	26	49	15	24	17	19	150

Table 6.3: Confusion matrix for the SCC clusters and predefined classes of the Iris data.

The fact that predefined classes 2 and 3 are spread over 4 SCC clusters each confirms the well-known property that they are non compact. To deal with such a situation, we believe, some appropriate new variables have to be produced from the original ones in such a way that the clusters become compact, in that new variable subspace (see Section 6.3.4).

Application of the principal clustering algorithm with the least-maximum criterion gives 7 clusters that are quite similar to the 6 clusters found with the least-squares criterion (see the confusion matrix in Table 6.3).

	Cl-r 1	Cl-r 2	Cl-r 3	Cl-r 4	Cl-r 5	Cl-r 6	Cl-r 7	Total
Class 1	19	31						50
Class 2				1	11	19	19	50
Class 3			27	12	7	1	3	50
Total	19	31	27	13	18	20	22	150

Table 6.4: Confusion (contingency) table between two Iris set partitions: by the classes pre-given and by the clusters found with the least-maximum principal clustering repeated.

6.2.2 How K-Means Parameters Should be Chosen

The user of the K-Means method faces, usually, problems in choosing the following five important kinds of parameter associated with the method:

1) preliminary transformation of the raw data X into matrix Y to be processed;

2) entity-to-center distance $d(x, c)$;

3) centroid concept;

4) number of clusters;

5) initial centers.

Traditionally, the parameters above are considered as completely independent except for the obvious equality of the numbers of clusters and centers. Yet, sometimes the user can know the number of clusters while being uncertain in some or all of the tentative centers.

The bilinear clustering model suggests that there is no independence anymore: the parameters are associated to the criterion for model fitting. The least-squares criterion implies the distance to be Euclidean squared while the centroid, the cluster center of gravity. The least-moduli criterion yields city-block distance and median vector as the centroid. The data standardization is determined by the two principles applied to the data scatter (equal variable contributions, **P1**, and minimality with respect to the origin, **P2**).

The number of clusters along with the tentative centers can be identified with the principal clustering procedure (algorithm SCC). Although the latter suggestion seems rather shaky, the correspondence among the former three items is based on the model. Actually, in a detailed setting presented in Table 6.5, the correspondence concerns the choices of criterion, data scatter, distance, centroid, and scale and shift parameters. The Chebyshev minmax criterion also is included since all the parameters can be derived from it.

Table 6.5 can be used for determining all six of the parameters when the user is able to choose at least one of them. If, for instance, the user prefers that the larger residual is to make larger contribution to the criterion of approximation, she/he should use the least-squares criterion along with all the parameters in its row. If, otherwise, the user prefers medians as the centroids, she/he is restricted, due to the bilinear model, with the least-moduli criterion along with the city-block distance, etc. When the user does not want to take into account the real distribution properties, just concerning the ranges of the variables only, the model dictates using all the parameters related to the minmax criterion.

However, the bilinear model cannot tell which of the fitting criteria to use; what it tells is the correspondence between otherwise independent parameters.

Analogous correspondences can be traced for the qualitative data when treated with the bilinear model. It involves interrelations between choice of the normalizing (scale) parameter, measure of association between subsets/categories, and

Criterion	Data Scatter	Metric	Centroid	Scale Parameter	Shift Parameter
Least Squares	Square	Euclidean	Average	Standard Deviation	Average
Least Moduli	Absolute Value	City-Block	Median	Absolute Deviation	Median
Minmax	Maximal Range	Chebyshev	Midrange	Half-range	Midrange

Table 6.5: Correspondence between parameters of cluster analysis due to the bilinear model.

contingency coefficient as presented in Table 6.6 (which refers to the least-squares fitting criterion only). The second column corresponds to the user-defined option indicating, for every particular nominal variable, what is considered as an 'equally contributing' item: the variable itself (Var) or every of its categories (Cat). Rows of the Table 6.6 correspond to the related parameters. If the user can identify her/his choice for any of them, the others are defined automatically by the bilinear model.

Standardizing Option	Category/ Variable	Normalizing Scale	Set-to-Set Measure	Contingency Coefficient
First	Cat	$\sqrt{p_v}$	RCP=	$X^2(S,k)$
	Var	$\sqrt{p_v(\#k-1)}$	$p(l/m)/p(l)-1$	$M(S,k)$
Second	Cat	1	ACP=	$P(k/S)$
	Var	$\sqrt{1-\sum p_v^2}$	$p(l/m)-p(l)$	$W(k/S)$

Table 6.6: Correspondences among various characteristics of the nominal/categorical data due to the bilinear model.

6.2.3   Discussion

In this section, a further elaboration in using the bilinear model has been described in the context of K-Means (moving-center) clustering. There are two major supplements beyond what the general analysis, in the previous section, suggests.

The first is an extension of the algorithm to the case when the user's knowledge of the situation is more vague than it is assumed usually: the user may know not all the tentative centroids or even none of them; she/he is also allowed to have no idea how many clusters there are and where they are. The sequential extraction of the clusters one by one, starting from the most extreme configurations with the principal clustering, facilitates finding a starting setting for K-Means. This procedure can be considered a model for typology making as a process based on extracting the most unusual, extreme configurations.

The second is a set of links between different clustering parameters emerged due to the bilinear model. There is no connection between choosing the distance measure and the centroid concept, in traditional clustering. There is a connection, in bilinear clustering (see Tables 6.5 and 6.6).

6.3   Contribution-Based Analysis of Partitions

6.3.1   Variable Weights

The concept of the "importance weight" of a variable as a term in an overall additive evaluation measure is not uncommon in data analysis. The bilinear model provides quite a natural set of salience weights, both partition-based and cluster-specific, due to the data scatter decomposition (see Corollary 6.1., 289).

In Table 6.7, the standard point values (means) c_{tk} are presented for the first cluster of Disorders data, along with corresponding relative contribution weights $w(k/t)$ from Corollary 6.1.. The farther c_{tk} is from zero (which is the grand mean here) the easier it is to separate the cluster from the other entities in terms of the variable k, which is reflected in the weight values.

The contribution weights can be employed in various partition analysis and interpretation problems, of which four, (a) concept learning, (b) feature selecting, (c) space transforming, and (d) knowledge discovery, will be considered in the subsequent sections.

Variable	Mean value (original scale)	Mean Value (standardized scale)	Contribution weight (%)
w1	4.64	1.77	5.03
w2	4.64	1.32	2.78
w3	4.73	1.30	2.68
w4	1.73	-1.41	3.18
w5	5.18	3.23	16.67
w6	2.27	-0.98	1.53
w7	0.82	-0.93	1.39
w8	0.09	-2.70	11.70
w9	6.00	3.39	18.34
w10	1.45	-1.59	4.05
w11	1.82	-1.14	2.07
w12	1.00	-1.14	2.07
w13	5.64	3.52	19.85
w14	2.55	-1.00	1.60
w15	3.09	-0.66	0.69
w16	2.36	0.02	0.00
w17	0.73	-2.00	6.40

Table 6.7: Cluster 1 described with 17 psychopathological variables: the central value in the original scale, the central value in the standardized scale, and the relative contribution, per cent.

6.3.2 Approximate Conjunctive Concepts

There exist many systems for learning a logical description of a subset of the entities (for references, see Fisher 1987, Michalski 1992, Wnek and Michalski 1994). The relative contribution weights introduced above give yet another way to find approximate conjunctive descriptions for every cluster independently. Here, only square scatter contribution weights will be considered; the absolute scatter contributions are treated analogously.

Based on the right column in Table 6.7, let us pick consecutively the features which contribute the most to cluster 1 in the Disorders data to form a conjunctive conceptual description of the cluster. Initially, let us take the range of the most salient variable, $w13$ (contribution 19.85%), within the cluster 1: it is interval $[5, 6]$, the boundary points included. Conceptual description $W : 5 \leq w13 \leq 6$ covers all 11 individuals belonging to class 1; however, there is one case which is false positive: individual 32 from class 3 satisfies condition W. This relates to what could be called "precision error", PE, of the concept W with regard to a class $S \subset I$, which is defined as the number of the elements from outside S satisfying W, related to the general number of the entities outside S (proportion of the false positives). To decrease $PE(W) = 1/33$, let us pick the next most contributing variable, w9 (contribution 18.34%), and consider the conjunctive concept formed by the within-cluster ranges of both, w13 and w9: $W : 5 \leq w13 \leq 6 \& w9 = 6$. Obviously, precision error of this combined category equals zero. Moreover, it is easy to see that the first term is not necessary; concept $W : w9 = 6$ corresponds to all 11 individuals from class 1 and no one else. This is an example of the situation when a less contributing variable (w9) gives a better conceptual description, which reflects the fact that the statistics-based contribution weights detect the tendencies of the logical relations rather than the exact patterns of them.

Let us describe a general algorithm for approximational conjunctive conceptual description $W(S)$ of a class $S \subset I$ (the data matrix is assumed square scatter standardized). The degree of approximation is characterized by the precision error $PE(W(S))$ which should be made less than a user-specified value ϵ. Another stopping criterion involved bounds the number of conjunctive terms in the concept $W(S)$ by a user-defined integer n.

ACCL (Approximate Conjunctive Concept Learning)

Step 0. Find the means, c_k, of the variables $k \in K$ within cluster S and consider list L of the variables ordered by decreasing their contribution weights, c_k^2. Let conjunction $W(S)$ be empty and $PE = 1$.

Step 1. Remove the first variable, x_k, from list L and consider combined concept $W' = W(S) \& m_{Sk} \leq x_k \leq M_{Sk}$ where m_{Sk} and M_{Sk} are minimum and maximum of x_k within S, respectively. Compute $PE(W')$.

Step 2. Take every conjunctive term W of W' (in the order of its joining to $W(S)$) and consider conjunctive concept W'' which is equal to W' with W removed. If $PE(W'') = PE(W')$, put $W' \leftarrow W''$ and begin Step 2 again. If $PE(W'') > PE(W')$ for every conjunctive term removed, check whether the number of the conjunctive terms in W' is greater than n or not. If not, define $W(S) \leftarrow W'$; if yes, take the conjunction from the preceding iteration as $W(S)$ and end.

Step 3. If $PE(W(S)) \leq \epsilon$ or $L = \emptyset$, end; otherwise, go to Step 1.

Let us illustrate the algorithm by applying it to class 3 of the Disorders data based on the variable weights presented in Table 6.8. Let $\epsilon = 1/33$, thus admitting no more than one other individual covered by the conjunctive concept sought. The maximum weight variable with regard to class 3 is w16 (contribution 27.43%). Its within-cluster range $W : 4 \leq w16 \leq 6$ covers 5 individuals in the other classes (one in class 4 and four in class 1), which makes $PE(W) = 5/33$. Adding the within-cluster range of the next mostly contributing variable, w8 (contribution 16.73%), we have $W : 4 \leq w16 \leq 6 \& 0 \leq w8 \leq 1$, which makes $PE(W) = 4/33$ since the previously covered individual from class 4, 44, has value w8=5 and does not satisfy the combined condition. Variable w16 cannot be removed from the concept (Step 2) since this makes precision error grow. Then, considering each of the next mostly contributing variables, w17, w3, w6, and w10, we can see that adding of none of them can decrease $PE(W)$. For example, the within-cluster range of w10 is $[0, 3]$ which is compatible with values of w10 for all the four individuals, 2, 9, 10, and 11, from class 1, satisfying W. However, the next contributing variable, w2, has its within-cluster range, $[0, 4]$, incompatible with the values of w2 for individuals 2, 10, and 11, which makes the concept $W : 4 \leq w16 \leq 6$ & $0 \leq w8 \leq 1$ & $0 \leq w2 \leq 4$ have $PE(W) = 1/33$. Moreover, w8 now can be removed from the concept at Step 2 of the algorithm ACCL. This leads to concept $W : 4 \leq w16 \leq 6$ & $0 \leq w2 \leq 4$ as a solution to the problem. Subsequent adding w9 (or w13) to W may reduce $PE(W)$ to zero; however, no two-variable conjunctive concept can have $PE(W)$ less than 1/33, for class 3.

Variable	Class 1	Class 2	Class 3	Class 4	General, $v(k)$
w1	5.03	6.71	0.29	0.23	2.44
w2	2.78	3.02	4.92	3.08	2.27
w3	2.68	19.31	9.59	0.02	5.41
w4	3.18	0.00	0.71	1.90	1.09
w5	16.67	3.68	1.32	3.21	5.06
w6	1.53	4.18	5.80	2.14	2.18
w7	1.39	1.92	2.22	2.14	1.27
w8	11.70	15.71	16.73	12.41	9.54
w9	18.34	5.11	0.77	3.66	5.70
w10	4.05	1.59	5.62	9.75	3.39
w11	2.07	0.99	3.65	18.78	3.97
w12	2.07	1.22	1.32	14.26	3.03
w13	19.85	7.05	0.18	8.09	7.00
w14	1.60	0.36	3.95	6.71	1.94
w15	0.69	0.98	0.91	7.82	1.64
w16	0.00	9.53	27.43	2.18	5.57
w17	6.40	18.64	14.61	3.61	7.29

Table 6.8: Relative contributions of 17 psychopathological variables to the original classes (columns 1 to 4) and to the data scatter according to the partition structure.

With $n = 2$, that is, with only two conjunctive terms permitted, the approximate conjunctive concepts found for all the clusterings in the data sets considered can be presented in Table 6.9.

	Disorder		Iris		Masterpieces	
	SCC clusters	Predefined classes	SCC clusters	Predefined classes	SCC clusters	Predefined classes
Number	4	4	6	3	4	3
Mean PE	0.01	0.01	0.06	0.11	0.00	0.00

Table 6.9: Mean precision error over all clusters in each of the six clusterings considered.

In the Iris data set, predefined classes can be described by the following concepts: $1 \leq w3 \leq 1.9$ (class 1, PE=0), $2 \leq w2 \leq 3.4$ & $3.3 \leq w3 \leq 5.1$ (class 2, PE=0.15), and $1.4 \leq w4 \leq 2.5$ & $4.5 \leq w3 \leq 6.9$ (class 3, PE=0.18). The relatively high level of precision error for two of the classes supports the conclusion that they are dispersed in the variable space.

Masterpiece classes are entirely separated with the following concepts: "InMon=No" (for Pushkin's novels), "LenDialogue $\geq$ 118.6" (for Dostoevski's), and "Nchar $\geq$ 4 & Presentat=Direct" (for Tolstoy's).

Let us now try to find conceptual descriptions of the Digit classes found by the Confusion table and presented in Table 5.7, p. 267. The four-class corrected partition of the integer digits found is $S = \{1 - 4 - 7, 3 - 5 - 9, 6 - 8 - 0, 2\}$. In the Digit data, the most contributing variables to the clusters are e7 and e1 (cluster 1), e5 and e7 (cluster 2), e5 and e2 (cluster 3), and e6 (cluster 4). It appears, the clusters are described, with no error, by the following conjunctive concepts: e7=0, e5=0 & e7=1, e5=1 & e2=1, and e6=0, respectively. Perhaps, this can be interpreted as an indication of the most confusing digit segments.

6.3.3 Selecting the Variables

The problem of reducing the space dimensionality through selecting a subset of the most informative variables (features) has attracted a considerable effort (for the latest references, see John, Kohavi & Pfleger 1994, Aho & Bankert 1995). Feature (variable) selection algorithms for learning a partition involve two major components: an evaluation function, which evaluates performance of every particular feature subset, and a search algorithm, which searches in the space of feature subsets. We'll focus on the search algorithms based on the so-called backward selection strategy starting with all the variables available and repeatedly removing some of them; usually, only one feature is removed at each step, which is not the case here because we use some "intermediate" selection rules (as PWS, CWS, and ACS introduced below). As to the evaluation functions, the following three kinds can be distinguished:

(1) numerical evaluation of the relationship between the variables selected and the classification to be learned;

(2) quality of a classifier involving the variables selected ("wrapper" models, Aho & Bankert 1995); in our case, the set of approximate conjunctive concepts designed for the clusters can stand as the classifier while the (average) precision error can be used as its evaluation function;

(3) evaluation of the difference between a clustering found with a clustering algorithm (SCC, in our case) and the classification to be learned.

The latter evaluation seems the hardest since if a feature subset is good in terms of an evaluation function of the third kind, it must be good in terms of the other kinds of evaluation functions, because the clustering algorithm employed can be considered both as a particular classifier itself and a device for revealing relationships between the feature subset and the clustering. This is why we prefer to use an evaluation function of the third kind. To measure the degree of difference between two partitions, there are quite a number of measures developed in the literature as described in Section 5.1.4. Here, it will be sufficient to take just the number (or proportion) of instances shifted from their "original" classes as the evaluation function (referred to as "shift error") to compare a partition with that one found at the preceding step. For example, shift error of SCC clustering of Iris data is 56/150, by Table 6.3, since there are only 26+49+19=94 entities unmoved, in the predefined classification.

A contribution-based searching algorithm can be suggested as consisting of one or several sequential search iterations. Each iteration starts with a feature (variable) set as its input, and it consists of the following three steps: (a) selection of a subset of the variables; (b) performing SCC clustering with the subset selected; (c) evaluation of the partition found. Depending on the evaluation result, the process ends, or the next iteration is performed (with the parameters of the selection procedure (a) changed). We suggest three version of the selection procedure depending on which information about the variables is employed:

Partition-Based Weight Selection (PWS)
Let us order the variables according to their partition-based contribution weights, $v(k)$, and select a number of the most contributing variables. The number can be determined with one of the following parameters: (a) a user-defined threshold number of the variables p; (b) a threshold of relative weight of each of the variables, t; (c) a threshold of the cumulative relative weight of the subset taken, ct.

For example, set of the Disorder variables, $\{w8, w17, w13, w9, w16, w3, w5\}$, would be selected if either $p = 7$ or $t = 5\%$ or $ct = 45\%$ (see the right column in Table 6.8).

Cluster-Specific Weights Selection (CWS)

For each cluster t, order the variables according to their cluster-specific relative contribution weights, $v(k/t)$, and select a subset of the most contributing variables from each of the orders. All the three kinds of parameters above are relevant in this. Then take the union of the subsets selected.

For example, subset $\{w3, w5, w8, w9, w11, w13, w16, w17\}$ is selected with threshold $t = 15\%$, according to Table 6.8.

Approximate Concept Based Selection (ACS)

Take all the variables occurring in the approximate conjunctive concepts derived with algorithm ACCL for the classification considered. The parameters here are the parameters of ACCL: the maximum number of conjunctive terms, n, and the maximum precision error, PE.

For example, subset $\{w2, w9, w11, w16, w17\}$ corresponds to the conjunctive concepts found for four Disorder classes (up to $PE = 1/33$) in Section 6.3.2.

Some results of the selection procedures above applied to SCC discovered clusters for each of the data sets considered are presented in Table 6.10.

Curiously, PWS results in the first row of Table 6.10, though departing from SCC clustering, approximate the classifications predefined: for Disorder data, the 10-variable subset gives SCC clustering coinciding with the four-class partition predefined. Selection of the variables for learning the predefined classifications on Disorder and Masterpieces data leads to similar results (see Table 6.11). Table 6.11 shows that the situation is quite different for Iris data where the predefined classes are not "compact" in the feature space.

6.3.4   Transforming the Variable Space

To improve performance of the contribution-based method in Iris-like situations, the feature set should be transformed. The rules suggested above for selection of the variables can be employed as a vehicle for transforming the variable space when combined with a special device, a generator of the variables, a subroutine producing new variables from the original ones. This can be considered what is called, in machine learning, a hypothesis-driven (actually, cluster-driven) constructive induction system (see Wnek & Michalski 1994 for references and review). Let us consider a generator of the variables that works iteratively; at the input of each iteration, a set of variables is considered available; then all pair-wise sums, differences, products and ratios of the variables (that is, columns of the data matrix) are

Selection procedure		Disorder, 4 SCC clusters	Iris, 6 SCC clusters
PWS	Parameter	$t = 3\%$	$p = 3$
	Subset	w8,w17,w13,w9,w16, w3,w5,w11,w10,w12	w3,w4,w1
	Shift error	1/44	39/150
PWS	Parameter	$t = 5\%$	
	Subset	w8,w17,w13,w9, w16,w3,w5	
	Shift error	2/44	
CWS	Parameter	$t = 10\%$	$t = 35\%$
	Subset	w3,w5,w8,w9,w11, w12,w13,w16,w17	w1,w2,w4
	Shift error	1/44	35/150
CWS	Parameter	$t = 15\%$	
	Subset	w3,w5,w8,w9, w11,w13,w16,w17	
	Shift error	2/44	
ACS	Parameter	$n = 2$	$n = 2$
	Subset	w1,w2,w9, w11,w16,w17	w1,w2,w3
	Shift error	2/44	35/150

Table 6.10: Results of selection procedures applied to SCC clusterings.

calculated; either the PWS or CWS or ACS procedure is applied to find a subset of the most contributing variables to be left for the next iteration.

Two iterations of this method have been applied to the set of three Iris variables, w2, w3, and w4 (since w1 does not participate in the approximate conjunctive concepts for the predefined classification). The results are presented in Table 6.12.

In Table 6.12, the best SCC clustering is based on the four ACS variables; it differs from the predefined classification by only 8 instances of class 3 joined to class 2. The variables are taken from the approximate conjunctive concepts made with $n = 2$: $w2/w3 \geq 1.77$ (class 1, PE=0), $0.33 \leq w2/(w3 * w4) \leq 0.76$ & $9.9 \leq w3 * w3 * w4 \leq 42.5$ (class 2, PE=0.02), and $7.5 \leq w3 * w4 \leq 15.87$ & $0.38 \leq w2/w3 \leq 0.63$ (class 3, PE=0.02).

In general, the latter three tables confirm the hypothesis that the three selection criteria considered must have different sensitivity and, thus, different performance: the best is the concept-driven selection procedure ACS while the partition-weight-

Selection procedure		Disorder, 4 classes	Iris, 3 classes
PWS	Parameter	$t = 3\%$	$p = 3$
	Subset	w8,w17,w13,w9, w16,w3,w5	w3,w4,w1
	Shift error	1/44	68/150
CWS	Parameter	$t = 15\%$	$p = 3$
	Subset	w3,w9,w11,w12, w13,w16,w17	w2,w3,w4
	Shift error	0	50/150
ACS	Parameter	$n = 2$	$n = 2$
	Subset	w2,w9,w11, w16,w17	w2,w3,w4
	Shift error	1/44	50/150

Table 6.11: Results of selection procedures applied to predefined classifications.

driven procedure PWS is the worst.

6.3.5   Knowledge Discovery

In our context, knowledge is a set of statements about the variables we deal with (for a review, see Frawley, Piatetsky-Shapiro, and Matheus 1992). The quantitative form of knowledge includes quantitative equations like a physical law, $y = F(x_1, ..., x_n)$, where $y, x_1, ..., x_n$ are the variables and F is an explicitly defined function. The qualitative form of knowledge is based upon logical rules, $A \rightarrow B$, (sometimes called productions) where A and B are some logical combinations of the variable categories.

Let us focus on the issues connected with derivation of the production rules. Any production, $A \rightarrow B$, can be considered an intensional statement which can be extensionally presented through a relation between two sets of the objects, S_A and S_B, satisfying logical conditions A and B, respectively. The set relation corresponding to implication $A \rightarrow B$ is nothing but inclusion $S_A \subseteq S_B$. Thus, the process of knowledge discovery in the form of productions may be thought of as a process of generating logical expressions A and B in such a way that $S_A \subseteq S_B$ holds. Since the empirical data often contain very superficial (having no explicit theoretical meaning) variables, it is appropriate to admit error in the relation $S_A \subseteq S_B$. The degree of exactness of the relation $S_A \subseteq S_B$ and implication $A \rightarrow B$ can be measured by the proportion of elements of S_A belonging to S_B,

PWS	Variables	$w2 - w3, w3 + w4, w3, w2/w3, w4, w2 - w4, w3 * w4$
	Number of clusters	4
	Shift error	34
CWS	Variables	$w2, w2/w3, w2/(w3 * w4), w2/w4,$
		$w2 - w4, w3 * w3 * w4, w3 * w4$
	Number of clusters	4
	Shift error	31
ACS	Variables	$w2/w3, w2/(w3 * w4), w3 * w3 * w4, w3 * w4$
	Number of clusters	3
	Shift error	8

Table 6.12: Performance of the procedures PWS, CWS, and ACS in constructive induction for learning the predefined classification of the Iris data.

$p(B/A) = |S_A \cap S_B|/|S_A|$. This is, actually, the (empirical) probability of B under condition A (conditional probability). The error of implication $A \rightarrow B$ is the complement of that to unity, $e(B/A) = 1 - p(B/A)$. Sometimes, knowledge is presented as the logical equivalence, $A \leftrightarrow B$, which is a conjunction of two implications, $A \rightarrow B$ and $B \rightarrow A$. The exactness of such a statement can be measured with the Jaccard match/mismatch coefficient based on $r(B, A) = |S_A \cap S_B|/|S_A \cup S_B|$ rather than with two coefficients $p(B/A)$ and $p(A/B)$ accompanying the constituent implications.

When the data set is not small so that there are hundreds of entities and dozens of variables, the number of potentially available expressions A and B whose extensions satisfy (or approximately satisfy) inclusion $S_A \subseteq S_B$ may become quite large. As experience shows, quite a number of them cannot be associated with any causal or other relations between the features of phenomena in question and, in this aspect, appear meaningless. Testing meaningfulness of automatically generated expressions is a hard, still unsolved, problem of artificial intelligence. It could be helpful if a strategy could be suggested leading to meaningful expressions only. A simple strategy of this kind is considered below based on the following two major restrictions:

(1) expressions A are generated using the variables belonging to a subset V_1 while another, nonoverlapping variable subset V_2 is used for generating Bs: the underlying assumption is that the user can specify a pair of subsets of the variables that are hypothetically causally connected;

(2) the logical expressions relate to the clusters of the entities found in V_1 or V_2 space.

The first restriction refers to meaningfulness of the relations to seek for while the second to that of the expressions A and B themselves. For example, a sociologist may say that demographic status of a person (variables V_1: sex, age, marital status, etc.) influences her/his vacation time habits (V_2: preferred activities). Then, a cluster of "aged women" may appear in a sample collected, which may be a subset of the individuals preferring traveling activity. This yields the production "Aged women prefer traveling as their vacation activity" discovered as a part of the general knowledge about the population.

With these restrictions made, the following strategy involving the approximate conjunctive concepts discussed can be suggested for knowledge discovery.

Production Discovery Algorithm

Step 1. Choose subsets of the variables, V_1 and V_2.

Step 2. Partition the entity set, I, with a clustering algorithm in the variable space V_1.

Step 3. For every cluster S_t, find corresponding approximate conjunctive concept A_t, in space V_1, and B_t, in space V_2.

Step 4. Evaluate exactness of each of the implications found, $A_t \to B_t$, $t = 1, ..., m$ and exclude those having degree of exactness less than a threshold, the rest represents the productions that express assumed relationship $V_1 \to V_2$. If the rest is empty, the hypothesis on that relationship is considered rejected (with respect to the data analyzed).

By construction of the concepts A_t and B_t, both subsets S_{A_t} and S_{B_t} (of the entities corresponding to concept A_t or B_t, respectively) include S_t. However, the fact that clustering is done by variables V_1 guarantees that conceptual description of the clusters in terms of V_1 will be more precise than that in terms of V_2. This yields that the subsets S_{A_t}, in general, will be "smaller" than S_{B_t}, thus providing a good exactness of the productions $A_t \to B_t$.

The method described can be adjusted to the situation when the partition is pregiven. In this case, at Step 2, transforming the variable spaces V_1 and V_2 must be done as described in the previous section.

Let us choose a subset of four Disorders variables characterizing the disorders most decisively as subset $V_1 = \{w9, w11, w16, w17\}$. Subset $V_2 = \{w1, w2, w5, w15\}$ is said to relate to the feelings of the patients. Let us apply the method above to derive what kind of feelings is implied by a particular mood, based on Disorders data along with the four classes of the mental disorders. Table 6.13 presents simplest concepts related to the classes found with the variables of V_1 (A concept) and V_2 (B concept). It can be seen that the latter group of variables admits rather large precision errors (amounting to 19, in class 3), which does not prevent the productions found, $A \to B$, from being quite exact.

As another example, let us turn to the problem of revealing interrelations between two

Cl	Concept A	PE	Concept B	PE	Production $A \to B$	Error
1	$w9 = 6$	0	$w1 \geq 3 \& w5 \geq 2$	6	$w9 = 6 \to$ $w1 \geq 3 \& w5 \geq 2$	0
2	$w17 = 6$	0	$w1 \leq 5 \& w5 \leq 1$	15	$w17 = 6 \to$ $w1 \leq 5 \& w5 \leq 1$	0
3	$4 \leq w16 \leq 6$	5	$w2 \leq 4$	19	$4 \leq w16 \leq 6 \to w2 \leq 4$	3
4	$w11 \geq 5$	2	$w15 \geq 4$	14	$w11 \geq 5 \to w15 \geq 4$	0

Table 6.13: Concepts and productions discovered for variable subsets V_1 and V_2 based on the original classes in Disorders data.

aspects of the author's style raised while discussing the Masterpieces data in Section 1.4.3: what is the relationship between linguistics ($V_1 = \{$LenSent, LenDial$\}$) and presentation ($V_2 = \{$NChar, InMon, Presentat$\}$)? We can see quite easily, that the author classes can be exactly characterized, in V_1, just by the intervals of LenDial: Pushkin (≤ 16.6), Dostoevski (≥ 118.6), and Tolstoy (between 30 and 58). In terms of V_2, the classes are exactly characterized by "InMon=No" (Pushkin), "InMon =Yes & Presentat=Thought" (Dostoevski), and "InMon =Yes & Presentat=Direct" (Tolstoy). Since all the errors are zero, we have found that each of the author clusters leads to a particular logical equivalence, as, for instance, "LenDial $\leq 16.6 \leftrightarrow$ InMon=No" (Pushkin). Up to the illustrative nature of the data, the knowledge found may serve as a base to look for supplementary perhaps more deep author's features to explain the equivalences.

6.3.6 Discussion

Yet another feature of the bilinear clustering model, variable contribution weights, has been employed, in this section, to develop various learning/interpreting tools for multivariate classification.

1. The cluster-specific weights derived may present a better tool for describing the clusters by conjunctive concepts than the partition-based weights, as the examples suggest. However, the weight as the difference (squared) between the within-cluster and the grand mean is a statistical concept related to the "majority", not all the objects in question. Thus, a less contributing variable may have a narrower interval to cover the cluster than a more contributing one, just because there is an entity in the cluster which is an outlier by the latter variable. Therefore, the weight should be utilized to preliminarily order the variables to be checked quite thoroughly with the precision error values.

2. In the problem of selecting the most informative variable subspace, we consider the problem of evaluation of the subspace as of crucial importance. Usually, evaluation is made based on the quality of a discrimination rule in the subspace: the better discrimination, the better subspace. This is a good principle when the subspace is sought for discriminating. However, we deal mostly with the clustering problems; this is why we suggest using the quality of a clustering, not discriminating, procedure for evaluating a variable subspace (the K-Means extended is taken as that clustering procedure). Again, the weights are used as an intermediate facility to make the variable selection process easier.

3. The variable selection procedures are suggested for use in transforming the variable space to make the supervised classes coherent in the space. To do so, the variables are propagated with simple arithmetical operations to allow for selecting the best in the breed. In the literature, there are much more sophisticated methods described. However the simple machinery suggested may work well too, as demonstrated with adjusting the variable space to the dispersed Iris classes.

4. Knowledge discovery from data is a hot subject in artificial intelligence. In the present author's opinion, the main problems in that are related to "meaningfulness" of the propositions generated, not to the generation itself. The classification context employed allows us to propose two principles leading to meaningful rule generating:

   1) the subject and predicate of the rule are sought within different subspaces that must be chosen by the user (restricting the variables);

   2) the extensional content of the proposition must be concentrated around homogeneous clusters found in the "subject" subspace (restricting the entities).

   However, all this seems quite speculative since there is no underlying theory of "meaningfulness" suggested.

6.4 Partitioning in Aggregable Tables

6.4.1 Row/Column Partitioning

There are two distinctive features of bilinear modeling for contingency data contrasting that from the case of processing of the general entity-to-variable tables: 1) the data entries are transformed into the flow indices that are referred to as RCP (relative change of probability) coefficients when the data relate to co-occurrences; the flow indices, not the original data, are employed as the inputs to the bilinear modeling; 2) the only fitting criterion used is the weighted least squares, with the weights being products of the marginal frequencies (probabilities).

Methods for row or column partitioning of contingency data based on the correspondence analysis constructions already have been considered in the literature. Two of heuristic clustering approaches were developed in detail: agglomerative clustering (see Jambu 1978, Greenacre 1988) and the moving centers (K-Means) method (Diday et al. 1979, Govaert 1980); these methods involve the chi-squared metric as the distance measure to reveal structural similarities between the rows.

In more precise terms, let us consider the problem of partitioning the rows of a contingency table $P = (p_{ij})$, $i \in I$, $j \in J$; for the sake of simplicity, the data are considered after they have been divided by the total flow p_{++}, which means that $p_{++} = 1$. Let $S = \{S_1, ..., S_m\}$ be a partition of the row set I accompanied with the cluster weights $p(t) = \sum_{i \in S_t} p_{i+}$ and centroid profiles $g_t = \{(p_{tj}/p(t)) : j \in J\}$ where $p_{tj} = \sum_{i \in S_t} p_{ij}$ ($t = 1, ..., m$). Then, at an agglomeration step, the Ward-like function

$$D(u, t) = (p(u)p(t)/(p(u) + p(t)))\chi^2(g_u, g_t) \qquad (6.13)$$

is considered to choose such a pair (u, t) for merging that minimizes $D(u, t)$ by $u, t = 1, ..., m, u \neq t$.

The chi-squared distance $d^2(g_s, g_t)$ is defined as in Section 2.2.3:

$$\chi^2(g_u, g_t) = \sum_{j \in J} p_{+j}(p_{uj}/p(u) - p_{tj}/p(t))^2. \qquad (6.14)$$

After clusters S_u and S_t are merged into cluster $S_u \cup S_t$, its weight and centroid profile can be calculated based on those for the merged clusters: $p(u \cup t) = p(u) + p(t)$, and $g_{u \cup t} = \{(p_{uj} + p_{tj})/p(u \cup t) : j \in J\}$.

As to the correspondence-analysis-wise K-Means method, it is completely defined with indication that the centroids are defined as $c(S_t) = g_t$ while the chi-squared distance between the profiles,

$$\chi^2(i, g_t) = \sum_{j \in J} p_{+j}(p_{ij}/p_{i+} - p_{tj}/p(t))^2,$$

is considered as the distance between a row-entity $i \in I$ and centroid g_t.

It appears that both of the methods can be considered as the algorithms of local optimization which fit a corresponding form of the bilinear model.

Let us remind the reader that the model represents a set of equations

$$q_{ij} = \sum_{h \in H} \mu_h v_{ih} w_{jh} + e_{ij} \qquad (6.15)$$

where $q_{ij} = p_{ij}/p_{i+}p_{+j} - 1$ are the flow indices, or RCP values for the original data $P = (p_{ij})$, $h \in H$ is an index of a cluster represented by two Boolean vectors, $v_h = (v_{ih})$ and $w_h = (w_{jh})$, corresponding to a subset of rows, $V_h = \{i : v_{ih} = 1\}$, and of columns, $W_h = \{j : w_{jh} = 1\}$, respectively, and μ_h is an intensity weight of the box-cluster $V_h \times W_h$. Values e_{ij} are, as usual, the residuals to be minimized, this time, with the weighted least-squares criterion:

$$L^2 = \sum_{i \in I} \sum_{j \in J} p_{i+}p_{+j}e_{ij}^2. \tag{6.16}$$

As it was proved in Section 4.5.4, the criterion provides the optimal μ_h to be equal to the flow index values calculated for the boxes:

$$\mu_h = q_{V_h W_h} = \frac{p_{V_h W_h}}{p_{V_h+}p_{+W_h}} - 1 \tag{6.17}$$

where $p_{V_h W_h} = \sum_{i \in V_h} \sum_{j \in W_h} p_{ij}$ is the aggregate flow (frequency) and the marginal aggregate flows (frequencies) are defined correspondingly.

To specify the model (6.15)–(6.16) for row partitioning, let us consider an m-class partition S on I and the trivial $|J|$-class singleton partition on J. The model involves boxes $S_t \times \{j\}$ and their intensities μ_{tj}. The equations (6.15), in this case, can be expressed as

$$q_{ij} = \sum_t \mu_{tj} v_{it} w_{jj} + e_{ij},$$

where $w_{jj} = 1$ for all $j \in J$. Substituting this into (6.16), one can have the following equivalent form of the model: minimize

$$L^2 = \sum_{i,j} p_i p_j (q_{ij} - \sum_t \mu_{tj} v_{it})^2 \tag{6.18}$$

over arbitrary μ_{tj} and Boolean v_{it}. No boxes here could overlap; thus, the optimal μ_{tj} are equal to corresponding flow index values: $\mu_{tj} = q_{tj} = p_{tj}/(p(t)p_{+j})$.

The following two statements put the methods discussed in the bilinear model framework.

Statement 6.5. *The agglomerative minimization algorithm for criterion (6.18) is equivalent to the agglomeration clustering algorithm based on function (6.13).*

Proof: Let us analyze the difference, D, between values of L^2 (6.18) for a partition S and for partition $S(u, t)$ obtained by merging its classes S_u and S_t. Obviously, the difference depends only on the changed classes:

$$D = \sum_{j \in J} [p(u)p_{+j}(p_{uj}/(p(u)p_{+j}) - 1)^2 + p(t)p_{+j}(p_{tj}/(p(t)p_{+j}) - 1)^2 -$$

$$(p(u) + p(t))p_{+j}((p_{uj} + p_{tj})/((p(u) + p(t))p_{+j}) - 1)^2].$$

After elementary transformations, this expression becomes

$$D = (p_{uj} + p_{tj})^2/((p(u) + p(t))p_{+j}) - p_{uj}^2/(p(u)p_{+j}) - p_{tj}^2/(p(t)p_{+j}) =$$

$p(u)p(t)/(p(u) + p(t)))d^2(g_u, g_t)$ which equals $D(u,t)$ in (6.13). □

The alternating minimization iterations involve the following groups of variables: μ_{tj}-values (real) and v_{it} values (Boolean).

Statement 6.6. *The alternating minimization procedure for criterion (6.18) is equivalent to the correspondence-analysis-wise K-Means algorithm.*

Proof: Let us consider the contents of an iteration in the alternating minimization algorithm. For a fixed partition $S = \{S_1, ..., S_m\}$ (v_{it}-values) the optimal μ_{tj} are q_{tj}. Then, with these μ_{tj} values the criterion in (6.18) can be rewritten in the following form (taking into account that the sought values v_{it} correspond to a partition S):

$$L^2 = \sum_{i,j} p_{i+}p_{+j}(q_{ij} - \sum_t \mu_{tj}v_{it})^2 = \sum_{t=1}^{m}\sum_{i \in S_t}\sum_j p_{i+}p_{+j}(q_{ij} - q_{tj})^2$$

The following easy-to-prove equality, $\sum_j p_ip_j(q_{ij} - q_{tj})^2 = p_{i+}\chi^2(i, g_t)$, leads to $L^2 = \sum_{t=1}^{m}\sum_{i \in S_t} p_{i+}\chi^2(i, g_t)$ which is obviously minimized by the minimal distance rule (in the moving-center algorithm). □

6.4.2 Bipartitioning

Let us extend the clustering techniques considered to the problem of bipartitioning, that is, simultaneous partitioning of both, row and column, sets.

The problem of bipartitioning is as follows: find a partition $S = \{S_1, ..., S_m\}$ of row set I and a partition $T = \{T_1, ..., T_l\}$ of column set J to present the data table structure in a best way. The bilinear clustering model for bipartitioning can be stated as the problem of finding $|I|$-dimensional mutually orthogonal Boolean vectors $v_1, ..., v_m$ and $|J|$-dimensional mutually orthogonal Boolean vectors $w_1, ..., w_l$ as well as real values μ_{tu} $(t = 1, ..., m; u = 1, ..., l)$ satisfying equations

$$q_{ij} = \sum_{t=1}^{m}\sum_{u=1}^{l} \mu_{tu}v_{it}w_{ju} + e_{ij} \qquad (6.19)$$

and minimizing the weighted least-squares criterion L^2 in (6.15).

Here, sought Boolean vectors correspond to the sought classes of partitions S and T as their indicator functions. The classes are $S_t = \{i : v_{it} = 1\}$, $t = 1, ..., m$, and $T_u = \{j : w_{ju} = 1\}$, $u = 1, ..., l$. The equations (6.19) can be considered as a special case of the model (6.15) with the box indices $h = (t, u)$. Since boxes $S_t \times T_u$ are not overlapping, the optimal values of μ_{tu} are equal to q_{tu}. The aggregate $m \times l$ class-to-class contingency data table $P(S, T)$ has entries $p_{tu} = \sum_{i \in S_t} \sum_{j \in T_u} p_{ij}$. The original data table can be considered as $P = P(I, J)$ corresponding to the trivial partitions of I and J with the singletons as their classes.

It can be proved the same way as in Statement 6.5. that the weighted chi-squared row-distance, $D(t, t') = (p_{t+}p_{t'+})/(p_{t+}+p_{t'+})\chi^2(t, t')$ (applied to the table $P(S, T)$), equals the increment of criterion L^2 after merging two S-classes S_t and $S_{t'}$. Dually, the weighted chi-square column-distance $D(u, u') = ((p_{+u}p_{+u'})/(p_{+u}+p_{+u'}))\chi^2(u, u')$ equals the increment of criterion L^2 after merging T-classes T_u and $T_{u'}$.

Let us describe the local search algorithms for fitting the model (6.19) with criterion (6.15) to modify both the agglomerative clustering and moving center methods for the bipartitioning problem.

Agglomerative Bipartitioning
Each iteration of the algorithm consists of the agglomeration step applied to the rows or to the columns. Finding minimal values of $D(t, t')$ (with regard to all $t \neq t'$) and $D(u, u')$ (by all $u \neq u'$) and then determining which of them is the least, the decision is made on the following points: a) what kind of classes, rows or columns, must be merged at the ongoing iteration; b) the pair of the classes to merge.

Evidently, after $|I| + |J| - 4$ iterations a 2×2 matrix $P(S, T)$ will be obtained: one-class partition S or T cannot be reached before since such a one-class partition implies $X^2(S, T) = 0$, which contradicts the "minimality of the increment" rule of the merging procedure. The whole process can be represented by two dendrograms (one for I, the other for J) with level values L^2 assigned to the internal nodes of both of the partitions to show the differences in the part of data explained at each step. An important feature is that these level values are measured in the same scale for both of the dendrograms. Each of the internal nodes corresponds to a table $P(S, T)$ generated by corresponding partitions S and T, and its level value is equal to L^2. All the leaves (the singleton classes) have zero as their level value while the root (corresponding to all the elements united) is assigned the initial value $X^2(I, J)$ as its level value.

In the alternating minimization approach, we deal with two partitions, S (of rows) and T (of columns), both having pre-fixed numbers of classes.

There are three groups of variables involved: v_{it} (corresponding to S), w_{ju} (corresponding to T), and μ_{tu}; however, the last group can be considered as having no independent meaning: the optimal μ_{tu} are functions q_{tu} of the aggregate table

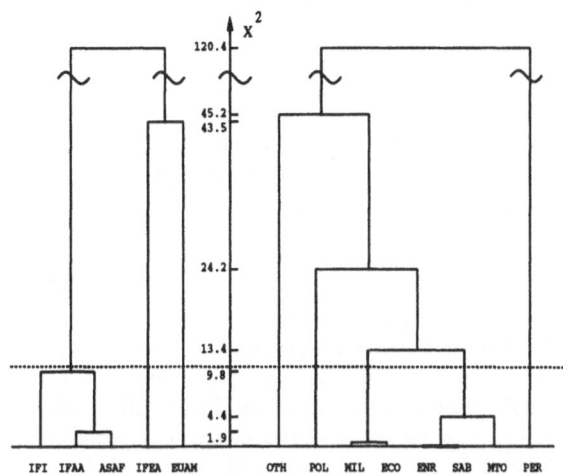

Figure 6.1: Hierarchical biclustering results for the Worries data.

$P(S, T)$ corresponding to the partitions.

> **Alternating Minimization Bipartitioning**
> Each iteration consists of two moving-center steps applied in succession to the rows (v-variables) and to the columns (w-variables). The iteration starts with the following elements fixed: (a) partitions S and T, (b) aggregate $m \times l$ contingency table $P(S, T)$ with the entries p_{tu}, and (c) the coefficients $\mu_{tu} = q_{tu}$ ($t = 1, ..., m; u = 1, ..., l$).
> The first step of the iteration: finding a suboptimal m-class row-partition by the table $P(I, T)$ using the ordinary alternating minimization (moving center) method, starting with S.
> The second step: for the found S, the table $P(S, J)$ is considered, and the moving center method is applied again, this time to the set of columns, starting from T.

Each step the value of the criterion is decreased by the nature of the algorithm. So the algorithm converges since the set of all pairs of the partitions is finite.

Bipartitioning, as well as partitioning, can be performed with the sequential fitting procedure based on sequential one-by-one extraction of the boxes with the algorithms described in Section 4.5.4, taking the residual values of the RCPs

$q_{ij} - \mu_h v_{ih} w_{jh}$ each time the box cluster h had been found. Any kind of a priori nonoverlapping requirement can be taken into account very easily due to the seriation nature of the box clustering procedure.

> An important feature of bipartitioning for aggregable data is that the bipartitioned data can be represented by the aggregate table $P(S, T)$, thus implying that the clusters may be considered the aggregate elements along with the aggregate interactions on their own.

Let us apply agglomerative bipartitioning to the Worries data (8 × 5 contingency table characterizing 8 kinds of worry depending on place of living/origin). The result is presented in Fig.6.1. We can see the structure of similarity between both the rows and columns reflected in the common scale of X^2 values. If we can sacrifice, say, 10% of the original data scatter (see dotted line in the picture), then the aggregate table Table 6.14 is quite sufficient for the subsequent analysis. The aggregation made shrinks all the classes below the dotted line. In this table, IAA is the category comprising original places/origins IFI, IFAA and ASAF. The rows MEC and ESM are aggregates of the rows MIL, ECO and ENR, SAB, MTO, respectively.

The Table 6.14 keeps a huge part of the different (large) flow index values of the original table, which follows from the meaning of X^2 as their average. The correspondence analysis display shows that the merged items have been quite close in that, also (see Fig.6.2).

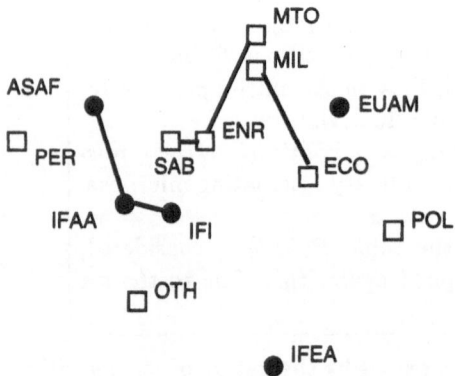

Figure 6.2: Displaying similar (merged at 10% level) row/column patterns in the Worries data.

	EUAM	IFEA	IAA
POL	118	28	45
MEC	229	30	129
ESM	263	52	182
PER	48	16	127
OTH	128	52	107

Table 6.14: The aggregate cross-classification of 1554 individuals by their worries and living places obtained by 10% cut of the original data scatter value.

6.4.3 Discussion

1. The correspondence analysis method aims at analysis of the interrelation between two sets, the rows and the columns. Analogously, the box clustering model and algorithm, though extended for any aggregable data, basically, are purported for the same goal. Aggregation of interacting objects, such as in mobility tables, also aims at the same problem: finding similar patterns of interrelations between output and input entities. Partitioning, though based on the same model, shifts the goal: it reveals classes of rows (or columns) that are similar just as they are (which, in this case, reflects similar pattern of interrelation with the columns). It appears, in particular, that the standard local search fitting algorithms are equivalent to the correspondence-analysis-wise clustering techniques developed earlier by analogy with K-Means and agglomerative algorithms.

2. Having the model allows quite easily extending the local search fitting algorithms for bipartitioning. Due to mathematical specifics of the model, it appears that bipartitioning is equivalent to shrinking the data table into an aggregate matrix with its row/columns being the clusters found and the entries just corresponding total transactions (frequencies). The aggregate data keeps a certain part of the flow index values of the original data (measured by the decrement of the chi-squared coefficient).

Chapter 7

Hierarchy as a Clustering Structure

FEATURES

• Directions for representing and comparing hierarchies are discussed.

• Clustering methods that are invariant under monotone dissimilarity transformations are analyzed.

• Most recent theories and methods concerning such concepts as ultrametric, tree metric, Robinson matrix, pyramid, and weak hierarchy are presented.

• A linear theory for binary hierarchy is proposed to allow decomposing the data entries, as well as covariances, by the clusters.

7.1   Representing Hierarchy

7.1.1   Rooted Labeled Tree

Hierarchies are represented usually by *trees*. Tree $T = (V, E)$ is a graph with a finite set V of its vertices called also *nodes* and a set E of the edges connecting some of the nodes pair-wise in such a manner that all the nodes are connected by paths but still no cycle occurs among the paths. There are a number of other characterizations for the tree: a) a connected graph with the number of edges equal to the number of nodes minus one, $|E| = |V| - 1$; b) a graph in which any two nodes are connected by one and only one path; c) a connected graph which loses that property with the deletion of any edge.

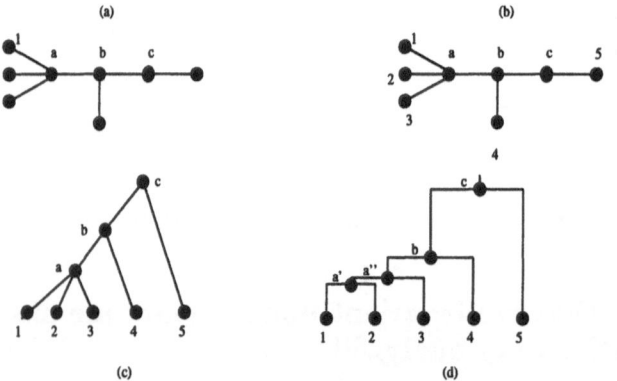

Figure 7.1: A tree (a) after its leaves have been labeled (b) and node c has been taken as its root (c); a binary tree (d).

Fig.7.1 (a) presents a tree with 8 nodes. There are three kinds of nodes in the trees: a) branching node which is an interior node adjacent to three or more other nodes (such as a and b in Fig.7.1 (a)); b) through-node which is adjacent to two other nodes (such as c in Fig.7.1 (a)); c) leaf or terminal (pendant) node which is adjacent to only one other node, such as 1 in Fig.7.1 (a). Usually, the terminal nodes correspond to the entities clusterized, which is presented by labeling the leaves by the entities (see Fig.7.1 (b) where labels 1 to 6 are assigned to the leaves). Such a tree, with all its leaves labeled one-to-one is referred to as a labeled tree. To represent a hierarchy, tree must be rooted. A *rooted tree* is a tree with an interior node singled out; the root gives a partial order to the tree: any node is preceded by the nodes belonging to the path connecting it with the root. A rooted tree can be drawn as a genealogy tree, with the root as the forefather (see Fig.7.1 (c) where labeled tree (b) is presented as a rooted tree with c taken as the root). For any node in the rooted tree (except for the root itself), there is only one path from the node to the root and, thus, the only adjacent node, called the parent, is on this path; all the other adjacent nodes are called children of the node. The leaves (terminal nodes) have no children. In Fig.7.1 (c), node a has b as its parent and leaves $1, 2, 3$ as children. A rooted labeled tree is referred to as a binary tree if any non-terminal node has exactly two children. The tree from Fig.7.1 (c), modified to become a binary one by splitting node a into two nodes, a' and a'', is presented in Fig.7.1 (d).

While the leaves represent entities/species/subsets classified, the tree itself corresponds to their combined classification structure. This can be explicated as follows.

Let us denote by I the leaf set of a rooted labeled tree T, and by I_n the set of leaves descending from a node n; such a subset I_n is referred to as a *node cluster*. One can say also that leaf i belongs to I_n if n lies on the path connecting i with the root.

For any through-node (below the root) n, its leaf set I_n coincides with the leaf set $I_{c(n)}$ of its only child $c(n)$; this means that these two nodes bear the same information so that one of them can be excluded without any loss of the classes. In the remainder, we consider the rooted tree concept with this additional constraint: such a tree contains no through-nodes (except for perhaps the root itself).

The set $I(T)$ of all the node clusters of a labeled rooted tree T satisfies the following properties:

H1. $\{i\} \in I(T)$ for any $i \in I$ (leaves are node clusters);

H2. $I \in I(T)$ (I corresponds to the root);

H3. for any $I_1, I_2 \in I(T)$, either they are nonoverlapping, $I_1 \cap I_2 = \emptyset$, or one of them includes the other, $I_1 \subseteq I_2$ or $I_2 \subseteq I_1$, which means that $I_1 \cap I_2 \in \{\emptyset, I_1, I_2\}$.

The properties H1. - H3. characterize the rooted tree completely; that is, if a set $S_W = \{S_w : w \in W\}$ of subsets $S_w \subseteq I$, $w \in W$, satisfies the properties, a rooted tree $T(S_W)$ can be defined in such a way that $I(T(S_W)) = S_W$. This tree $T(S_W)$ must have set S_W as the set of its nodes, singletons $\{i\}$, $i \in I$ as the leaves and I as the root; an edge connects subsets S_w and $S_{w'}$ if one of them includes the other and no other subset from S_W can be put between them (by inclusion). No cycle can occur since condition H3. is satisfied. Moreover, any node cluster I_{S_w}, obviously, coincides with S_w (after the leaves $\{i\}$ are identified with the corresponding elements $i \in I$).

Let us refer to any set of subsets, S_W, as a *set tree* if it satisfies conditions H1. to H3. We have proven the following statement.

Statement 7.1. *A set of subsets S_W is the set of all node clusters of a rooted labeled tree iff it is a set tree.*

The root cluster I and singletons $\{i\}$ are called *trivial* node clusters. In Fig.7.1 (c), the nontrivial clusters are 1-2-3 and 1-2-3-4.

7.1.2   Indexed Tree and Ultrametric

A classification tree is usually supplied with additional information of either kind: (a) an index (weight) function or (b) a character function, depending on which kind of data has been used in drawing the tree: dissimilarity or entity-to-category data.

The index function reflects degree of internal dissimilarity associated with the node clusters. Sometimes a two-leaf class can have a larger within dissimilarity than a bigger class which is farther from the root. The *index function* of a rooted tree is a function $w(n)$ defined on the set of its nodes and satisfying the following conditions: (a) $w(i) = 0$ for all leaves $i \in I$; (b) $w(n) < w(n')$ if n is a subordinate of n', that is, $I_n \subset I_{n'}$. Reverse scaling of the index would show the similarity, which can be defined quite analogously. The index function can be easily implemented in the drawn presentation of the rooted trees (see Fig.7.2 where, obviously, cluster 1-2 has much greater internal dissimilarity than cluster 3-4-5-6-7).

An *indexed tree* T_w is a rooted tree T with an associated index function w.

There are some other representations of the indexed trees: weight nested partitions and ultrametrics.

A set $S_W = \{S_w : w \in W \subset [0, a]\}$ of partitions of I is called a *weight nested partition* if $S_w \subset S_{w'}$ when $w < w'$ (a is a positive real, usually, 1 or 100). Values w, in this case, can be referred to as the weight levels of the nested partition. Any classification tree produces a nested partition by splitting its levels corresponding to the interior nodes. For example, the non-trivial partitions in the nested partition corresponding to the indexed tree in Fig.7.2 are: $\{1, 2, 3 -$

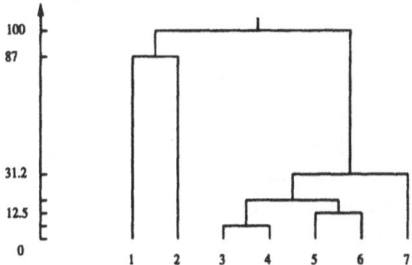

Figure 7.2: An example of indexed tree.

$4, 5, 6, 7\}(w = 6.25), \{1, 2, 3 - 4, 5 - 6, 7\}(w = 12.5), \{1, 2, 3 - 4 - 5 - 6, 7\}(w = 18.75), \{1, 2, 3 - 4 - 5 - 6 - 7\}(w = 31.25)$ and $\{1 - 2, 3 - 4 - 5 - 6 - 7\}(w = 87.5)$. Obviously, the tree can be easily reconstructed by its nested partition: the adjacent partitions show the structure of branching in the corresponding nodes while the levels provide for the index function values.

Since the weights, actually, correspond to the node clusters merged, the indexed tree can be also represented by the set of corresponding node clusters I_n accompanied with weights $w(n)$, which will be called *weighted set tree*.

For any indexed tree, a dissimilarity measure $D = (d(i, j))$ on the set of the entities can be defined as $d(i, j) = w(n[i, j])$ where $n[i, j]$ is the minimum node cluster being the ancestor for both $i, j \in I$. Such a measure is special: it is an ultrametric.

A dissimilarity matrix $d = (d_{ij})$, $i, j \in I$, is called an *ultrametric* if it satisfies a strengthened triangle inequality: for any $i, j, k \in I$

$$d_{ij} \le \max(d_{ik}, d_{jk}). \tag{7.1}$$

There is an analogous concept for the similarity measure: *ultrasimilarity* is a similarity matrix $b = (b_{ij})$, $i, j \in I$, satisfying the reverse inequality

$$b_{ij} \ge \min(b_{ik}, b_{jk}). \tag{7.2}$$

Condition (7.2) can be considered as an extension of the transitivity property for conventional binary relations to the case of fuzzy relations (represented by similarity matrices); this is why this property is called sometimes *fuzzy transitivity*.

Obviously, a similarity matrix b is an ultrasimilarity if and only if there exists an ultrametric d such that $b_{ij} = c - d_{ij}$ for any $i, j \in I$, where c is a constant. To find such an ultrametric, it is sufficient to take $c = \max_{i,j \in I} b_{ij}$ and let $d_{ij} = c - b_{ij}$.

An intuitive meaning of the ultrametric inequality: for any entities $i, j, k \in I$, two of the three distances, $d(i, j), d(i, k)$, and $d(j, k)$, have the same value while the third one is less than or equal to that. This implies that any threshold graph defined as $G_\pi = \{(i, j) : d(i, j) \leq \pi\}$ is an equivalence relation on I. Indeed, G_π is transitive: $d(i, j) \leq \pi$ and $d(j, k) \leq \pi$ imply $d(i, k) \leq \pi$ since $d(i, k)$ may not be greater than the maximum of the two former values.

To prove that the tree-defined dissimilarity D is an ultrametric, let us take arbitrary $i, j, k \in I$ and consider $d(i, k)$ and $d(j, k)$. If $d(i, k) = d(j, k)$, then both pairs i, k and j, k belong to the same minimal node cluster I_n, which implies $d(i, j) \leq d(i, k) = d(j, k)$ corresponding to (7.1). If the equality does not hold so that, for instance, $d(i, k) < d(j, k)$, then i, k belong to $I_{n'}$ and j, k belong to $I_{n''}$ for some nodes n', n''. Inclusion $I_{n'} \subset I_{n''}$ holds since the node clusters have their intersection containing k and, thus, nonempty. This implies that $i, j \in I_{n''}$ and $d(i, j) = d(j, k)$, which completes the proof.

Obviously, the threshold graphs G_π for D form a nested set of the equivalence relations that correspond to the nested partition generated by the tree.

We have proven the following statement.

Statement 7.2. *For any indexed tree T_w, its nested partition, weighted set tree, ultrametric, and set of the threshold equivalence graphs G_w, $w \in W$, are equivalent representations.*

The concept of ultrametric, in the modern classification research, appeared simultaneously in papers by Hartigan 1967, Jardine, Jardine, and Sibson 1967 and Johnson 1967 (a detailed account of these and some subsequent developments is given by Barthélemy and Guenoche 1991, Section 3.5). Yet, it was introduced some earlier by R. Baire 1905: see F. Hausdorff 1957, p. 116-117, where an ultrametric is referred to as a *Baire space*.

7.1.3   Hierarchy and Additive Structure

Let $S_W = \{S_w : w \in W\}$ be a weighted set tree, which means that S_W satisfies H1. to H3. and a positive weight λ_w is assigned to any $w \in W$. Let us consider corresponding *additive hierarchy structure* as a similarity measure $B = (b_{ij})$, $i, j \in I$, defined as $b_{ij} = \sum_{w \in W} \lambda_w s_{iw} s_{jw}$ where $s_w = (s_{iw})$ is the indicator function: $s_{iw} = 1$ for $i \in S_w$ and $s_{iw} = 0$, otherwise. The following result can be found in Bock 1974, Bandelt and Dress 1990.

Statement 7.3. *A similarity measure is an ultrasimilarity if and only if it can be represented by a uniquely defined additive hierarchy structure.*

Proof: Let $B = (b_{ij})$ be an ultrasimilarity matrix and set $S_W = \{S_1, ..., S_M\}$ consist of all different classes of the equivalence threshold graphs $G_\pi = \{(i, j) : b(i, j) \geq \pi\}$ for arbitrary thresholds π. Every $S_n \in S_W$ can be assigned the value $w(n)$ being the maximum π such that S_n is a class of G_π (there is no ambiguity in that definition since there is only a finite number of similarity values b_{ij} and thus a finite number of different threshold graphs). Obviously, set S_W along with function $w(n)$ is a weighted set tree. Let us define now, for any n, $n = 1, ..., M$, $\lambda_n = w(n) - w(a(n)) > 0$ where $a(n)$ is index of the minimal cluster $S_{a(n)}$ properly including S_n. If there is no such cluster, then $S_n = I$; in this case, let $\lambda_n = 0$. Let us prove that $b_{ij} = \sum_{n=1}^{M} \lambda_n s_{in} s_{jn}$ where s_n is the Boolean indicator vector of S_n. Indeed, let $S_{n[i,j]}$ be the minimal class containing both elements i, j. Then, the classes $S_{n_1}, ..., S_{n_m}$ properly including $S_{n[i,j]}$ form a chain by inclusion which makes $\sum_{k=1}^{m} \lambda_{n_k} + \lambda_{n[i,j]} = w(n[i, j]) = b_{ij}$. Conversely, let $b_{ij} = \sum_{w \in W} \lambda_w s_{iw} s_{jw}$ for a set tree $S_W = \{S_w : w \in W\}$. Let $S_{n(i,k)}, S_{n(j,k)}$ be the minimal set clusters containing both i, k and j, k, respectively. Since these clusters are overlapping (k belongs to both), one includes the other, say, $S_{n(i,k)} \subseteq S_{n(j,k)}$. Thus, $b_{ij} \geq b_{jk} = \min(b_{ik}, b_{jk})$, which proves the statement. □

Ultrasimilarity $B = (b_{ij})$, $i, j \in I$, will be referred to as *resolved* if no three different elements $i, j, k \in I$ exist for which all the values b_{ij}, b_{ik}, and b_{jk} are coinciding.

Corollary 7.1. *Ultrasimilarity $B = (b_{ij})$ corresponds to a binary tree iff it is resolved.*

Proof: Indeed, the fact that $b_{ij} = b_{ik} = b_{jk}$ means that in the corresponding weighted tree $S_{n(i,k)} = S_{n(j,k)}$ and there are nonoverlapping paths from the corresponding interior vertex n to the leaves i, j and ki, which can be only when n is adjacent to four nodes, at least. □

7.1.4 Nest Indicator Function

This concept is applicable only to the binary hierarchies. For a nonsingleton cluster $S_w = S_{w1} \cup S_{w2}$ ($w, w1, w2 \in W$) of a binary set tree S_W, let us define its three-value *nest indicator function* ϕ_w as follows:

$$\phi_{iw} = \begin{cases} a_w & \text{if } i \in S_{w1} \\ -b_w & \text{if } i \in S_{w2} \\ 0 & \text{if } i \notin S_w \end{cases}$$

where the positive reals a_w and b_w are selected to satisfy the following two conditions: (1) vector ϕ_w is centered; (2) vector ϕ_w has its Euclidean norm equal to 1. To be more precise, let us denote the cardinalities of clusters S_{w1} and S_{w2} by N_1 and N_2, respectively. Obviously, $N_1 + N_2 = N_w$ where N_w is cardinality of S_w. Then, (1) means that $N_1 a_w - N_2 b_w = 0$ while (2), $N_1 a_w^2 + N_2 b_w^2 = 1$. These two equations lead to the following uniquely defined values of a_w and b_w:

$$a_w = \sqrt{\frac{N_2}{N_1 N_w}} = \sqrt{\frac{1}{N_1} - \frac{1}{N_w}}, \quad b_w = \sqrt{\frac{N_1}{N_2 N_w}} = \sqrt{\frac{1}{N_2} - \frac{1}{N_w}}.$$

Statement 7.4. *The set of vectors $\phi_w, w \in W$, defined by a binary set tree $S_W = \{S_w : w \in W\}$ is an orthogonal basis of the $(N-1)$-dimensional Euclidean space of the centered N-dimensional vectors.*

Proof: Let us prove that the vectors ϕ_w, $w \in W$, are mutually orthogonal; that is, $\sum_{i \in I} \phi_{iw} \phi_{iw'} = 0$ for different $w, w' \in W$. If $S_w \cap S_{w'} = \emptyset$, then $\phi_{iw} \phi_{iw'} = 0$ for any $i \in I$ since either $i \notin S_w$ or $i \notin S_{w'}$. Otherwise, one of the sets includes the other, say, $S_{w'} \subset S_w$, which implies that $S_{w'}$ is included in one of the children sets S_{w1}, S_{w2} (such that S_w is their union), for instance, $S_{w'} \subseteq S_{w1}$. Then, $\phi_{iw} = a_w$ for any $i \in S_{w'}$, which implies that $\sum_{i \in I} \phi_{iw} \phi_{iw'} = a_w \sum_{i \in I} \phi_{iw'} = 0$, since vector $\phi_{w'}$ is centered. Thus, the orthogonality is proven. The number of nontrivial vertices $w \in W$ in a binary set tree and, therefore, the number of mutually orthogonal vectors ϕ_w, $w \in W$, is equal to $N - 1$ which is exactly the dimension of the space of all the centered N-dimensional vectors. $\quad\square$

This statement guarantees that any entity-to-variable matrix can be decomposed by any binary hierarchy, provided that the variables have been centered preliminarily.

Can the similarity data be decomposed by a binary tree? In general, no: there are $N(N-1)/2$ arbitrary similarity entries while only $N-1$ binary tree nodes. However, the ultrasimilarity data can.

Let us consider a similarity matrix $P_w = \phi_w \phi_w^T = (\phi_{iw} \phi_{jw})$ corresponding to a node of a binary tree S_W. Non-zero entries of such a matrix are within the subset S_w in such a way that $\phi_{iw} \phi_{jw} = -1/N_w$ (negative) when i and j are split in the children of S_w, and (i, j) entry equals $\frac{1}{N_k} - \frac{1}{N_w}$ (positive) when $i, j \in S_{wk}$ $(k = 1, 2)$. This is, obviously, the projector matrix onto the linear subspace $L_u(S_w)$ corresponding to partition of S_w in two classes, S_{w1} and S_{w2} (see Section 5.1.4). Since matrices P_w have all their rows centered, we have to deal with the row-centered ultrasimilarity matrices, as well, which necessitates considering the diagonal similarities that so far did not matter. Columns of such a matrix will be centered too due to symmetry. Thus, let ultrasimilarity matrix $S = (s_{ij})$ satisfy equalities $s_{ii} = -\sum_{j \neq i} s_{ij}$, $i \in I$ (the latter formula can be considered a rule for determining appropriate values of the diagonal entries). Evidently, the matrices of this form, corresponding to a set tree, form a linear subspace of the linear space of all symmetric $N \times N$ matrices considered as N^2-dimensional vectors.

Corollary 7.2. *The projection matrices* $P_w = (\phi_{iw}\phi_{jw})$, $i, j \in I$, *defined by a binary set tree* S_W *are mutually orthogonal and form a basis of the subspace of all the row-centered ultrasimilarity matrices corresponding to the tree* S_W.

Proof: The orthogonality follows from the Statement 7.4.; thus $N - 1$ of the matrices are enough to determine all the $N - 1$ ultrasimilarity values. $\square$

The nest indicator functions were considered by Benzécri 1973, but only in a specific context of the Correspondence analysis theory.

7.1.5 Edge-Weighted Tree and Tree Metric

There is an important generalization of the indexed tree concept related to the situation when the weights are assigned to the tree edges rather than to the nodes. Such a tree can be referred to as an *(edge) weighted tree*. The tree distance $d(i, j)$ is defined as the sum of the weights of the edges belonging to the unique path, $T(i, j)$, between i and j. The distance can be related to some latent attributes differing at the entities involved (see Sattah and Tversky 1977, Corter and Tversky 1986). The concept is widely employed in molecular biology as a model of evolutionary history for the entities being currently living species: the number of mutation changes between the species has been supposedly accumulated during their descent from the ancestors represented by the internal nodes (as in Fig.1.4, p. 13). Since any (node) weighted tree can be transformed into an edge weighted tree (keeping the property that, for any $i, j \in I$, the total distances between i or j and the minimal node $n[i, j]$ containing both i and j are equal to $w(n[i, j])/2$), the ultrametric is a tree distance. In general, the tree distance is characterized by the so-called *four-point condition* coined first by Zaretsky 1965 and supplemented by Smolensky 1969, though the most frequent reference is Buneman 1971 who gave the statement in its current formulation: for any $i, j, k, l \in I$ two of the three values, $d(i, j) + d(k, l), d(i, k) + d(j, l)$, and $d(i, l) + d(j, k)$, coincide while the third one is less than or equal to that common value:

$$d(i, j) + d(k, l) \leq \max(d(i, k) + d(j, l), d(i, l) + d(j, k)). \tag{7.3}$$

The proof, actually, can be easily derived from the picture in Fig.7.3 presenting the general pattern of the tree paths pair-wisely joining the four vertices involved in (7.3). Indeed, all of the sums in (7.3) include $d(i, x) + d(j, x) + d(k, y) + d(l, y)$ while two of them contain also $2d(x, y)$ (see Mirkin and Rodin 1984, Barthélemy and Guenoche 1991).

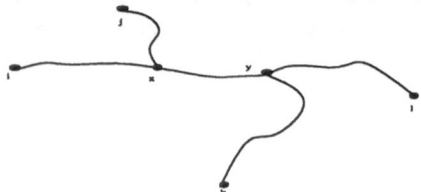

Figure 7.3: Four-point pattern in an edge weighted tree.

The four-point condition resembles that of the ultrametric, which is employed in the following construction, due to Farris, Kluge and Eckart 1970 (see also Leclerc 1995): let us pick an arbitrary $c \in I$ and define yet another distance

$$d_c(i, j) = MM + d_{ij} - d_{ic} - d_{jc} \ (i, j \in I - \{c\}) \tag{7.4}$$

where $MM > 0$ is chosen to make all the values of d_c non-negative; for instance, $MM = 2\max_{i,j \in I} d_{ij}$ makes it for sure.

Statement 7.5. *For a dissimilarity d on I, the following properties are equivalent:*

(1) d is a tree metric;

(2) d_c is an ultrametric for any $c \in I$;

(3) d_c is an ultrametric for some $c \in I$.

Proof: Let us take (7.3) with l substituted by any $c \in I$ and subtract $d_{ic} + d_{jc} + d_{kc} - MM$ from both parts of the inequality, which gives the ultrametric inequality for d_c immediately. This proves that (1) implies (2). To prove that (3) implies (1), let us assume that d_c is an ultrametric for some c and consider some $i, j, k, l \in I - \{c\}$. Then, obviously, $d_{ij} + d_{kl} = d_c(i, j) + d_c(k, l) + C$, $d_{ik} + d_{jl} = d_c(i, k) + d_c(j, l) + C$, and $d_{il} + d_{jk} = d_c(i, l) + d_c(j, k) + C$ where $C = d_{ic} + d_{jc} + d_{kc} + d_{lc} - 2MM$. Since d_c as an ultrametric satisfies the four-point condition so does d, which completes the proof since implication (2) → (3) is trivial. $\square$

In the sequel, the derived metric $d_c(i, j)$ will be referred to as *FKE-transform* of the original dissimilarity d_{ij}.

Correspondence between the tree representations of a tree metric d and its FKE-transform d_c is illustrated in Fig.7.4. One way, $d \rightarrow d_c$, is simple: tree (a) yields its tree metric (b) which is FKE-transformed into an ultrametric (c) corresponding to

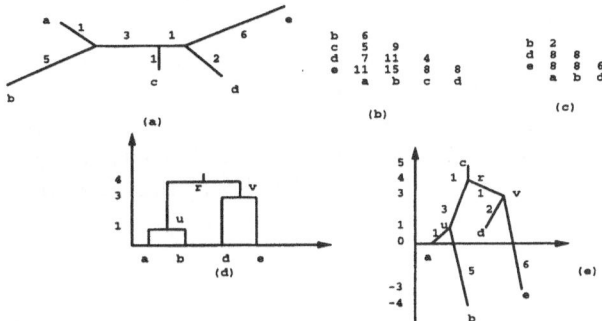

Figure 7.4: Weighted tree (a) and its tree metric (b) transformed into ultrametric (c) and indexed hierarchy (d) with FKE-transformation; (e) is the tree resulting with the inverse transformation.

hierarchy (d). The other way is not that obvious, although not complicated. It is based on the inverse version of the equation (7.4), $d_{ij} = d_c(i, j) + d_{ic} + d_{jc} - MM$. For the indexed tree derived from ultrametric d_c, let us define new index values for all its leaves: $w(i) = MM/2 - d_{ic}$ $(i \in I - \{c\})$; then let us add the leaf c to the tree, as joined by an edge to the root r, and define $w(c) = MM/2$; the other index values remain unchanged. In the new tree T_c (see Fig.7.4 (e)), the edge weights are defined as $d(u, v) = w(u) - w(v)$ for all its edges $\{u, v\}$ with u being the parent of v in the original rooted tree (see Fig.7.4 (e)). This means that the length of every interior edge remains invariant while each of the terminal edges (leading to a leaf, i) is subtracted by the value $w(i)$ defined. The only thing left is to prove that the tree found gives exactly the original tree metric.

Statement 7.6. *For tree T_c obtained from the ultrametric d_c, its tree metric coincides with the original metric d.*

Proof: Let us start with the distances $d(i, c)$, $i \in I - \{c\}$. By definition, $d(i, c) = w(c) - w(i) = MM/2 - (MM/2 - d(i, c)) = d_{ic}$. Now, for any $i, j \neq c$, $d(i, j) = d(i, c) + d(j, c) - 2d(u, c)$ where u is the node at the path between i and j, $T(i, j)$, where c joins $T(i, j)$. This implies $d(i, j) = d_{ic} + d_{jc} - 2(w(c) - w(u)) = d_{ij}$ since $w(u) = d_c(i, j)/2 = [MM + d_{ij} - d_{ic} - d_{jc}]/2$ because corresponding node cluster I_u is the minimal one to contain both i and j in the rooted hierarchy.   □

Construction of the tree T_c involves ultrametric d_c and the weight function w defined for every node of the tree. This gives rise to yet another characterization of the tree metric, using the so-called star-dissimilarity concept (Carroll 1976). A dissimilarity matrix $D = (d(i, j))$ is referred to as *star dissimilarity* measure if, for any different $i, j \in I$, $d(i, j) = w_i + w_j$ where $g = (w_i)$ is a vector (perhaps, nonpositive) defined for $i \in I$. The name of the concept is based on a graphical representation of the distance D in terms of a star, which is, basically, a root directly connected to all the leaves $i \in I$ with w_i as the corresponding edge weights. In the example of Fig.7.4, the star vector is $g = (0, -4, 5, 1, -3)$ thus containing two negative rays, which is reflected in Fig.7.4 (e).

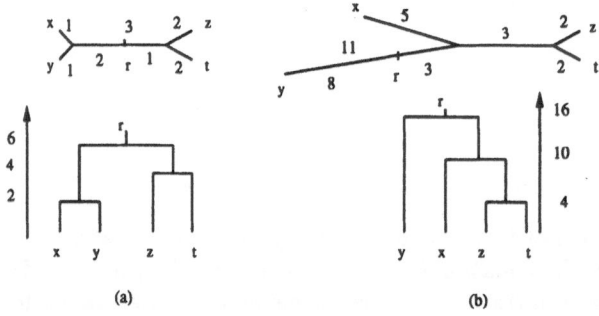

Figure 7.5: Two topologically equivalent trees generate different ultrametrics depending on the weights and root r location.

Statement 7.7. *A tree distance $d = (d_{ij})$ can be decomposed into the sum of a star dissimilarity and an ultrametric in infinitely many ways.*

The proof follows from the construction above (and can be found in Barthélemy and Guenoche 1991, p. 114-115). These authors note also that there is no simple relationship between the tree distances and corresponding ultrametrics; the quantitative information contained in the tree distances can be related to different ultrametric hierarchies. To illustrate that, let us consider an example from Barthélemy and Guenoche 1991, p. 116-117 (see Fig.7.5). We can see that the edge-weighted trees in (a) and (b) are topologically equivalent, though the roots and weights of the edges are different. But the weighted hierarchies representing corresponding tree metrics (turned out ultrametrics!) are very different topologically. This raises the following problem: find a description and a diagnostic tool for all the ultrametrics generated by the same edge-weighted tree. The notion of a T-split described below might be a convenient concept for that.

7.1.6   T-Splits

Let T be an unrooted edge-weighted tree with leaf set I. Let us refer to a biclass partition $S = \{A, B\}$ of I as a T-split if it is obtained by splitting the tree by one of its edges: eliminating an edge makes the resulting graph have two connected components, A (or B) is just the subset of the leaves belonging to one of them. Eliminating leaf edges we produce the trivial splits consisting of a singleton and all-the-others subset. In Fig.7.6, there are only two nontrivial splits: $S_1 = \{a - b, c - d - e\}$ and $S_2 = \{a - b - c, d - e\}$.

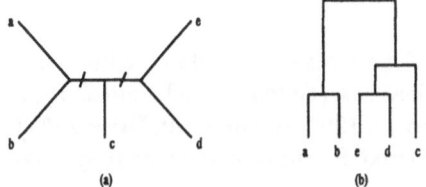

Figure 7.6: Hierarchy (b) produced by the left nontrivial split of tree (a).

Obviously, the split concept can be considered as a symmetrized version of the node cluster: all the node clusters are classes of the T-splits, though some of the T-split classes are not node-clusters. Let us create a rooted version of T putting the root in the middle of a tree edge, thus, taking corresponding split as the root divergence (in Fig.7.6 (b) a rooted tree is shown as based on the first of the nontrivial splits, S_1). Then, all the T-splits will be biclass partitions consisting of a node cluster and its complement in I; note that, within the T-split concept, there is no opportunity to distinguish the clusters from non-clusters. It seems quite evident that the set S_T of all T-splits (for a T fixed) can be characterized by a symmetrized version of the characteristic H1-H3 of the node clusters, p. 331:

G1. All the trivial splits belong to S_T.

G2. For any two different splits $S = \{A, B\}$ and $U = \{C, D\}$ from S_T, exactly one of the intersections, $A \cap C, A \cap D, B \cap C, B \cap D$, is empty.

Obviously, G1 is equivalent to H1 while G2 is equivalent to H3; there is no symmetrical match to H2. To make the equivalence between G2 and H3 more clear, let us show how the node clusters can be reconstructed from the splits. To do that, we need to find a rooted-tree-point of view on the splits. This can be done taking a split, $S = \{A, B\}$, as the root-specified one. All the other splits can be sorted depending on which of the classes A or B includes a class from a split considered (the fact that such an inclusion occurs is guaranteed by G2). The classes included are to be considered as the node clusters; H3 for them follows from G2 (Fig.7.6 illustrates this procedure). We have proven the following statement.

Statement 7.8. *A set of biclass partitions is a T-split set (for a tree T) if and only if it satisfies conditions G1 and G2.*

7.1.7 Neighbors Relation

Another concept for structural representation of the trees employed in the literature is a quaternary "neighbors" relation on the leaf set (Colonius and Shulze 1981, Bandelt and Dress 1986, Barthélemy and Guenoche 1991, Bandelt and Dress 1994). Traditionally, the neighbors relation is considered for unrooted trees only. We consider it for both rooted and unrooted trees.

Let T be a rooted tree with leaf set I and $S \in I(T)$. Then, for any $i, j, k, l \in I$ such that $i, j \in S$ while $k, l \notin S$, we say that the quadruple (i, j, k, l) belongs to the *rooted neighbors* relation $\varphi_T \subset I^4$; symbol $ij\|kl$ is usually employed as a synonym to $(i, j, k, l) \in \varphi_T$.

Obviously, the rooted neighbors relation satisfies the following properties, for any $i, j, k, l \in I$:

(a) antisymmetric comparability: one and only one of $ij\|kl$, $ik\|jl$, $il\|jk$ holds;

(b) symmetry: $ij\|kl$ implies $ji\|kl$ and $ij\|lk$;

(c) substitution: for any $m \in I$, $ij\|kl$ implies $ij\|km$ or $im\|kl$.

In fact, these conditions are also sufficient.

Statement 7.9. *A quaternary relation $\varphi \subset I^4$ is a rooted neighbors relation if* *and only if it satisfies the properties (a) through (c) above.*

Proof: We only need to prove that if $\varphi \subset I^4$ satisfy (a) to (c), a tree can be uniquely defined. Let us define a subset S be a cluster if it satisfies inclusion $S \times S \times (I - S) \times (I - S) \subset \varphi$. Let us prove that if clusters S_1 and S_2 are overlapping, one of them is a part of the other. If, conversely, there are $i \in S_1 \cap S_2, j \in S_1 - S_2, k \in S_2 - S_1$, and $l \in \bar{S}_1 \cap \bar{S}_2$, then both $ij\|kl$ and $ik\|jl$, which contradicts (a); thus, the clusters form a rooted tree T. To show that $\varphi_T = \varphi$, let us prove, initially, that φ is transitive, that is, 1) $ij\|kl$ and $im\|kl$ imply $jm\|kl$, and 2) $ij\|kl$ and $ij\|km$ imply $ij\|lm$. Indeed, $ij\|kl$ implies, by (c) and (b), $jm\|kl$ or $ij\|km$. Analogously, $im\|kl$ implies $jm\|kl$ or $im\|kj$. Suppose $jm\|kl$ is not true. Then both $ij\|km$ and $im\|kj$ hold simultaneously, which contradicts (a) and proves (1); (2) is proved analogously. The transitivity is proved. Now, let $ij\|kl$ and S be the set of all the leaves m satisfying condition $im\|kl$. Let us show that S is a cluster as defined above in the proof; that is, for any $m', m'' \in S$ and $l', l'' \notin S$, $m'm''\|l'l''$. Indeed, $m'm''\|kl$ by the property 1) of transitivity of φ since $im'\|kl$ and $im''\|kl$ by definition of S; $l', l'' \notin S$ are easy to put in the condition since $im\|kl$ implies, by (c), $il'\|kl$ or $im\|kl'$ where the former is not true by definition of S. $\qquad\square$

The unrooted neighbors relation is defined, for an unrooted tree T, by its splits: $ij\|kl$ if $i, j \in A$ and $k, l \in B$ for a T-split $S = \{A, B\}$. A most important feature of the unrooted neighbors relation is that it can be recovered from the tree metric without knowing the tree itself (Colonius and Schulze 1981, Bandelt and Dress 1986). This follows from the following fact:

Statement 7.10. *For any unrooted tree T and corresponding tree metric $d = (d_{ij})$, $ij\|kl$ if and only if*

$$d_{ij} + d_{kl} < d_{ik} + d_{jl} = d_{il} + d_{jk}.$$

The proof, actually, is quite obvious from the picture shown in Fig.7.3.

A *score function* $\sigma(i, j)$ is associated with any neighbors relation; it is defined as

$$\sigma(i, j) = |\{(k, l) : ij\|kl\}| \tag{7.5}$$

It can be easily seen that, for a rooted neighbors relation, $\sigma(i, j) = |I - S[i, j]|^2$ where $S[i, j]$ is the minimal node cluster containing both i and j. This implies that, in this case, the score function is an ultrasimilarity. For the unrooted neighbor relations, the score function is far more complex. Still, it can be proven that the score function for any unrooted neighbor relation satisfies the reverse four-point condition. More precisely, function $\nu(i, j) = MM - \sigma(i, j)$ (where $MM > \max \sigma(i, j)$) is a tree metric corresponding to the same tree T (Colonius and Schulze 1981, Bandelt and Dress 1986).

7.1.8   Character Rooted Trees

A *character rooted tree* is a rooted tree along with some characters assigned one-to-one to all its nodes (root excluded). The characters are taken from a set C; their assignment may be considered as made to the edges rather than to the nodes since, in every rooted tree, there is one-to-one correspondence between the nodes and edges: each node is adjacent to one and only one edge belonging to the path between the node and root.

Every node in a character rooted tree is characterized by the sequence of the characters along the path from root to the node. This sequence (set) of the characters is used as an identification key of the objects in corresponding node cluster. For example, when characters present some categories or predicates defined for an entity set (like in Fig.1.1, 3.2, and 3.9) the character tree represents the so-called *decision tree* or a *concept* (in terms of machine learning discipline, see Breiman et al. 1984 and Michalski 1992): each node relates to a subset of the entities defined by conjunction of the categories along the path from root to the node. In another applied area, biological taxonomy, the leaves represent species, interior nodes, the higher taxa or ancestral organisms, and the characters correspond to biological parameters (usually, binary "present-absent" attributes).

Another way of representing such a character rooted tree is through the rectangular "leaf-to-character" Boolean matrix $F = (f_{ic})$, $i \in I$, $c \in C$, where $f_{ic} = 1$ iff character c belongs to the character sequence corresponding to the leaf i. There is a simple correspondence between this kind of matrices and character rooted trees (Estabrook, Johnson, and McMorris 1975, Gusfield 1991). Let us define a subset $F_c = \{i : f_{ic} = 1\}$ of the leaves covered by the character c.

Statement 7.11. *Matrix F corresponds to a character rooted tree if and only if set $\{F_c : c \in C\}$ is a set tree, that is, for any $c, c' \in C$, subsets $F_c, F_{c'}$ are disjoint or one contains the other.*

7.1.9   Comparing Hierarchies

Above, four basic representation formats of the hierarchies have been considered: graph (rooted and unrooted tree), set (node clusters, neighbors relation, T-splits), nested partition, and matrix (tree metric, ultrametric, and nest indicator matrix). Comparing the hierarchies depends much on the representation format. Let us consider similarity/dissimilarity measures accordingly.

Graph

The most known dissimilarity measure between the hierarchies in terms of the labeled trees was proposed by Robinson 1971. The measure counts the smallest number of steps involving an exchange between two branches in a tree necessary to transform one of the trees into the other.

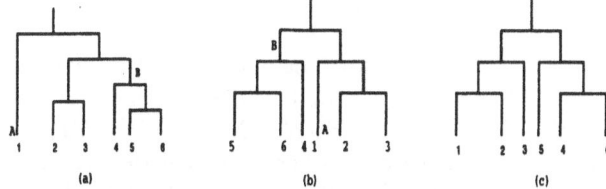

(a)   (b)   (c)

Figure 7.7: Exchange between nodes A and B produces (b) from (a); then (c) is obtained with two pairs of the leaves exchanged: 1 and 3, and 4 and 5.

In Fig.7.7, three exchanges between the branches are needed to transform tree (a) into tree (c); (b) represents the result of exchange between nodes A and B, after which just simple switch of 1 and 3, and 4 and 5 gives (c). This makes the distance equal to 3. This seemingly simple dissimilarity measure is, in fact, quite complicated because of irregularities involved in the exchanges. To see the irregularities, the exchange step should be reformulated in terms of the node clusters. We leave this as an exercise to the reader. Other measures defined so far in terms of the number of some elementary graph-transforming operations needed to transform one of the trees in the other, also become quite complicated when translated into set terms.

Set

When presented by their node clusters, the trees can be compared using any measure discussed in Section 4.1.2. The mismatch coefficient (sometimes referred to as a partition metric [Steel 1993]) has become most popular. It counts the number of node clusters that are present in one and only one of the trees. Obviously, only non-trivial clusters corresponding to the branching internal nodes may give the difference. The trees (a) and (c) in Fig.7.7 have their nontrivial node cluster sets 2-3, 5-6, 4-5-6, 2-3-4-5-6, and 1-2, 1-2-3, 4-6, 4-5-6, respectively, which leads to the mismatch number equal to 6. Having the maximum number of nontrivial clusters in both of the trees equal to $2(N-2) = 8$, we get the relative mismatch $6/8 = 3/4$ while the match coefficient is $1 - 3/4 = 1/4$. The Jaccard match coefficient seems quite relevant here; it equals 1/7 as there is only one cluster, 456, coinciding in both of the trees, out of the seven different clusters available.

The difference between neighbors relations, measured by the mismatch coefficient, is considered sometimes as relevant to the evolutionary considerations; Estabrook, McMorris and Meachem 1985 analyzed that for unrooted trees (quartet distance). It seems, the rooted neighbors relation is a simpler structure which also can be used for the purpose.

For a set tree $S_W = \{S_w : w \in W\}$, the rooted neighbors relation can be presented as $\nu = \cup_{w \in W} S_w \times S_w \times (I - S_w) \times (I - S_w)$, which implies that the number of quadruples in ν is equal to $|\nu| = \sum_{w \in W} N_w^2 [(N - N_w)^2 - (N - NP_w)^2]$ where $N = |I|$, $N_w = |S_w|$, and NP_w are the total number of the entities (leaves), the cardinality of S_w and its parent, respectively. For two neighbors relations, $\nu = \cup_{w \in W} S_w \times S_w \times (I - S_w) \times (I - S_w)$ and $\mu = \cup_{v \in V} T_v \times T_v \times (I - T_v) \times (I - T_v)$, their intersection equals, obviously,

$$\nu \cap \mu = \cup_{(w,v) \in W \times V} (S_w \cap T_v) \times (S_w \cap T_v) \times (I - (S_w \cup T_v)) \times (I - (S_w \cup T_v)),$$

which, regretfully, does not lead to as simple counting formula for $|\nu \cap \mu|$ as for $|\nu|$. This can be seen with the two trees, (a) and (c), in Figure 7.7.

To find $|\nu|$ for tree (a), we start with the maximum cluster $\alpha = 2 - 3 - 4 - 5 - 6$ which supplies quadruples $(i, j, 1, 1)$ in ν, with their amount equal to $5^2 1^2 = 25$. Next cluster, $\beta = 4 - 5 - 6$ counts for $3^2(3^2 - 1^2) = 72$ quadruples since we must exclude quadruples like $(4, 5, 1, 1)$ as already taken into account in relation to the cluster α. The next cluster $\gamma = 5 - 6$ adds $2^2(4^2 - 3^2) = 28$ of the quadruples; the subtracted number concerns quadruples like $(5, 6, 1, 2)$ already counted with cluster β. The last cluster, $\delta = 2 - 3$ adds $2^2(4^2 - 1) = 60$ where subtracted are the quadruples, like $(2, 3, 1, 1)$, already counted with the parent cluster α. Thus, $|\nu| = 25 + 72 + 28 + 60 = 185$. Analogously, the cardinality of the neighbors relation for tree (c) is equal 218, as the sum of two times 81 related to clusters $a = 4 - 5 - 6$ and $b = 1 - 2 - 3$, and of two times 28 related to smaller clusters $c = 4 - 6$ and $d = 1 - 2$. What about the intersection of these neighbors relations? The quadruple subsets generated by the clusters from tree (c) and α from (a) are as follows: $4 - 5 - 6 \times 4 - 5 - 6 \times 1 \times 1$ corresponding to $\alpha \cap a$, $2 - 3 \times 2 - 3 \times 1 \times 1$ corresponding to $\alpha \cap b$, $4 - 6 \times 4 - 6 \times 1 \times 1$ for $\alpha \cap c$, $2 \times 2 \times 1 \times 1$ for $\alpha \cap d$. Obviously, the latter two subsets are absorbed by the former two and do not contribute to the intersection. Analogously, two nonempty intersections of cluster β with the clusters from (c) are $4 - 6 \times 4 - 6 \times 1 - 2 - 3 \times 1 - 2 - 3$ and $4 - 6 \times 4 - 6 \times 1 \times 1$ (absorbed). Cluster γ gives $5 - 6 \times 5 - 6 \times 1 - 2 - 3 \times 1 - 2 - 3$ and $6 \times 6 \times 1 - 2 - 3 \times 1 - 2 - 3$ (absorbed). Cluster δ gives $2 - 3 \times 2 - 3 \times 4 - 5 - 6 \times 4 - 5 - 6$ and $2 \times 2 \times 4 - 5 - 6 \times 4 - 5 - 6$ (absorbed). Two maximal subsets related to cluster α contribute to $|\nu \cap \mu|$ with $3^2 1^2 = 9$ and $2^2 1^2 = 4$, respectively. With the quadruples related to unity subtracted, subset $4 - 6 \times 4 - 6 \times 1 - 2 - 3 \times 1 - 2 - 3$ for β gives $2^2(3^2 - 1) = 32$. With quadruples like $(6, 6, 1, 2)$ already counted at the previous step, the unabsorbed subset $5 - 6 \times 5 - 6 \times 1 - 2 - 3 \times 1 - 2 - 3$ (for γ) adds $3(3^2 - 1) = 24$. The unabsorbed subset for δ adds $2^2 3^2 = 36$ raising the total number of the common quadruples to 105.

The mismatch number equals $218 + 185 - 2 \times 105 = 193$; the relative mismatch (divided by the maximum number of the quadruples $6^4 = 1296$) equals 0.15. This is an underestimate: the cause is that no tree can produce as many clusters as 1296 or even approximate the figure. Jaccard mismatch coefficient seems more appropriate here; it equals $193/(218+185-105) = 0.65$, which shows quite a difference.

In substantive research, even more strange measures can emerge, such as that one in Mirkin, Muchnik, Smith 1995: it counts the number of co-occurrences of such nodes s in S and t in T that node cluster I_t partly overlaps both node cluster of s and of each of its children; the maximal pairs (s, t) among them, proved to correspond to "gene duplications", appear to be counted four times each.

Nested Partition

The hierarchies considered as nested partitions can, virtually, lead to an enormous number of coefficients based on the partition-to-partition indices. Fowlkes and Mallows 1983 suggested calculating a series of index values (m, c_m) where m is an integer between 2 and N-1 and c_m is a similarity index between m-class partitions taken from the compared hierarchies. This allows to analyze which parts of the hierarchies, say, upper (m is small) or lower (m is large) ones, are nearer to or farther from each other.

Matrix

Matrix representation of the hierarchies allows for using the matrix correlation coefficients (started by Sokal and Rohlf 1962). This subject has received no extensive analysis yet. Some may consider the matrix correlation coefficient as a measure of goodness-of-fit for comparing the original similarity/dissimilarity data with the distance matrix corresponding to the found hierarchy. However, such a measure has tremendously low capability to differentiate between good and bad solutions since, ordinarily, the correlation coefficient values are quite close to 1, in this case, because, on average, the pattern of changes in ultrametric or tree metric values is very similar to that in the raw matrix.

As a whole, the study of dissimilarity between hierarchies seems still underdone. There have been made some attempts of abstract algebraic considerations (Monjardet 1981, Leclerc 1985) and of analysis of the probabilistic distribution for some of the indices under a hypothesis on the distribution of the trees (usually, the uniform one) (Lapointe and Legendre 1992, Steel 1993).

7.1.10   Discussion

The most customary representations of hierarchical classification are the rooted labeled tree, set tree, and ultrametric. However, unrooted labeled trees, splits and tree metrics, currently, have become quite popular in some substantive studies, especially in molecular evolution and psychology research. Nest indicator bases and neighbors relations are examples of somewhat more strange representations of the hierarchical classification.

An overview presented is an attempt to have a more or less complete list of the representations along with characterizations and interconnection results. More systematic theory of these and, perhaps, other representations of the hierarchical classification is a subject for future research.

7.2   Monotone Equivariant Methods

7.2.1   Monotone Equivariance and Threshold Graphs

Any hierarchical clustering method dealing with dissimilarity matrices $d = (d_{ij})$ can be considered as a map F defined on the set D of all dissimilarity matrices d into the set of all ultrametrics U. Jardine and Sibson 1971 suggested a set of requirements (axioms) for a "universally acceptable" clustering method $F : D \to U$. One of the axioms will be investigated in this section.

Let us consider, out of all the clustering methods, $F : D \to U$, those that are invariant with regard to monotone transformations of the between-entity dissimilarities. Actually, Jardine and Sibson 1971 claimed that only such methods should be considered as appropriate ones, rejecting thus such powerful clustering techniques as based on the square-error criterion which is one of the major concerns in this book. Although monotone-invariance seems neither universal nor necessary, it is a nice property, and it is interesting to learn what the clustering techniques are which satisfy the requirement. The study was undertaken by M. Janowitz and his collaborators (see, for example, Janowitz 1978, Janowitz and Stinebrickner 1993, and Janowitz and Wille 1995).

First, let us specify the monotone transformations. Let us consider the set of all non-negative reals $[0, \infty)$; a mapping θ on this set will be referred to as *isotone* if $\theta(0) = 0$ and $a \leq b \to \theta(a) \leq \theta(b)$. (Note that, usually, the isotone mappings are not required to be 0-preserving, but here we confine ourselves with this narrower concept only). A one-to-one isotone mapping of $[0, \infty)$ onto itself is referred to as an *order automorphism* on $[0, \infty)$.

Second, let us specify the concept of cluster method: this is any mapping F of the dissimilarity matrices $d \in D$ into ultrametrics $F(d) \in U$; the set D consists of all the dissimilarities $d : I \times I \to [0, \infty)$ such that $d_{ii} = 0$, and $U \subset D$, of all the ultrametrics defined at the same N-entity set I.

Now, we define a cluster method F as being *θ-compatible* if $F(\theta d) = \theta F(d)$

for any dissimilarity $d \in D$, where $\theta d(i,j) = \theta(d(i,j))$. A cluster method is called *monotone-equivariant* if it is θ-compatible for every order automorphism θ on $[0, \infty)$.

Intuitively, the monotone-equivariance must guarantee that the method reflects only structural, not quantitative, properties of the dissimilarities. To put this intuition in a formal setting, let us separate the quantitative and structural information contained in a dissimilarity measure $d \in D$. Let $h(d) = \{h_0, h_1, h_2, ..., h_n, h_{n+1}\}$ be the image of d, that is, the set of different d-values, naturally ordered: $0 = h_0 < h_1 < h_2 < ... < h_{n+1}$. This is the quantitative information. The set of all threshold graphs $TG(d) = \{G_{h_0}, G_{h_1}, ..., G_{h_{n+1}}\}$ where $G_h = \{(i,j) : d_{ij} \leq h\}$ represents the structural information. Obviously, $\Delta \subseteq G_{h_0} \subset G_{h_1} \subset ... \subset G_{h_{n+1}} = I \times I$ where $\Delta = \{(i,i) : i \in I\}$. Since the maximum dissimilarity always gives $I \times I$ as the corresponding threshold graph, it will be convenient to eliminate it from our consideration. In the sequel, $h(d) = \{h_0, h_1, h_2, ..., h_n\}$ where h_n is the second-maximum of the dissimilarities, and, respectively, $TG(d)$ does not contain $I \times I$ (unless $d(i,j) = 0$ for all $i, j \in I$, which is the zero metric), $TG(d) = \{G_{h_0}, G_{h_1}, ..., G_{h_n}\}$.

Statement 7.12. *A cluster method F is monotone-equivariant if and only if $h(F(d)) \subseteq h(d)$ and $TG(F(d))$ depends only on $TG(d)$.*

Proof: Let F be monotone-equivariant and $d_1, d_2 \in D$ have their threshold graph sets coinciding, $TG(d_1) = TG(d_2)$, while their images may be different. However, since both of the image value sets are ordered by the relation $<$, there are infinitely many order automorphisms θ mapping one image set onto the other, meaning that $d_1 = \theta d_2$. Thus, $F(d_1) = F(\theta d_2) = \theta F(d_2)$ which implies that $F(d_1)$ and $F(d_2)$ also have coinciding threshold graph sets, $TG(F(d_1)) = TG(F(d_2))$. Moreover, if $a \in F(d)$ and $a \notin h(d)$, then any order automorphism θ' which has all $h_k \in h(d)$ fixed, $\theta'(h_k) = h_k$, but a moved, $\theta'(a) \neq a$, satisfies equation $d = \theta' d$ followed by $F(d) = F(\theta' d) = \theta' F(d)$, which implies that $\theta'(a) = a$. This contradiction implies that $h(F(d)) \subseteq h(d)$. The theorem is proved since the reverse statement is trivial. $\square$

So, any monotone-equivariant cluster method F works as a device collapsing the set of the original distance values $h(d)$ (numbering up to $(N-1)N/2$) into at most $N - 1$ ultrametric values $h(F(d))$ and generating threshold graphs that are equivalence relations (partitions). It can be claimed, also, that the monotone-equivariant methods, intrinsically, are graph-theoretical since they are based on processing the set of threshold graphs.

A simple example of such a method is the single-linkage method: it takes the values $h \in h(d)$ corresponding to the minimum spanning tree and generates the equivalence relations $F_h = \epsilon(G_h)$ where $\epsilon(G)$ is defined as the transitive closure of G, that is, the set of pairs (i,j) such that i and j belong to the same connected component of G. Curiously, F_h, in this method, depends only on corresponding

G_h, although, in general, designing F_h may require all the threshold graphs from $TG(d)$. Such a method, with F_h depending only on G_h, is called a *flat* cluster method (Jardine and Sibson 1971, Janowitz 1978).

7.2.2   Isotone Cluster Methods

Let us look at the monotone-equivariant cluster methods, in relation to more rough (that is, not necessarily one-to-one) isotone mappings ξ of $[0, \infty)$. For an arbitrary isotone ξ, let us look upon the class $F(\xi)$ of all the ξ-compatible monotone-equivariant methods F. Such a class of cluster methods will be referred to as an *isotone* class.

Before going into further analysis, let us present here some examples of the isotone classes we deal with. Any flat method is generated by a two-valued isotone mapping ξ_h, which is equal to 0 for any $x \le h$ and a constant k for $x > h$. Indeed, $\xi_h d$ equals zero, for every $(i, j) \in G_h$, and k, otherwise. This implies that the only nontrivial threshold equivalence for $\xi_h F(d)$ is F_h, thus completely corresponding to what was defined above as flat methods. Two other examples are the so-called *divisive* and *agglomerative* methods. In divisive methods, any resulting threshold equivalence depends only on the threshold graphs from $TG(d)$ corresponding to the same and larger thresholds. Dually, the threshold equivalences obtained with an agglomerative method depend only on the threshold graphs from $TG(d)$ corresponding to the same and smaller thresholds. This terminology matches the standard notions since the traditional agglomerative methods generate partitions joining smaller clusters (having smaller within dissimilarity values) while the divisive ones generate partitions splitting larger clusters in the smaller ones, thus decreasing the within-cluster distances. Yet a vast majority of cluster agglomerative/divisive techniques are not monotone-equivariant. The fact that the two classes defined are isotone will be proved later.

There are also three isotone classes of somewhat more artificial kind: *0-flat*, *0-divisive*, and *0-stable* cluster methods which differ from corresponding classes of flat, divisive and general monotone-equivariant methods, respectively, in that they admit a particular kind of dependency on the pattern of zeros in the dissimilarity matrix d provided by G_0: F_0 depends only upon G_0 while the other threshold equivalences F_h depend on their respective arguments, G_h (0-flat), G_{h_k} for $h_k \ge h$ (0-divisive), and all $TG(d)$ (0-stable), supplemented by G_0. More precise definitions along with the proofs of isotonicity will be given within the Statement to follow.

Let us define $x, y \in [0, \infty)$ as ξ–equivalent if $\xi(x) = \xi(y)$. Obviously, ξ-equivalence classes are either proper intervals (on which ξ is constant) or singletons (belonging to the intervals where ξ makes a one-to-one mapping). Thus, the maximal intervals where ξ is constant along with the maximal intervals on which ξ

is a one-to-one mapping cover all the real line, except for the boundary points of the intervals in the case when they are open (an open interval does not contain its extreme point(s)). This means that the maximal intervals along with the boundary singletons give a partition of $[0, \infty)$ called *kernel* of ξ and denoted $ker(\xi)$.

The kernel consists of at most a countable number of intervals. Each of the intervals is either a (boundary) single point or a proper interval on which ξ is constant or one-to-one. Denote these three types of intervals, respectively, by the symbols s, c, and o. If we think of a word in the alphabet $\{s, c, o\}$ as a mapping from the intervals of $ker(\xi)$ into $\{s, c, o\}$, this produces a representation of $ker(\xi)$ as a finite or countably infinite word in that alphabet.

Since, obviously, $ker(\xi_1) = ker(\xi_2)$ implies $F(\xi_1) = F(\xi_2)$, any isotone class $F(\xi)$, actually, is determined by the word corresponding to $ker(\xi)$. This allows us to change the denotation of the isotone classes from $F(\xi)$ to $F(w)$ where w belongs to the set $W(s, c, o)$ of all finite and countably infinite words in the alphabet $\{s, c, o\}$ (the empty word included). Actually, as it is proved below, there is no need to consider the infinite words, for all of them are equivalent to some two- or three-letter words.

7.2.3 Classes of Isotone Methods

Now we are ready to formulate the major result (Janowitz and Wille 1995).

Statement 7.13. *There are only seven different isotone classes of cluster methods, $F(o)$, $F(cc)$, $F(scc)$, $F(sc)$, $F(co)$, $F(oc)$, and $F(sco)$.*

Proof: First, let us note that, by definition of the kernel, the following sub-words are illegal: ss, so, os, and oo. Moreover, no proper word can end with s since $[0, \infty)$ has no largest element.

This implies that the possible 1-letter, 2-letter and 3-letter words are: c (constant), o (morphism), sc, cc, co, oc, scc, sco, csc, ccc, cco, coc, occ, and oco.

Obviously, $F(o)$ consists of all the monotone-equivariant methods since, actually, the isotone mappings with kernel o are one-to-one mappings very much resembling the order automorphisms. Moreover, $F(c) = F(o)$. Indeed, there is only one constant isotone map ξ such that $\xi(0) = 0$: it is the zero mapping, $\xi(x) = 0$ for every $x \in [0, \infty)$. But every monotone-equivariant cluster method satisfies $F(0) = 0$ where 0 denotes the zero metric (equal to zero for every $i, j \in I$), thus belonging to $F(c)$.

When $ker(\xi) = cc$, there is an $h > 0$ such that $\xi(x) = 0$ for $0 \le x \le h$ and $\xi(x) = k$ for $x > h$ where k is a constant. This means that ξd has G_h as the only proper threshold graph. Thus, to be ξ-compatible, a cluster method F must have its result depending on G_h only, that is, be flat. Moreover, if $ker(\xi) = cw_1cw_2$ where $w_1, w_2 \in W(s, c, o)$, then for any dissimilarity d, the w_1 and w_2 portions of ξ can be shifted by an order automorphism so that they miss the image $h(d)$, which means again that the only threshold graph the cluster method F deals with is G_h for an arbitrary $h > 0$. This proves that $F(cw_1cw_2) = F(cc)$.

Let us prove now that all the other words containing two or more occurrences of c are equivalent to scc. Let us, first, consider the most general form containing scc, $ker(\xi) = sw_1cw_2cw_3$ where $w_1, w_2, w_3 \in W(s, c, o)$. Again, for any dissimilarity d, we can apply an order automorphism θ so that w_1, w_2, w_3 miss all values in the image $h(d)$. This means that we may choose ξ so that ξd has only two different threshold graphs, G_0 and G_h, and G_0 cannot be eliminated since $\xi(h) > 0$ for every $h > 0$ by the formula of $ker(\xi)$. Thus, $F \in F(sw_1cw_2cw_3)$ acts so that every threshold graph F_h of its image $F(d)$ depends only on G_0 and G_h. This class of cluster methods is referred to as *0-flat*. Analogous argument shows that $F(ow_1cw_2cw_3) = F(scc)$ thus proving that only 0-flat methods are compatible with the mappings having their kernels $sw_1cw_2cw_3$ or $ow_1cw_2cw_3$.

We have proven that all the methods which are compatible with isotone mappings having more than one c in their kernels, are flat (if there is no prefix to c) or 0-flat (if there is a prefix to c). This implies also that there is no need to consider (countably) infinite words: obviously, the infinite words must contain an infinite number of occurrences of c (intervals on which ξ is constant), which has been proven to be equivalent to having two cs only. There are only five possibilities remaining: sc, co, oc, sco, and oco (since ss, oo, so, and os are forbidden). These possibilities are considered in turn:

1) If $ker(\xi) = sc$, then $\xi(0) = 0$ and $\xi(x) = k$ for any $x > 0$ where k is a constant. That means that each of ξd and $\xi F(d)$ have, respectively, G_0 or F_0 as the only proper threshold graph, which implies that F_0 depends on G_0 only. As to the other threshold equivalences F_h in $h(F(d))$, the resulting ultrametric $\xi F(d)$ does not depend on them at all, which means that $F(sc)$ consists of all the cluster methods having F_0 depending on G_0 only while F_h may depend on all the set of threshold graphs $TG(d)$. These methods are called *0-stable*.

2) If $ker(\xi) = co$, then there is an $h > 0$ such that $\xi(x) = 0$ and all $\xi(h_k)$ are different for $h_k > h$ ($h_k \in h(d)$). This shows that $F \in F(\xi)$ must have its threshold equivalences F_h depending on all G_{h_k} for $h_k \geq h$. This kind of method is called *divisive*.

3) If $ker(\xi) = oc$, then the situation is dual to that in 2): $F \in F(oc)$ must have its threshold equivalences F_h depending on all G_{h_k} for $h_k \leq h$. This kind of method is called *agglomerative*.

4) If $ker(\xi) = sco$, then the situation is much as in 2). If ξ is chosen to have all the image of d mapped into a constant, which means that there is only one proper threshold graph G_0 for ξd, then F_0 of $F(d)$ must depend on that. In a general situation, ξ maps 0 in 0, collapses some lower levels of positive d values, and maps the larger h_k values in different values $\xi(h_k)$. That means that F_h may depend only on G_0 and G_{h_k} for all $h_k \geq h$. Such a method is called 0-*divisive*.

5) If $ker(\xi) = oco$, then $F(\xi) = F(scc)$. Indeed, it is obvious that if $F(\xi_1) = F(\xi_2)$ then $F(\xi_2\xi_1)$ is the same. Let ξ_2 have its c-interval being the end part of the first o-interval of ξ_1, then $ker(\xi_2\xi_1) = occo$, which proves that $F(oco) = F(occo) = F(scc)$ as is proved above. □

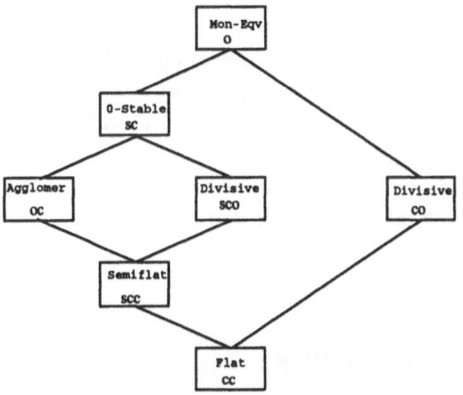

Figure 7.8: Seven classes of isotone-equivalent monotone-equivariant clustering methods presented in the set-theoretic inclusion order.

The classes of the monotone-equivariant cluster methods are presented in Fig.7.8 according to the set-theoretic inclusion. This follows from the dependencies between the structures of $TG(d)$ and $TG(F(d))$ revealed: for example, flat methods are 0-flat since dependence of F_h on only G_h is a special case of more general dependence of F_h on F_0 and F_h, which itself is a special case of more general dependencies in agglomerative and 0-divisive methods. The "meets" in the diagram in Fig.7.8 correspond to the intersections of the corresponding classes, which again can be proved by analyzing the structural dependencies.

7.2.4   Discussion

Another example of axiomatic analysis is presented. The monotone equivariance requirement is a nice mathematical property, though a cluster method should not be considered as being at a disadvantage if it does not satisfy it.

Basically, the requirement picks out the methods based on processing the threshold graphs on the entity set. All the most interesting methods involving averaging or more complicated arithmetic operations applied to the dissimilarity data fail to satisfy the requirement and drop out. Still, in some situations, when the dissimilarities are known up to their ordering only, using the arithmetic-driven methods may seem meaningless, and this is a case when the monotone equivariant methods are welcome.

By further strengthening the requirement, we arrive at clustering methods that are equivalent to each other with regard to an isotone mapping on the real line. Such a mapping may collapse some different distances into the same value, and the methods may not distinguish those distances anymore. It happens that there are only seven isotone equivalence classes of monotone-equivariant methods. This kind of result seems to belong to meta-classification studies; it illustrates both the gains and limits of the axiomatic approach.

7.3   Ultrametrics and Tree Metrics

7.3.1   Ultrametric and Minimum Spanning Trees

An *ultrametric* is a distance matrix $D = (d_{ij})$, $i, j \in I$, satisfying a strengthened triangle inequality for any $i, j, k \in I$:

$$d_{ij} \leq \max(d_{ik}, d_{jk}). \tag{7.6}$$

A wide set of flat clustering techniques can be proposed, based on relations between the ultrametrics and minimum spanning trees (Gower and Ross 1969, Leclerc 1981). Let us recall that a tree $T = (I, E)$ is referred to as a spanning tree of length $l(T) = \sum_{ij \in E} d_{ij}$ where d is a dissimilarity measure between the entities. $T(i, j)$ denotes the only path in T joining its nodes $i, j \in I$. Here, I is set of all the nodes, not just leaves only. A minimum (maximum) spanning tree mT (MT) has minimum (maximum) length $l(T)$. A simple algorithm for constructing a minimum (maximum) spanning tree has been described in Section 2.2.4. What is important here is that the minimum (maximum) spanning tree depends only on the order of the dissimilarities, not their quantitative values. This is especially readily seen from the following characteristic property of the minimum and maximum spanning trees (Leclerc 1981).

Statement 7.14. *Spanning tree T is an mT if and only if, for any $i, j, k, l \in I$, $(i, j) \in T(k, l)$ implies $d_{ij} \leq d_{kl}$, and, dually, T is an MT if and only if, for any $i, j, k, l \in I$, $(i, j) \in T(k, l)$ implies $d_{ij} \geq d_{kl}$.*

Proof: Let $(i, j) \in mT(k, l)$; then $d_{ij} \leq d_{kl}$ since if $d_{ij} > d_{kl}$ then putting (k, l) instead of (i, j) in mT would make its length lesser, which contradicts the definition of mT. If, conversely, T is a spanning tree satisfying the condition, then its length must be minimum since its edges cannot be substituted by other ones with the total length decreased. □

Any spanning tree contains exactly $N - 1$ edges while any ultrametric has at most $N - 1$ different values; this gives us a hint that ultrametrics may be somehow encoded with spanning trees. An explicit definition follows.

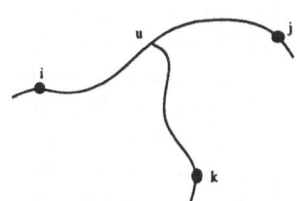

Figure 7.9: A general pattern of joining three leaves in a labeled tree.

For any spanning tree $T = (I, E)$ and dissimilarity d, let us construct a dissimilarity d_T:

$$d_T(i, j) = \max\{d_{i'j'} : (i', j') \in T(i, j)\}$$

where $T(i, j)$ is the unique path between the vertices i and j in tree T. As can be easily seen in Fig.7.9, the paths $T(i, k)$ and $T(j, k)$ cover path $T(i, j)$, which implies $\max[d_T(i, k), d_T(j, k)] \geq d_T(i, j)$ and proves that d_T is an ultrametric. Moreover, method $F(d) = d_T$ is flat since all classes of the equivalence threshold graphs defined by d_T are connected components of corresponding threshold graphs defined by d. Although the method derives an ultrametric distance matrix, not the hierarchy itself, its construction does not present any difficulties: the spanning tree T employed gives all the necessary information. The different metric values $d_1 < d_2 < ... < d_n$ are the node cluster indices while, for every d_h, $h = 1, ..., n$, the node clusters are just connected components of the d_h-threshold graph obtained from T.

Obviously,

$$d_{mT} \leq d \leq d_{MT}$$

where mT and MT are minimum and maximum spanning trees defined for d, respectively. Moreover, d_{mT} is the maximum ultrametric satisfying inequality $d_{mT} \leq d$ since $d_{mT}(i, j) = d_{ij}$ for all edges in mT and there are no other values involved in d_{mT}. Regretfully, no dual statement holds about d_{MT}.

The minimum spanning tree is associated with single linkage clustering.

Statement 7.15. *For any minimum spanning tree, mT, defined for a given dissimilarity d, the mT-ultrametric corresponds to a hierarchy found with the single linkage (nearest neighbor) method.*

Proof: We'll prove that every cluster found with the single linkage corresponds to a connected component of the minimum spanning tree mT, and the nearest-neighbor dissimilarity between two single-linkage clusters $d(S, T) = \min_{i \in S, j \in T} d_{ij}$ can be realized with an edge (i^*, j^*) belonging to mT. Since the minimal dissimilarity value $d_{i_0 j_0}$ is present in mT for a pair $i_0, j_0 \in I$, the cluster $\{i_0, j_0\}$ can be found with the single linkage and it is a connected component of mT. Then, let some two clusters S and T be connected components in T while their joining pair $(i^*, j^*) \in S \times T$ is not an edge of mT. This implies that $d_{i^* j^*} = d_{i' j'}$ where (i', j') is the only edge of mT connecting S and T. Indeed, $d_{i^* j^*} < d_{i' j'}$ contradicts minimality of mT and $d_{i^* j^*} > d_{i' j'}$ is impossible by the definition of the single linkage. Thus, the pair (i', j') can be taken in the single linkage method for merging S and T, which proves the statement. $\square$

In Fig.7.10, the mT- and MT-ultrametrics ((b) and (e), respectively) are shown for the Primates dissimilarity data (a). Corresponding hierarchies are presented in (d) and (g). The results found are much different: chain (b) versus star (e), and the hierarchies (d) and (g) are reversely nested. This is caused by the fact that the $N - 1$ entries determining mT are almost nonoverlapping with the $N - 1$ entries determining MT; this can and does produce as great a difference in the results as possible. The peculiarity of the methods that they involve just $N - 1$ dissimilarity entries rather than all $N(N - 1)/2$ of them results in a twofold effect: (1) a nice

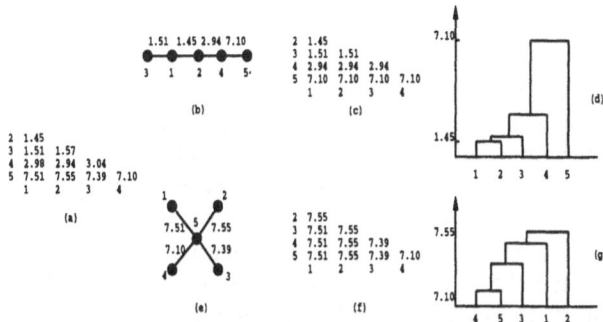

Figure 7.10: The minimum (b, c, d) and maximum (e, f, g) spanning tree generated hierarchies for Primates data (a).

mathematical theory, (2) a poor application capability.

An analogous theory can be developed for the case of the rectangular dissimilarity data $d = (d_{ij})$, $i \in I$, $j \in J$, where $I \cap J = \emptyset$; some details can be seen in De Soete et al. 1984 including the following extension of the ultrametric inequality:

$$d_{ij} \leq \max(d_{i'j}, d_{ij'}, d_{i'j'}))$$

for all $i, i' \in I$, $j, j' \in J$.

7.3.2 Tree Metric and Its Adjustment

A dissimilarity $d = (d_{ij})$ is a tree metric if it satisfies four-point condition (7.3):

$$d_{ij} + d_{kl} \leq \max(d_{ik} + d_{jl}, d_{il} + d_{jk}). \tag{7.7}$$

How can one derive a weighted tree representing a given tree metric? There have been developed several approaches to the problem as based on: (1) definition, (2) FKE-transform reduction to the ultrametric, and (3) sequentially finding "immediate" neighbors. Let us discuss, in brief, all the three approaches.

Most straightforward and perhaps the simplest, definition-based approach relies on the fact that, for any $i, j \in I$, the tree metric distance from another entity $k \in I$ to the only path $T(i, j)$ joining i and j in T is equal to $d(k, u) = (d_{ik} + d_{jk} - d_{ij})/2$, which is obvious from the picture in Fig.7.9. Thus, we may start drawing the tree from any pair i, j joined by an edge of length d_{ij} considered as a starting draft of the path $T(i, j)$ between i and j in T. To join an arbitrary k with this "current path draft", we determine its distance from that, $d(k, u)$, which allows us putting u on the path draft after its location is determined uniquely with the distances from i and j, $d(u, i) = d_{ik} - d(k, u)$ and $d(u, j) = d_{jk} - d(k, u)$. Continuing this process, we arrive at a tree sought (see Waterman et al. 1977). A drawback of this approach is that it cannot be applied to arbitrary dissimilarity data since values $d(k, u)$ based on different pairs i, j are inconsistent with each other and with distances from the other entities to $T(i, j)$. In this aspect, the other two approaches seem somewhat better since they lead to a uniquely defined tree even when the data does not satisfy the four-point condition. However, the meaning of the tree found remains unclear and heuristic.

Let us discuss the FKE-transform based approach. Selecting an arbitrary $c \in I$ and defining a derivative ultrametric $d_c(i, j) = d_{ij} - d_{ic} - d_{jc}$ (see Statement 7.5.), we can see that, actually, the tree metric has at most $2N - 3$ different values. Indeed, ultrametric d_c has at most $N - 2$ values since it is defined on set $I - \{c\}$, and there are $N - 1$ values d_{ic}, $i \neq c$; these $2N - 3$ values determine the tree metric completely due to the equations $d_{ij} = d_c(i, j) + d_{ic} + d_{jc}$. This allows us to use a minimum spanning tree mT_c of the ultrametric d_c to store all the information on the tree metric (Bandelt 1990); the restoring formula:

$$d_{ij} = \max_{(k,l) \in mT_c(i,j)} d_c(k, l) + d_{ic} + d_{jc} \qquad (7.8)$$

is based on Statement 7.14.

The same formula can be utilized when an arbitrary dissimilarity measure $d = (d_{ij})$ is given to approximate it (in an intuitive sense) by a tree metric. To do that, let us:

1) choose a $c \in I$;

2) calculate dissimilarity $d_c(i, j) = MM + d_{ij} - d_{ic} - d_{jc}$ for every $i, j \in I - \{c\}$ (where MM is a large number, say, $MM = 2 \max_{i,j \in I} d_{ij}$);

3) determine a minimum spanning tree mT_c for d_c;

4) calculate a tree dissimilarity with formula (7.8).

A nice property of the procedure is that the dissimilarity found satisfies the four-point condition (7.7); thus, the procedure restores the given d when it is a tree metric. Yet there are also some drawbacks: (1) each $c \in I$ chosen may lead to a different result, (2) the dissimilarity found may have some negative values (Leclerc 1995). A more direct way for storing and restoring the tree metric based just on a minimum spanning tree constructed for the tree metric itself has been suggested recently by Leclerc 1995 (based on a method elaborated by Critchley 1994 for a particular case).

Sattah and Tversky 1977 developed an approach based on the following observation. A pair of leaves of a tree are called *immediate neighbors* if they are connected through a single interior node; the interior connecting node will be referred to as the *marginal* node. It is obvious that, in any tree, there is at least one pair of immediate neighbors. Let us assume that we always can identify a pair of immediate neighbors by the tree metric. This gives rise to the following algorithm for designing an underlying tree, which is referred to usually as the *neighbor joining (NJ) algorithm* (Saitou and Nei 1987).

Neighbor Joining Algorithm
1. Pick a pair of immediate neighbors, i and j.
2. Form a new node u with its distances $d_{u,k} = (d_{ik} + d_{jk} - d_{ij})/2$, $k \in I - \{i, j\}$, put it in I and remove i and j (after deleting i and j, u becomes a leaf).
3. If there are still some entities of I unremoved, go to step 1 (with the data reduced); otherwise end.

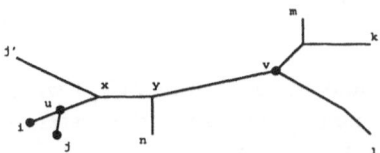

Figure 7.11: After removal of the immediate neighbors i and j, u remains a leaf.

The process can be illustrated with Fig.7.11. If the underlying tree is not binary, for example, the distance between u and x on Fig.7.11 is zero, it will be manifested in the fact that all the distances defined for x will be the same as for u: $d(u, k) = d(x, k)$ for any $k \in I - \{i, j, j'\}$. The distances $d(i, u)$ and $d(j, u)$ are evaluated based on the same considerations: say, $d(i, u) = (d_{ik} + d_{ij} - d_{kj})/2$ for any $k \in I - \{i, j\}$ when the distance d is a tree metric.

To perform the first step of NJ algorithm, a procedure for determining a pair of immediate neighbors is needed. Sattah and Tversky 1977 pointed out that the immediate neighbors i and j must satisfy inequality

$$d_{ij} + d_{kl} < d_{ik} + d_{jl}$$

for every pair $\{k, l\} \neq \{i, j\}$; they proposed to count, for every pair i, j, the number of the pairs $\{k, l\}$ satisfying the inequality above (which is just the score function $\sigma(i, j)$ (7.5)). Obviously, when the dissimilarity is a tree metric, any pair which has received the maximum number of counts (maximum $\sigma(i, j)$) must be immediate neighbors; this principle is utilized in their version of the NJ algorithm (Sattah and Tversky 1977).

Gascuel 1994 demonstrated that the other neighbor-joining algorithms suggested by Saitou and Nei 1987 and Studier and Keppler 1988, actually, are based on maximization of a quantitative form of Sattah and Tversky's criterion:

$$D(i, j) = \sum_{k, l \in I - \{i, j\}} (d_{ik} + d_{jl} + d_{il} + d_{jk} - 2d_{ij} - 2d_{kl})/2.$$

We suggest using, for the same purpose, a centrality index of the path $T(i, j)$ between $i, j \in I$ defined as

$$c(i, j) = \sum_{k \in I} d(k, T(i, j))$$

where $d(k, T(i, j))$ is the distance between $k \in I$ and the node u in $T(i, j)$ where k joins the path (see Fig.7.9). For the points i and j themselves the distances $d(i, T(i, j))$ and $d(j, T(i, j))$ can be defined as equal to $d(i, j)/2$.

With elementary arithmetic, it can be proven that the centrality and Gascuel's indices are linearly related:

$$(N - 1)c(i, j) = \sum_{i, j \in I} d_{ij} + D(i, j).$$

The centrality index can be calculated with a simpler formula derived from the equality $d(k, u) = (d_{ik} + d_{jk} - d_{ij})/2$:

$$c(i, j) = (d_{i+} + d_{j+} - (N - 2)d_{ij})/2$$

where $d_{i+} = \sum_{k \in I} d_{ik}$ for any $i \in I$. This formula allows much simpler calculation since it is based on pairs of the entities, not quadruples of them, as the other ones.

The NJ algorithm is correct if the claim that the maximum of the centrality index (or, equivalently, of Saitou-Nei or Studier-Keppler index) corresponds to a pair of immediate neighbors is correct; though the proofs published by the latter authors do not seem to be quite clear, the fact seems quite plausible.

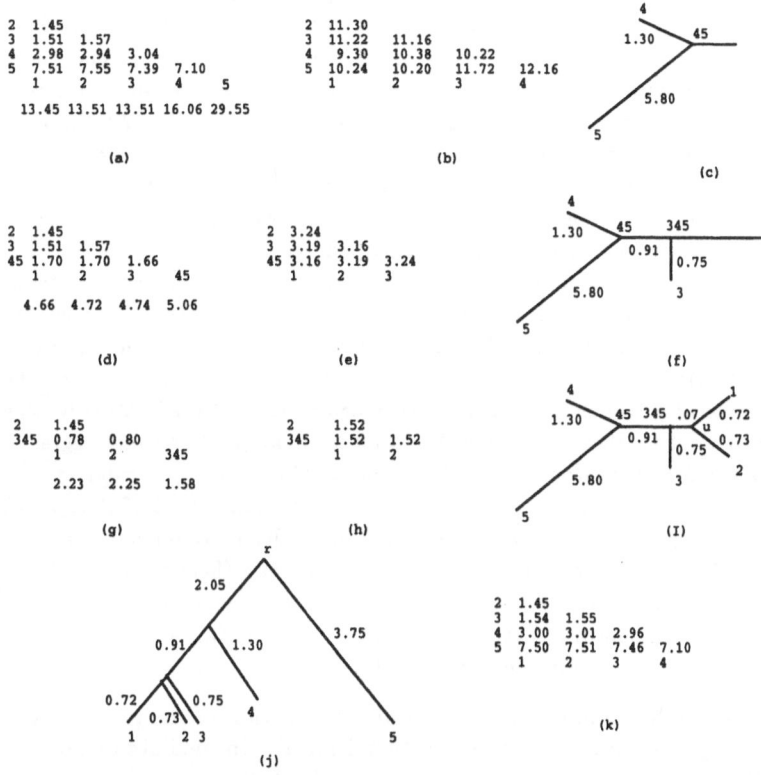

(a)
```
2  1.45
3  1.51  1.57
4  2.98  2.94  3.04
5  7.51  7.55  7.39  7.10
   1     2     3     4     5
13.45 13.51 13.51 16.06 29.55
```

(b)
```
2  11.30
3  11.22  11.16
4   9.30  10.38  10.22
5  10.24  10.20  11.72  12.16
    1      2      3      4
```

(c)
```
      4
   1.30     45
            5.80
   5
```

(d)
```
2   1.45
3   1.51  1.57
45  1.70  1.70  1.66
    1     2     3     45
4.66  4.72  4.74  5.06
```

(e)
```
2   3.24
3   3.19  3.16
45  3.16  3.19  3.24
    1     2     3
```

(f)
```
      4      45    345
   1.30      0.91
                   0.75
             5.80   3
   5
```

(g)
```
2    1.45
345  0.78  0.80
     1     2     345
2.23  2.25  1.58
```

(h)
```
2    1.52
345  1.52  1.52
     1     2
```

(I)
```
      4    45  345  .07    1
                          u  0.72
   1.30    0.91          0.73
                  0.75      0.73
           5.80   3         2
   5
```

(j)
```
          r
   2.05
      0.91   1.30        3.75
  0.72  0.75   4
     0.73
   1    2 3                 5
```

(k)
```
2  1.45
3  1.54  1.55
4  3.00  3.01  2.96
5  7.50  7.51  7.46  7.10
   1     2     3     4
```

Figure 7.12: NJ-algorithm applied to Primates data.

In Fig.7.12, the sequential iterations of the NJ algorithm applied to the 5 by 5 Primates dissimilarity matrix are presented. The matrix is presented along with the summary distances d_i, $i = 1,...,5$ in (a); corresponding centrality index values are in matrix (b). We can see that the maximum equals $c(4,5) = 12.16$; this allows us to agglomerate the species 4 and 5 into an interior (ancestral species) node 45, as shown in (c). The distances between 45 and its constituents are calculated based on the formula mentioned above, $d(i,u) = (d_{ik} - d_{jk} + d_{ij})/2$, where u is the agglomerate of i and j and $k \in I$ is arbitrary. The latter formula holds when d is a tree metric. When the dissimilarity is arbitrary, there is no preference for choice of $k \neq i,j$. The standard practice, in this case, is to average the estimates by $k \neq i,j$, which leads to

$$d(i,u) = [\sum_{k \neq i,j}(d_{ik} - d_{jk})/(N - 2) + d_{ij}]/2,$$

and an analogous formula is utilized to estimate $d(j, u)$. In our case, $\sum_{k \neq 4,5}(d_{4k} - d_{5k})/3 = (4.53 + 4.61 + 4.35)/3 = 4.50$. This results in the estimates $d(4, 45) = (4.50 + 7.10)/2 = 5.80$ and $d(5, 45) = (-4.50 + 7.10)/2 = 1.30$ (see 7.12 (c)). It remains to recalculate dissimilarities between the new node 45 and the others using formula $d(u, k) = (d(i, k) + d(j, k) - d(i, j))/2$ (see the resulting distances in Fig.7.12 (d)). At the next iteration, we can see that the centrality index (e) has two maxima: $c(1, 2) = c(45, 3) = 3.24$. Actually, this can be used for parallel merging 1 with 2 and 45 with 3 since such a result clearly indicates that both pairs are immediate neighbors (in the case when d is a tree metric). However, let us do it sequentially: merge 3 and 45 along with the subsequent estimation of the edges obtained (Fig.7.12 (f)). Once again, we start with an aggregate data (g), this time having all the centrality index values equal (h), which clearly indicates that all three nodes, 345, 1, and 2, must be merged into the same aggregate node u (see Fig.7.12 (i)). The tree is found, but there is no information to choose the root location. In molecular biology, supplementary information is utilized. Usually, this is done by adding to I yet another species which is known to have quite a small degree of the evolutionary kinship to the species in I. The node of joining this particular species to the tree constructed for I, indicates the location of the "ancestral" root of the species in I. In our case, the species 5, Rhesus monkey, is, obviously, more distant from the others, thus indicating the root location for the other four species (between 3 and 4), as well as for all the set of five (between 4 and 5). A corresponding rooted tree (with the appropriate edge lengths) is shown in Fig.7.12 (j). The tree metric (k), in this particular case, does not look too different from the original data.

It should be noted, that due to the sequential organization of the process considered, the following regularity holds: the farther from the immediate neighbors, the larger the differences between the original and derived dissimilarities.

7.3.3 Discussion

There are three major questions about each of the two classes of metrics considered, ultrametric and tree metric:

(1) Whether a given dissimilarity metric belongs to a class (recognition problem);

(2) If yes, how a corresponding tree can be reconstructed; and how all the set of corresponding trees can be characterized (reconstruction problem);

(3) If no, how the given data set can be approximated in a class of metrics considered (approximation problem).

Addressing the two former questions is not an issue; a handful of possible approaches to that have been described above. The methods of recognition and reconstruction for both ultrametric and tree metric seem quite satisfactory computationally. Moreover, the tree reconstructed is uniquely defined. On the other hand, we have presented nothing but heuristic methods for the approximation problem. The problem of approximation with least-squares or least-moduli criterion has been proven computationally hard (Day 1987), though minimizing maximum-moduli criterion may be computationally easier (Agarwala et al. 1995). Eventually, substantive research may suggest other forms of approximation criteria.

Yet there is another aspect of the analysis: the relationship between the two kinds of metrics provided by FKE-transformation. There is something mysterious in that. The ultrametric inequality holds up to monotone transformations of the distance: for any non-decreasing real function ξ, distance ξd is an ultrametric iff d is, which is not true when d is a tree metric; the four-point condition holds up to any non-decreasing linear (not arbitrary monotone) function, $\xi(x) = ax + b$, where $a \geq 0$. This means that the topology of a rooted tree can be reconstructed in a unique way under any non-decreasing transformation ξd of the ultrametric d while the unrooted tree reconstructed from a tree metric d is invariant only with regard to linear functions ξ. The question is this: what makes the switching between the ordinal and cardinal kinds of data so easy?

7.4 Split Decomposition Theory

7.4.1 Split Metrics and Canonical Decomposition

Let $d = (d_{ij})$, $i, j \in I$, be a tree metric on an N-element set I where $N \geq 5$. As we have stated already, for any $i, j, k, l \in I$, inequality in the four-point condition,

$$d_{ij} + d_{kl} < \max(d_{ik} + d_{jl}, d_{il} + d_{jk}),$$

indicates that vertices i and j can be separated from k and l in the corresponding weighted tree by eliminating an edge. To fix the split edge location, we need information on the whole T-split $S = \{A, B\}$. Moreover, the weight of the split edge can be estimated with the following "four-point" dissimilarity function

$$d(ij, kl) = [\max(d_{ij} + d_{kl}, d_{ik} + d_{jl}, d_{il} + d_{jk}) - d_{ij} - d_{kl}]/2, \qquad (7.9)$$

applied to every $i, j \in A$ and $k, l \in B$. The weight equals

$$d(A, B) = \min_{i,j \in A, k,l \in B} d(ij, kl) \qquad (7.10)$$

which can be easily seen in Fig.7.3.

The T-splits $S = \{A, B\}$ can be recovered from d as the biclass partitions of I such that for every $i, j \in A$ and $k, l \in B$ the strong four-point inequality above is true.

When d is an observed symmetrical dissimilarity function, not an exact tree metric, this method can be generalized in various ways. The following discussion presents the results found by Bandelt and Dress 1992 based on use of formula (7.10) applied to all possible pairs of nonoverlapping subsets $A, B \subset I$, biclass partitions included.

Obviously, $d(A, B) \geq 0$ since $d(ij, kl) \geq 0$ by its definition in (7.9). The pairs $\{A, B\}$ with $d(A, B) > 0$ will be called d-splits (with added adjective "partial" when $A \cup B \neq I$). Since some of i, j and k, l may be coincident, in general, $d(\{i, j\}, \{k, l\})$ may be less than $d(ij, kl)$. However, these values coincide when d satisfies the triangle axiom because, in this case, $d(ij, kl)$ cannot be larger than any $d(i_1 i_2, k_1 k_2)$ with $i_1, i_2 \in \{i, j\}$ and $k_1, k_2 \in \{k, l\}$. For example, $d_{ik} + d_{jl} - d_{ij} - d_{kl} \leq d_{ik} + d_{il} - d_{kl} = 2d(ii, kl) \leq 2d_{ik} = 2d(ii, kk)$.

Let us denote the set of all d-splits by $S(d)$. For a partial split $\{\{i, j\}, \{k, l\}\}$, let us denote $S(d)(ij, kl)$ the set of all d-splits $\{A, B\} \in S(d)$ extending it, that is, such that $\{i, j\}$ is included in one of the classes A, B while $\{k, l\}$ in the other. Since, obviously, at least one of the values $d(ij, kl), d(ik, jl)$, and $d(il, jk)$ is zero, the set of d-splits must be compatible in that, for every quadruple $i, j, k, l \in I$, $S(d)(ij, kl) \cap S(d)(ik, jl) \cap S(d)(il, jk) = \emptyset$. This latter equation will be referred to as the condition of *weak compatibility* (which extends the previous condition of mere compatibility held for T-splits). A dissimilarity function d may have $S(d)$ empty, which means that $d(A, B) = 0$ for every split $\{A, B\}$; such a d is called *split-prime*. The simple mismatch coefficient between all 8 subsets of a three-element set (or Hamming distance between all 3-digit Boolean vectors, which is the same) is an example of the split-prime metric.

Any split $S = \{A, B\}$ can be associated with a so-called *split metric* d_{AB} defined so that $d_{AB}(i, j) = 1$ whenever i and j belong to different classes of the split, and $d_{AB}(i, j) = 0$ when i, j are in the same class, A or B. Actually, this is dissimilarity $1 - s_{ij}$ associated with the equivalence indicator matrix for partition S, p. 230; another term used here underscores its metric application.

Since a set of d-splits, $S(d)$, obviously, does not change when d is multiplied or/and added with a real, that is, $S(d) = S(\alpha d + \beta)$, for any real β and positive α, in the rest of this section, we consider only nonnegative dissimilarities with zero diagonal, $d_{ij} \geq d_{ii} = 0$ for every $i, j \in I$. It appears (Bandelt and Dress 1992) that the set $S(d)$ has remarkable properties:

Properties of d-splits

1. For each d-split $S = \{A, B\}$, there exist $i, j, k, l \in I$ (some of i, j or/and k, l may coincide) defined by condition $d(A, B) = d(ij, kl)$ such that $S(d)(ij, kl) = \{S\}$; that is, S is the only element of $S(d)(ij, kl)$.

2. The following *canonical* decomposition holds:

$$d = \sum_{\{A,B\} \in S(d)} d(A, B) d_{AB} + d^0 \qquad (7.11)$$

where residue d^0 is a split-prime nonnegative dissimilarity function (which is a metric if d is a metric).

3. All the elements of the canonical decomposition (d-split metrics, d_{AB}, and the split-prime residue, d^0) are linearly independent in $N(N-1)/2$-dimensional space of the dissimilarity functions on I.

4. Any set of splits of I is a set of d-splits for a dissimilarity function d if and only if it is weak compatible.

5. A dissimilarity d has its split-prime residue d^0 vanished, $d^0 = 0$, if and only if d satisfies the following *five-point* condition: for every $i, j, k, l, u \in I$,

$$d(ij, kl) \leq d(iu, kl) + d(ij, ku).$$

These statements show that the number of d-splits is not larger than $N(N-1)/2$, and $d^0 = 0$ when the bound is exact. To find all the d-splits, the following recursive procedure can be carried out.

Finding d-splits

Suppose that all partial d-splits have been found for a proper k-element subset $X \subset I$. Then, pick any $u \in I - X$ and check whether $\{X, \{u\}\}$ or $\{A \cup \{u\}, B\}$ or $\{A, B \cup \{u\}\}$ is a partial split of $X \cup \{u\}$, for any partial d-split $\{A, B\}$ of X. In this way, all partial d-splits of $X \cup \{u\}$ will be found. The values $d(A, B)$ are updated at each iteration.

After some $O(N)$ iterations all the d-splits S along with their $d(S)$ will be found. Since every iteration step involves some $O(k^3 k(k-1)/2) = O(k^5)$ comparisons (where $k = |X|$), the complexity of the whole procedure is $O(N^6)$, and it remains to find out whether any procedure of lesser complexity exists.

Let us apply the procedure to Primates 5×5 distance data. It can be seen rather easily that there are 10 d-splits listed in Table 7.1. Curiously, the number of d-splits

Split	Index
1	0.695
2	0.705
3	0.695
4	1.245
5	5.725
1-2	0.03
1-3	0.01
2-4	0.04
3-5	0.09
4-5	0.96

Table 7.1: Set of all d-splits for the Primates distance data; each presented with its smaller class and split index value.

found is exactly $N(N-1)/2 = 10$ for $N = 5$, which means that there is no residue in decomposition of the metric through the splits. For instance, $d_{34} = 3.04 = 0.695 + 1.245 + 0.01 + 0.04 + 0.09 + 0.96$ where terms correspond to splits 3, 4, 1-3, 2-4, 3-5, and 4-5 from Table 7.1, respectively.

No simple visual representation for the canonical decomposition has been found yet. However, an approximate tree representation has been proposed by Bandelt and Dress 1994 via finding a compatible subset of splits. A greedy procedure works as follows.

Greedy Split-Based Tree
Order the splits in order of decreasing split indices. Pick them in this order checking whether the current is compatible with those previously selected. If yes, add it to the collection; if no, drop it out, and take the next one.

Applied to the data in Table 7.1 already ordered bottom-up, the algorithm leads to a collection consisting of only two nontrivial splits, 1-2 and 4-5 (the others are noncompatible with split 4-5 taken first). The trivial splits are compatible always and may be omitted from the procedure. This gives the tree presented in Fig.7.13 along with corresponding tree metric. The metric is obviously subdominant to the original metric, which reflects

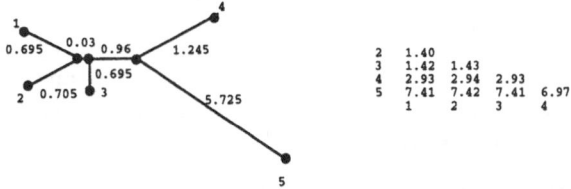

2	1.40			
3	1.42	1.43		
4	2.93	2.94	2.93	
5	7.41	7.42	7.41	6.97
	1	2	3	4

Figure 7.13: The weighted tree and its tree metric produced by the "greedy" splits for Primates data.

the fact that three d-splits have been excluded from the tree.

7.4.2 Mathematical Properties

The statements 1 to 4 in the box on p. 365 will be proved in this section to substantiate the algorithm described.

To prove item 1, let us demonstrate that the following holds.

Statement 7.16. *For any partial d-split $\{A_0, B_0\}$*

$$\sum_{\{A,B\} \in S(d)(A_0, B_0)} d(A, B) \leq d(A_0, B_0).$$

Proof: Let us prove that, for any $u \neq i, j, k, l$,

$$d(\{i, j, u\}, \{k, l\}) + d(\{i, j\}, \{k, l, u\}) \leq d(ij, kl). \tag{7.12}$$

Let the inequality fail; that is, $d(\{i_1, i_2, u\}, \{k_1, k_2\}) + d(\{i_1, i_2\}, \{k_1, k_2, u\}) > d(i_1 i_2, k_1 k_2)$ for some i_1, i_2, k_1, k_2, u. The fact that $d(i_1 i_2, k_1 k_2) > 0$ implies that all the three quantities are positive since no single item on the left can be larger than $d(i_1 i_2, k_1 k_2)$, by definition.

Let $\{p, q\} = \{1, 2\}$ so that $d(i_1 u, k_1 k_2) = (d_{i_1 k_q} + d_{u k_p} - d_{i_1 u} - d_{k_1 k_2})/2$. By the assumption,

$$d(i_1 u, k_1 k_2) + d(i_1 i_2, u k_p) \geq d(\{i_1, i_2, u\}, \{k_1, k_2\}) + d(\{i_1, i_2\}, \{k_1, k_2, u\}) > d(i_1 i_2, k_1 k_2),$$

which means that

$$d_{i_1 k_q} - d_{i_1 u} + \max(d_{i_1 u} + d_{i_2 k_p}, d_{i_1 k_p} + d_{i_2 u}) > \max(d_{i_1 k_1} + d_{i_2 k_2}, d_{i_1 k_2} + d_{i_2 k_1}).$$

This strict inequality may only hold when $\max(d_{i_1 u} + d_{i_2 k_p}, d_{i_1 k_p} + d_{i_2 u}) \neq d_{i_1 u} + d_{i_2 k_p}$, that is, when $d_{i_1 u} + d_{i_2 k_p} < d_{i_1 k_p} + d_{i_2 u}$ for $p = 1, 2$. With i_1 and i_2 in the argument above interchanged, the same items appear in the reverse strict inequality, which is impossible, thus proving the inequality.

Let now $S = \{A', B'\}$ be a partial split extending $\{\{i, j\}, \{k, l\}\}$ (that is, $i, j \in A'$ and $k, l \in B'$) and satisfying $d(A', B') = d(ij, kl)$. For any $u \notin A' \cup B'$,

$$d(A' \cup \{u\}, B') + d(A', B' \cup \{u\}) \leq d(A', B')$$

by (7.12) since $d(A' \cup \{u\}, B') \leq d(\{i, j, u\}, \{k, l\})$ and $d(A', B' \cup \{u\}) \leq d(\{i, j\}, \{k, l, u\})$. This means that the statement is proven for the situation when partial split $\{A_0, B_0\}$ takes into account all but one element of I.

Let us now make inductive assumption that the statement holds when $|I - (A_0 \cup B_0)| = n$ and prove it for $|I - (A_0 \cup B_0)| = n + 1$. In the latter case, obviously, $\sum_{\{A,B\} \in S(d)(A_0, B_0)} d(A, B) = \sum_{\{A,B\} \in S(d)(A_0 \cup \{u\}, B_0)} d(A, B) +$

$\sum_{\{A,B\} \in S(d)(A_0, B_0 \cup \{u\})} d(A, B) \leq d(A_0 \cup \{u\}, B_0) + d(A_0, B_0 \cup \{u\}) \leq d(ij, kl),$

which proves the statement.   □

Statement 7.17. *For every d-split $S = \{A, B\}$, there exist $i, j, k, l \in I$ (some of i, j or/and k, l may coincide), defined by condition $d(A, B) = d(ij, kl)$ such that $S(d)(ij, kl) = \{S\}$; that is, S is the only element of $S(d)(ij, kl)$.*

Proof: We have

$$d(A, B) \leq \sum_{\{A', B'\} \in S(d)(ij, kl)} d(A', B') \leq d(ij, kl) = d(A, B).$$

This shows that $d(A', B') = 0$ unless $\{A', B'\} = \{A, B\}$.   □

The first claim in the box is proven. To prove the second one, let us state the following.

Statement 7.18. *For a d-split $\{A, B\}$ and $\lambda \leq d(A, B)$, function*

$$d' = d - \lambda d_{AB}$$

is nonnegative (satisfying the triangle inequality if d does) and $d'(A', B') = d(A', B')$ for all d-splits $\{A', B'\} \neq \{A, B\}$ while $d'(A, B) = d(A, B) - \lambda$.

Proof: To prove that d' is nonnegative, note that $d'_{ij} = d_{ij}$ when $i, j \in A$ or $i, j \in B$. For $i \in A$ and $j \in B$, we have $d'_{ij} = d_{ij} - \lambda = d(ii, jj) - \lambda \geq d(A, B) - \lambda \geq 0$.

When d satisfies the triangle inequality, $d_{ij} \leq d_{ik} + d_{kj}$, two kinds of situations are possible: (a) all i, j, k belong to one of the classes, say, A, (b) two of the entities, say, i, j belong to one of the classes, A, while $k \in B$. Case (a) is trivial since $d' = d$ within the classes. In the second case, $d'_{ij} = d_{ij}$ while $d'_{ik} = d_{ik} - \lambda$ and $d'_{jk} = d_{jk} - \lambda$. Since $\lambda \leq d(A, B) \leq d(ij, kk) = (d_{ik} + d_{jk} - d_{ij})/2$, we have $d_{ij} \leq (d_{ik} - \lambda) + (d_{jk} - \lambda)$, which proves the triangle inequality for d'.

Let us prove that $d'(ik, jl) = d(ik, jl)$ for all disjoint pairs of sets $\{i, k\}$ and $\{j, l\}$ such that $\{A, B\} \notin S(d)(ik, jl)$. If either of A or B contains at least three of i, j, k, l, then the equality is trivial. When A contains only two of i, j, k, l, say, $i, k \in A$ (and thus $j, l \in B$),

$$d_{ik} + d_{jl} \leq \max(d_{ij} + d_{kl} - 2\lambda, d_{il} + d_{jk} - 2\lambda)$$

since $\lambda \leq d(A, B) \leq d(ik, jl) = [\max(d_{ij} + d_{kl}, d_{il} + d_{jk}) - d_{ik} - d_{jl}]/2$.

This implies that $d'(ik, jl) = [\max(d_{ij} + d_{kl} - 2\lambda, d_{ik} + d_{jl}, d_{il} + d_{jk} - 2\lambda) - d_{ik} - d_{jl} + 2\lambda]/2 = [\max(d_{ij} + d_{kl} - 2\lambda, d_{il} + d_{jk} - 2\lambda) - d_{ik} - d_{jl} + 2\lambda]/2 = [\max(d_{ij} + d_{kl}, d_{il} + d_{jk}) - d_{ik} - d_{jl}]/2 = d(ik, jl)$, as required.

When $\{A, B\} \in S(d)(ij, kl)$, obviously, $d'(ij, kl) = [\max(d_{ik} + d_{jl} - 2\lambda, d_{il} + d_{jk} - 2\lambda) - d_{ij} - d_{kl}]/2 = d(ij, kl) - \lambda$. This implies that $d'(A, B) = d(A, B) - \lambda$ since $\{A, B\}$ is the only d-split to extend $\{i, j\}, \{k, l\}$.

It remains to show that $d'(A', B') = d(A', B')$ for all the other d-splits. Choose $i, k \in A'$ and $j, l \in B'$ such that $d(A', B') = d(ik, jl)$. Since $\{A, B\} \notin S(d)(ik, jl)$ by the Statement 7.17. above, that implies that $d(A', B') = d(ik, jl) = d'(ik, jl) \geq d'(A', B')$. To prove the reverse inequality, assume that $d'(A', B') = d'(tu, vw)$. Then, $d(A', B') \leq d(tu, vw) = d'(tu, vw) = d'(A, B)$ when $\{A, B\} \notin S(d)(tu, vw)$. Otherwise, if $\{A, B\} \in S(d)(tu, vw)$, then $d(A', B') \leq d(A', B') + d(A, B) - \lambda \leq d(tu, vw) - \lambda = d'(tu, vw)$, which completes the proof. □

The statement proven shows that the set of d-splits can be contracted by one-by-one eliminating the splits sequentially via subtractions $d' = d - d(A, B)d_{AB}$ since the only difference in the split structure of d' and d is that $\{A, B\}$ is not a d'-split. This shows that, actually, the second claim has been proven and the following theorem holds.

Statement 7.19. *The canonical decomposition (7.11) holds along with the residue d^0 being a split-prime nonnegative dissimilarity function (which is a metric if d is a metric).*

Let us now analyze the concept of weak compatibility.

Statement 7.20. *For any collection S of weakly compatible splits of I and positive λ_S chosen for every $S \in \mathcal{S}$, dissimilarity function $d = \sum_{S \in \mathcal{S}} \lambda_S d_S$ is a metric having its d-split set equal to $\mathcal{S}$ along with $d(S) = \lambda_S$ for every $S \in \mathcal{S}$.*

Proof: Let us consider a split $S = \{A, B\} \in \mathcal{S}$ and pick $i, j \in A$ and $k, l \in B$ such that $d(A, B) = d(ij, kl)$. Let us prove that $d(A, B) > 0$, that is, $\{A, B\}$ is a d-split. By weak compatibility, there is no split in $\mathcal{S}$ extending, say, $\{i, l\}, \{j, k\}$. Thus, we can partition $\mathcal{S}$ into three parts, S_1 collecting all those $S \in \mathcal{S}$ that extend $\{i, j\}, \{k, l\}$, S_2 consisting of all those $S \in \mathcal{S}$ that extend $\{i, k\}, \{j, l\}$, and $S_3 = \mathcal{S}$ $-(S_1 \cup S_2)$. Since all the splits from S_3 equally contribute to each of the four-point distance sums involving i, j, k, l, we have

$$d(ij, kl) = [\max(d_{ik} + d_{jl}, d_{il} + d_{jk}) - d_{ij} - d_{kl}]/2 =$$

$$\max(\textstyle\sum_{S \in S_1} \lambda_S, \sum_{S \in S_1 \cup S_2} \lambda_S) - \sum_{S \in S_2} \lambda_S = \sum_{S \in S_1} \lambda_S \geq \lambda_{\{A,B\}} > 0,$$

which proves that $\{A, B\}$ is a d-split. Thus, the set of d-splits includes $\mathcal{S}$. Canonically decomposing d, we have $d = d^0 + \sum_{d-splits\,S} d(S)d_S \geq \sum_{S \in \mathcal{S}} d(S)d_S \geq \sum_{S \in \mathcal{S}} \lambda_S d_S = d$, which implies $d^0 = 0$ and $d(S) = \lambda_S$ for $S \in \mathcal{S}$, or $d_S = 0$, otherwise. $\qquad\square$

Statement 7.21. *The split metrics d_S for any collection $\mathcal{S}$ of weakly compatible splits of I are linearly independent, thus, $\mathcal{S}$ cannot contain more than $N(N-1)/2$ splits.*

Proof: Assume that $\sum_{S \in \mathcal{S}} \lambda_S d_S = 0$ for some real λ_S, $S \in \mathcal{S}$. Partition $\mathcal{S}$ into three classes, S_1, S_2 and S_3, with S_1 corresponding to positive, S_2 negative, and S_3 zero lambdas. Then consider metric d given by either of the equal expressions

$$\sum_{S \in S_1} \lambda_S d_S = \sum_{S \in S_2} (-\lambda_S)d_S.$$

The previous statement implies that $S_1 = S_2$ which can be true only if both of the sets are empty. Thus $\mathcal{S}=S_3$ and all $\lambda_S = 0$, which proves that the split metrics d_S, $S \in \mathcal{S}$, are linearly independent. The dimensionality of the set of N by N symmetric matrices with zero diagonal equals, obviously, $N(N-1)/2$, and the statement is proven. $\qquad\square$

To complete the proof of the first four properties of d-splits listed in the box, it remains to prove that the split-prime residue is linearly independent from the d-split metrics.

Statement 7.22. *The split-prime residue d^0 in canonical decomposition (7.11) is linearly independent from the all d-split metrics.*

Proof: Let d^0 be a linear combination of d-split metrics d_S, $d^0 = \sum_S \lambda_S d_S$. Let us denote the set of all d-splits S with nonnegative $\lambda_S \geq 0$ by S_1, and the set of all d-splits with negative $\lambda_S < 0$, by S_2. Consider metric

$$d' = \sum_{S \in S_1} d(S)d_S + \sum_{S \in S_2} (d(S) + \lambda_S)d_S.$$

For any split S of I, $d'(S) = d(S)$ if $S \in S_1$, $d'(S) = d(S) + \lambda_S$ if $S \in S_2$, and $d'(S) = 0$, otherwise.

On the other hand, $d' = d - \sum_{S \in S_1} \lambda_S d_S$. This implies that $d'(S) = d(S) - \lambda_S$ if $S \in S_1$ and $d'(S) = d(S)$ if $S \in S_2$. Comparing the equations for $d'(S)$, we conclude that all $\lambda_S = 0$, which proves the statement. $\square$

7.4.3 Weak Clusters and Weak Hierarchy

The concept of weak cluster has been considered in Section 3.2.1, p. 130. Let us put it here in the following way. For a distance matrix $d = (d_{ij})$, let us define a three-argument function: $d(i, j/k) = \max(d_{ij}, d_{ik}, d_{jk}) - d_{ij}$. A subset $A \subset I$ is a *weak cluster* if, for any $i, j \in A$ and $k \notin A$, $d(i, j/k) > 0$. The minimum value $d(A) = \min_{k \notin A} \min_{i,j \in A} d(i, j/k)$ is called the *isolation index* of A; obviously, A is a weak cluster if and only if $d(A) > 0$.

The isolation index can be presented as $d(A) = min_{k \notin A} d(k, A)$ where $d(k, A) = \min_{i,j \in A} d(i, j/k)$ is a linkage function which is obviously monotone (as defined in Section 4.2.5). Thus, the isolation index is a quasi-convex set function, that is, it satisfies inequality $d(A \cap B) \geq \min(d(A), d(B))$ when $A \cap B \neq \emptyset$. Obviously, every entity i forms a weak cluster on its own since $d(i, i/k) > 0$ when $k \neq i$ (we do not consider the trivial zero distance here).

Statement 7.23. *If weak clusters A and B have nonempty intersection $A \cap B$, then it is a weak cluster also.*

Proof: Indeed, $d(A \cap B) \geq \min(d(A), d(B)) > 0$. $\square$

This implies that, actually, the intersection of any number of weak clusters is a weak cluster if not empty.

Statement 7.24. *If S_1, S_2 and S_3 are weak clusters, then $S_1 \cap S_2 \cap S_3 = S_k \cap S_l$ for some $k, l \in \{1, 2, 3\}$.*

Proof: Let $S_1 \cap S_2 \cap S_3$ not be equal to any of $S_1 \cap S_2$, $S_1 \cap S_3$, $S_2 \cap S_3$. Then there exists $x_i \in S_j \cap S_k - S_i$ for $\{i, j, k\} = \{1, 2, 3\}$ (see Fig.7.14). Let, without loss of generality, $d_{x_1 x_2} \leq d_{x_2 x_3} \leq d_{x_1 x_3}$. Then $d(x_1, x_3/x_2) \leq 0$, which contradicts the assumption that S_2 is a weak cluster. $\square$

Let us remind the reader the concept of weak hierarchy. A *weak hierarchy* is a set S_W of subsets S_w, $w \in W$, satisfying the following two properties: (1) for any $S_1, S_2 \in S_W$, $S_1 \cap S_2 \neq \emptyset$ implies $S_1 \cap S_2 \in S_W$; (2) for all $S_1, S_2, S_3 \in S_W$, $S_1 \cap S_2 \cap S_3 = S_k \cap S_l$ for some $k, l \in \{1, 2, 3\}$.

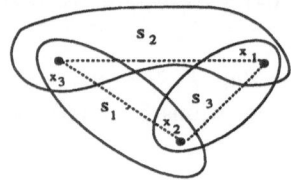

Figure 7.14: A three subset pattern which does not fit in the weak hierarchy definition.

The properties proven mean that the set of all weak clusters, for a given dissimilarity d, is a weak hierarchy. There exists yet another characteristic of the weak hierarchies, in terms of the closure operator associated. For any subset $A \subset I$, let us denote its S_w-closure, that is, the intersection of all the $S_w \in S_W$ containing A, through $\langle A \rangle$. Then, it can be proved that any weak cluster is, really, the closure of a pair of the entities (Jamison-Waldner 1982, Bandelt and Dress 1989). The set trees also have this property. Howewer, for the weak hierarchies, it is a characteristic.

Statement 7.25. *An intersection-closed set of subsets, S_W, is a weak hierarchy if and only if, for any cluster $A \in S_W$, there exists a pair of the entities, $i, j \in A$ (perhaps coinciding) such that $A = \langle i, j \rangle$.*

Proof: Let S_W be a weak hierarchy while there exists $A \in S_M$ such that $A \neq \langle i, j \rangle$, for any $i, j \in A$. Let us take a pair $i, j \in A$ such that $\langle i, j \rangle$ is maximal. Let $k \in A - \langle i, j \rangle$. Then $i \notin \langle k, j \rangle$ and $j \notin \langle i, k \rangle$ since $\langle i, j \rangle$ is maximal. This contradicts property 2 in the definition of weak hierarchy. Conversely, if the nonempty intersection of S_1, S_2, and S_3 from S_W is not equal to either $S_1 \cap S_2$ or $S_1 \cap S_3$ or $S_2 \cap S_3$, then the latter three intersections contain elements x_3, x_2, and x_1, respectively, not belonging to $S_1 \cap S_2 \cap S_3$. This implies that $\langle x_1, x_2, x_3 \rangle \neq \langle x_i, x_j \rangle$ for all $i, j \in \{1, 2, 3\}$, which contradicts the assumed property. $\square$

Due to the statement, all the weak clusters are closures of subsets consisting of one or two of the entities, which implies that number of the weak clusters cannot exceed the total number of those, $N + N(N - 1)/2 = N(N + 1)/2$.

The concepts of weak clusters and d-splits are quite interrelated.

Let us employ a given dissimilarity $d = (d_{ij})$ as a d-split generator while weak clusters will be defined in set $I_c = I - \{c\}$ where c is an arbitrary element of I.

Let us recall that FKE-transform (7.4) is a dissimilarity function defined on I_c:

$$d_c(i,j) = MM + d_{ij} - d_{ic} - d_{jc} \; (i,j \in I - \{c\})$$

where $MM > 0$ is chosen to make all the values of d_c positive. Analogously, an FKE-transform similarity function can be defined as:

$$a_c(i,j) = d_{ic} + d_{jc} - d_{ij}$$

thus satisfying equation $a_c = MM - d_c$.

The concept of a d_c-weak cluster as a subset $A \subseteq I_c$ satisfying inequality

$$d_c(A) = \min_{i,j \in A, k \notin A} d_c(i,j/k) > 0$$

where $d_c(i,j/k) = \max(d_c(i,j), d_c(i,k), d_c(j,k)) - d_c(i,j)$, can be easily reformulated in terms of a_c. A subset $A \subset I_c$ is a weak a_c-cluster if and only if, for every $i,j \in A$ and $k \in I_c - A$,

$$a_c(i,j/k) = a_c(i,j) - \min(a_c(i,j), a_c(i,k), a_c(j,k)) > 0.$$

The minimum value of $a_c(i,j/k)$ in the condition above will be referred to as the *similarity isolation index* $a_c(A)$ of A. Obviously, $a_c(A) = d_c(A)$.

A major observation made by Bandelt and Dress 1992 is the following correspondence between d-splits and weak d_c-clusters.

Statement 7.26. *In every d-split $\{A, B\}$, one of the classes – that one not containing c – is a weak d_c-cluster. A biclass partition $\{A, B\}$ is a d-split if and only if A is a weak d_c-cluster for every $c \in B$, and $d(A,B)$ equals the minimum of $d_c(A)$, $c \in B$.*

Proof: Let us consider $d(ij, kc) = \max(d_{ij} + d_{kc}, d_{ik} + d_{jc}, d_{ic} + d_{jk}) - (d_{ij} + d_{kc})$. After adding $MM - d_{jc} - d_{ic} - d_{kc}$ to every term, we have: $d(ij, kc) = \max[d_c(i,j), d_c(i,k), d_c(j,k)] - d_c(i,j) = d_c(i,j/k)$. Thus A is a weak d_c-cluster if and only if $\{A, \{k, c\}\}$ is a partial d-split for every $k \in I - A - \{c\}$. □

Let us extend the concept of additive structure to the weak hierarchy. Let us consider an arbitrary non-negative function $\lambda(S_w)$ for the clusters S_w belonging to a given weak hierarchy $S_W(c)$ on the set I_c, which is assumed positive for the irreducible clusters (not being intersections of the other clusters). The cluster indicator function s_w is defined, as usual, with $s_{iw} = 1$ if $i \in S_w$ and $s_{iw} = 0$ if $i \notin S_w$. A matrix indicator function is defined as $s_w^c(i,j) = s_{iw}s_{jw}$, being equal to unity if both i and j belong to S_w. Curiously, this function is the FKE-transform similarity function for the split metric of the corresponding split $\{S_w, I - S_w\}$:

$$s_w^c(i,j) = (d_{S_w,I-S_w}(i,c) + d_{S_w,I-S_w}(j,c) - d_{S_w,I-S_w}(i,j))/2,$$

which is easily proven by checking out all the possibilities: (a) both i, j belong to S_w; (b) one of i, j belongs to S_w; (c) both of i, j belong to $I - S_w$. Then the additive (S_w, λ)-structure is defined as matrix $a^c = \sum_{S_w \in S_W(c)} \lambda(S_w) s_w^c$. A similarity matrix is called a *weak hierarchy additive structure* if it is a (S_w, λ)-structure for a weak hierarchy S_W and a function λ.

The following fact holds (see proposition 6 in Bandelt and Dress 1992).

Statement 7.27. *A metric $d = (d_{ij})$ $(i, j \in I)$ is d-split decomposable if and only if its FKE-transform similarity matrix a_c is a weak hierarchy additive structure for all $c \in I$.*

Proof: Let $S(c)$ be the set of all $A \subseteq I - \{c\}$ for which $\{A, I - A\}$ is a d-split. Since metric d is d-split decomposable, it can be presented as $d = \sum_{A \in S(c)} d(A, I - A) d_{A,I-A}$. Applying the FKE-transformation to both parts of the equality, we have $a_c = \sum_{A \in S(c)} d(A, I - A) s_A^c$. Conversely, let $a_c = \sum_{A \in S(c)} \lambda(A) s_A^c$ for some $c \in I$. It implies that $d_{ic} = a_c(i,i) = \sum_{i \in A \in S(c)} \lambda(A)$. The equation can be rewritten as follows: $d_{ij} - \sum_{A \in S(c)} \lambda(A) d_{A,I-A} = (d_{ic} - \sum_{i \in A \in S(c)} \lambda(A)) + (d_{jc} - \sum_{j \in A \in S(c)} \lambda(A))$ for any $i, j \in I_c$, which implies that $d_{i,j} = \sum_{A \in S(c)} \lambda(A) d_{A,I-A}(i,j)$, or $d = \sum_{A \in S(c)} \lambda(A) d_{A,I-A}$. Thus, $\lambda(A) = d(A, I - A)$ for any $A \in S(c)$ by the previous statement. □

We can note that the weak clusters are uniquely defined up to any monotone transformation of the similarity/dissimilarity function while the d-splits would change under a nonlinear transformation of the distances, which is again connected somehow with FKE-transformation (see p. 363).

More on the theory of weak clusters and hierarchies can be found in Bandelt and Dress 1989, 1992.

7.4.4 Discussion

The contents of this section is an outline of a theory developed, mainly, by Bandelt and Dress, on the tree-like structure information which can be extracted from any nonnegative dissimilarity function d. It appears, there is quite a structure out there.

1. There exists a uniquely defined set of biclass partitions $\{A, B\}$ (d-splits) along with weights $d(A, B)$ such that corresponding matrix indicator functions, d_{AB} (d-split metrics), are linearly independent as $N \times N$ dimensional vectors

and, moreover, form the weighted summary dissimilarity function d' which is super-dominated by d, that is, $d' \le d$.

2. The weight $d(A, B)$ of a split $\{A, B\}$ is equal to the weight of the edge splitting A and B in the tree representing d when d is a tree metric.

3. The splits $\{A, B\}$ with positive weights can be found sequentially, one-by-one, sequentially subtracting the corresponding metric $d(A, B)d_{AB}$ from d after a split $\{A, B\}$ is found, which makes a theoretically determined alternative to the heuristic subtraction process in the framework of a sequential fitting clustering strategy SEFIT.

4. The biclass splits, in general, do not form a hierarchy; they do form a restricted structure (weak compatibility) which relates to the weak hierarchy formed by classes of the splits that are weak clusters of a corresponding FKE-transform similarity a_c of d. The weak hierarchy is a nice generalization of the hierarchy concept in that every cluster can be considered as the intersection of all the clusters containing some pair of the entities, but, in contrast to the ordinary hierarchy, the number of parents of a cluster in a weak hierarchy may be more than one.

5. When the residual $d - d'$ is zero, the FKE-transform similarity matrix a_c is an additive cluster structure which can be effectively decomposed into a weighted sum of uniquely defined cluster indicator matrices (following the split decomposition of d).

7.5 Pyramids and Robinson Matrices

7.5.1 Pyramids

Let $P = (i_1, i_2, ..., i_N)$ be an ordering of I. A subset $S \subset I$ is called P-compatible if it can be represented as an interval of P, which means that if $iPjPk$ and $i, k \in S$, then $j \in S$. A *pyramid* (Diday 1986) is a set of subsets $S_W = \{S_w \subseteq I : w \in W\}$ satisfying the following conditions:

1. $\{i\} \in S_W$ for any $i \in I$ (singleton clusters);

2. $I \in S_W$ (universal cluster);

3. For any S_w and S_v $(v, w \in W)$, either $S_v \cap S_w = \emptyset$ or $S_v \cap S_w \in S_W$;

4. There exists an ordering P such that S_w is P-compatible for any $w \in W$.

The definition immediately implies that any set hierarchy is a pyramid. Any pyramid is a weak hierarchy; this follows from the properties of intervals: the intersection of three intervals, if nonempty, must be the intersection of two of them (see Fig.7.16 where the intersection of the intervals having a, b and c as their endpoints is, obviously, the intersection of the b and c intervals).

Figure 7.15: Twenty protein indexed pyramid according to Smith and Smith 1990.

According to Smith and Smith 1990, the scores of the twenty amino acids in problems of finding the best alignment of protein sequences can be represented by the picture in Fig.7.15 which is nothing but a pyramid; the order of the amino acids is compatible with all the clusters. Curiously, no node in the pyramid has more than two parents. This is not a coincidence.

Statement 7.28. *No cluster $S \in S_W$ in a pyramid S_W has more than two parents.*

Proof: Let S have three parents, S_1, S_2, and S_3, at least. Obviously, $S = S_1 \cap S_2 \cap S_3$. Then, there must be a pair among the parents which gives S as their overlap since S_W is a weak hierarchy. Let it be S_1 and S_2 so that $S = S_1 \cap S_2$. It can be, in terms of Fig.7.16, $S_1 = [a, b]$ and $S_2 = [c, d]$ so that $S = [c, b]$. The other interval, $S_3 = [e, f]$, must have $e \leq c$ and $f \geq b$, to include S (here, the real-line relation symbols, $\leq$ and $\geq$, are used to denote the corresponding pyramidal ordering P).

Moreover, relation $a < e$ would imply that interval set $[e, b]$ belongs to S_W, which contradicts the assumption that $S_1 = [a, b]$ is a parent of S since $[e, b] \subseteq [a, b]$. Thus, $e \leq a$ and, therefore, $S_1 \subseteq S_3$, which contradicts the assumption that S_3 is a parent of S. $\square$

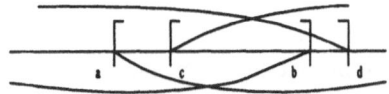

Figure 7.16: Four real line intervals represented with their left (a, c) and right (b, d) end-points.

An *indexed pyramid* is a pair (S_W, h) where S_W is a pyramid and h is an index function satisfying the following conditions:

1. $S \subset T$ implies $h(S) < h(T)$ for any $S, T \in S_W$;

2. $h(S) = 0$ if and only if $S = \{i\}$ for some $i \in I$.

The two-parent property proven guarantees that any indexed pyramid can be presented graphically by a dendrogram (having its singletons ordered according to the pyramid ordering) as in Fig.7.15.

Diday and Bertrand 1986 also considered a wider class of index functions: a set function h is referred to as a *weak index* for pyramid S_W if it satisfies the following conditions:

1. $S \subset T$ implies $h(S) \leq h(T)$ for any $S, T \in S_W$;

2. $h(S) = 0$ if and only if $S = \{i\}$ for some $i \in I$;

3. $S \subset T$ and $h(S) = h(T)$ implies that there exist clusters $S_1, S_2 \in S_W$ (different from S) such that $S = S_1 \cap S_2$.

Let us consider a pyramid having $\{a, b, c\}$, $\{b, c, d\}$, and $\{b, c\}$ as its only nontrivial clusters. With its index function equal to 1 at the two-element cluster and 2 at the three-element clusters, the pyramid can be represented as in Fig.7.17 (1). With weak index

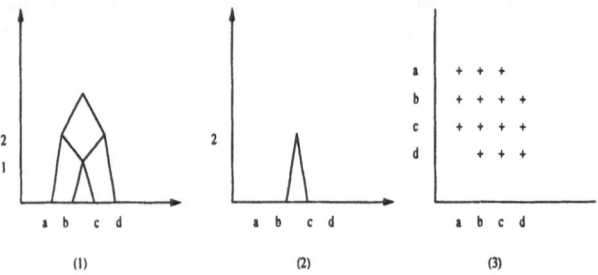

Figure 7.17: Three subsets, a-b-c, b-c-d, and b-c, represented as an indexed pyramid (1); when all the index values are 2 (weak index), the pyramid cannot be drawn corectly (2), however, it can be visualized with a matrix pattern (3).

function equal to 2 for each of the clusters, no such picture can be drawn: in Fig.7.17 (2), if both a and d would be joined with the top of cluster $\{b, c\}$, no three-element cluster could be separated; in this case, the pyramid can be represented with just the weak index function highlighted by crosses in Fig.7.17 (3).

Obviously, if h is a (weak) pyramid index function, fh is also a (weak) pyramid index function, for any monotone function f.

An indexed pyramid defines a dissimilarity index for every pair of entities: $d(i, j) = h(S_{ij})$ where S_{ij} is the smallest cluster containing both i and j ($i, j \in I$). For example, $d(C, Y) = 1$ and $d(F, Y) = 3$ by the pyramid in Fig.7.15 indexed (though in the opposite direction) by its score function. It turns out, the pyramidal dissimilarity indices are closely related to the so-called Robinson matrices (called linear, in Mirkin and Rodin 1984).

A dissimilarity measure $d = (d_{ij})$ is referred to as having Robinson form with respect to an ordering P of I, if

$$d_{ij} \geq \max(d_{ik}, d_{kj})$$

whenever $iPkPj$. A Robinson similarity measure $a = (a_{ij})$ is defined analogously, by the inequality $a_{ij} \leq \min(a_{ik}, a_{kj})$ whenever $iPkPj$. The entries in matrix d (or a), subject to the row/column ordering P, never decrease (or increase) when moving away from a main diagonal entry within any row or column. Actually, since the matrix is symmetrical, even more may be said: the entries in matrix d (or a), subject to the row/column ordering P, never increase (respectively, decrease) when moving towards the main diagonal entries from any non-diagonal entry (i, j) within its row i and column j border. This means that $d_{kl} \leq d_{ij}$ when $iPkPlPj$, which follows from the inequalities $d_{kl} \leq d_{kj}$ and $d_{kj} \leq d_{ij}$ implied by the definition of the Robinson form.

Let us introduce also a stronger kind of Robinson dissimilarity measure. Let $d = (d_{ij})$ be a Robinson form matrix with respect to the natural order of its row/columns. We say that the *strong inequality condition* holds if $d_{kj} < \min(d_{ij}, d_{kl})$ whenever $\max(d_{ij}, d_{kl}) < d_{il}$, $1 \leq i < k < j < l \leq N$. A dissimilarity measure d will be referred to as *pyramidal* if I can be reordered in such a way that d becomes a Robinson matrix satisfying the strong inequality condition.

Obviously, for any monotone function ξ, matrices d and ξd simultaneously have or do not have Robinson (pyramidal) form with respect to an order P.

Obviously, the chain $C_P = \{(i_k, i_k + 1) : k = 1, ..., N - 1\}$ along the ordering $P = (i_1...i_N)$ for a Robinson matrix is a minimum spanning tree (MST) in the corresponding weighted graph. This can be utilized for recognition: given a dissimilarity matrix, is it of Robinson form or not? It is sufficient to find all the minimum spanning trees that are chains, and check the basic inequality in the Robinson form definition with respect to the ordering corresponding to such a chain. The matrix is not Robinson if the test fails, for instance, because there are no chains among the minimum spanning trees.

Yet some other ways can be utilized to address the problem. One is based on the fact that any dissimilarity matrix is Robinson if and only if all its threshold graphs have their maximal cliques compatible with the same ordering P (Roberts 1979, Gross and Fulkerson 1965). Finding maximal cliques is, in general, an NP-hard problem; however, it is a simple problem for interval threshold graphs (see Booth and Lueker 1976 who also provide a linear-time algorithm to analyze whether a graph is an interval one or not [A graph $G = (V, E)$ is called an interval graph if its vertices can be represented as real line intervals in such a way that the intervals are overlapping if and only if the corresponding vertices are joined by an edge in E.]). Another way is based on "threshold" subsets (balls) $B(i, h) = \{k : d_{ik} \leq h\}$ ($i \in I$ and h is a real); $B(i, h)$ is just the set consisting of i and the vertices adjacent to i in the h-threshold graph G_h for dissimilarity $d = (d_{ij})$. The balls $B(i, h)$ are P-compatible if and only if matrix d is Robinson with respect to P (Kupershtoh and Mirkin 1971). Based on this, the problem becomes one of checking whether the incidence matrix for the set of all balls (whose cardinality, obviously, does not exceed N^2) can be reordered in such a way that every column has the consecutive ones property (every $B(i, h)$ is an interval of the order obtained); this can be resolved with the PQ-tree algorithm of Booth and Lueker 1976.

It appears that the Robinson or pyramidal matrices are exactly the pyramidal indices corresponding to weak index functions or to index functions, respectively. The former result belongs to E. Diday who introduced the concept of pyramid itself

(Diday 1986, Diday and Bertrand 1986); the latter statement seems new, though some related concepts and results can be found in Durand and Fichet 1988, Diatta and Fichet 1994.

To prove the statements, let us associate a pyramid with any given Robinson matrix d. For the sake of simplicity, the dissimilarity d is assumed even, so that $d_{ik} = d_{jk}$ for all $k \in I$ when $d_{ij} = 0$. Let S_h be a threshold clique, which means that S_h is a maximal clique of the h-threshold graph $G_h = \{(i, j) : d_{ij} \leq h\}$. Obviously, if d is Robinson with respect to an ordering P, then S_h is P-compatible, as well as the intersection $S_{h_1} \cap S_{h_2}$ (if not empty) of threshold cliques. Consider the set S_{Pd} of all subsets of the form S_h or $S_{h_1} \cap S_{h_2}$ for some h, h_1 and h_2. The overlap $S_{h_1} \cap S_{h_2}$ may be not a threshold clique, as can be seen from the following example of a Robinson matrix d:

$$d = \begin{pmatrix} 0 & 1 & 5 & 5 & 5 & 5 \\ 1 & 0 & 1 & 2 & 3 & 4 \\ 5 & 1 & 0 & 2 & 2 & 3 \\ 5 & 2 & 2 & 0 & 1 & 2 \\ 5 & 3 & 2 & 1 & 0 & 1 \\ 5 & 4 & 3 & 2 & 1 & 0 \end{pmatrix}$$

Sets $S_2 = \{2, 3, 4\}$ and $T_2 = \{3, 4, 5\}$ are maximal cliques of the 2-threshold graph while their intersection $S_2 \cap T_2 = \{3, 4\}$ is not a maximal clique in G_h, for any $h > 0$.

On the other hand, any non-empty intersection of the threshold clique intersections is a threshold clique intersection itself as is clearly seen in Fig.7.16. In the real line representing the ordering P, we can distinguish four intervals S_h labeled by their endpoints a, c (the "clique" sets are to the right of them) and b and d (the intervals are to the left); their intersections are the line segments $[a, b]$ and $[c, d]$ having themselves segment $[c, b]$ as their overlap. However, the overlap $[c, b]$ is the intersection of the primary interval sets labeled by c and b. This proves that the set S_{Pd} is a pyramid.

Let us show that set function $h_d(S) = \max_{i,j \in S} d_{ij}$ is a weak index function for the pyramid S_{Pd}. The first two items in the definition are obvious; the third follows from the fact that S_{Pd} contains only the threshold cliques S_h and their pair-wise intersections.

It can be seen also that when the strong inequality condition holds for a Robinson matrix d rearranged according to the ordering P, the weak index h_d is an index function for the pyramid S_{Pd} thus satisfying the property that $h(S) < h(T)$ whenever $S \subset T$ for $S, T \in S_{Pd}$. Indeed, strict inclusion $S \subset T$ can hold in one of the following two cases: (1) $S = S_{h_1}$ and $T = S_{h_2}$; (2) $S = S_{h_1} \cap S_{h_2}$ and $T = S_{h_1}$ or $T = S_{h_2}$. In the first case, $h_1 < h_2$ since the opposite inequality would contradict the maximality of S and T as cliques, which means that $h_d(S) < h_d(T)$. In the second case, let i and j be the first and the last elements of S_{h_1} in the ordering P while k and l are those of S_{h_2}. Obviously, $i < k < j < l$ since the overlap S is not empty. Let, say, $h_1 = \max(h_1, h_2)$. Then $d_{il} > h_1$ since, in the opposite case, the vertex l would be adjacent to all the elements of S_{h_1} in G_{h_1}, which contradicts the assumption that S_{h_1} is a maximal clique in G_{h_1}. This implies that both d_{ij} and d_{kl} are smaller than d_{il} since $(i, j) \in G_{h_1}$ and $(k, l) \in G_{h_2}$. Thus, due to the strong inequality condition, $d_{kj} < \min(h_1, h_2)$, and therefore, all the dissimilarities within S are smaller than $\min(h_1, h_2)$, which implies $h_d(S) < \min(h_d(S_1), h_d(S_2))$, and proves that h_d is an index function.

We have proven a part in the following statement.

Statement 7.29. *A dissimilarity matrix $d = (d_{ij})$ is Robinson or pyramidal (with respect to an order P) if and only if it corresponds to a pyramid (with respect to the same order) indexed with a weak index function or an index function, respectively.*

Proof: It remains to prove that a weakly indexed or an indexed hierarchy corresponds to a Robinson or pyramidal dissimilarity, respectively. Let S_W be a pyramid weakly indexed with index function h (without any loss of generality, we assume that the entities are numbered in the corresponding P order). Define a corresponding dissimilarity d by the following rule: for any $i, j \in I$, $d(i, j) = h(S(i, j))$ where $S(i, j)$ is the minimum cluster containing both i and j. To prove that the dissimilarity satisfies the Robinson condition (with regard to the pyramidal ordering P), $d(i, k) \geq max(d(i, j), d(j, k))$ whenever $i < j < k$, assume that it is not true and, thus, $d(i, k) < d(i, j)$ for some i, j, k such that $i < j < k$. But $S(i, k)$ must contain j implying that $S(i, j) \subseteq S(i, k)$ and in turn $h(S(i, j)) \leq h(S(i, k))$ and $d(i, j) \leq d(i, k)$. This proves that d is a Robinson matrix.

Assume now that h is an index function and $\max(d(i, j), d(k, l)) < d(i, l)$ where $i < k < j < l$. The fact that $d(i, j) < d(i, l)$ implies there exists a cluster, S_1, such that $d(i, j) = h(S_1)$, containing both i, j (and, therefore, k) but not l. Analogously, there exists a cluster, S_2, such that $d(k, l) = h(S_2)$, containing both k, l (and, therefore, j) but not i. The intersection $S = S_1 \cap S_2$ thus contains both j and k but neither i nor l, which implies that $h(S)$ is less than each of $h(S_1)$ and $h(S_2)$. Thus, $d(j, k) < \min(d(i, j), d(k, l))$ since $d(j, k) \leq h(S)$. The statement is proven.   □

7.5.2 Least-Squares Fitting

There exist agglomerative versions of pyramidal clustering (Diday and Bertrand 1986, Gaul and Shader 1994) involving expanding an ordering with respect to clusters merged, starting from the initial setting: the singletons with no order prescribed. The idea of the algorithm can be illustrated with an example from

Diday and Bertrand 1986, p. 40, shown in Fig.7.18. Initially, the nearest neighbors a and b are joined and put together in the order sought. Then, the second nearest neighbors b, c must be joined, which places c to the right of b in the resulting order. In the hierarchy, joining b and c (without a involved) is forbidden, which creates a different picture.

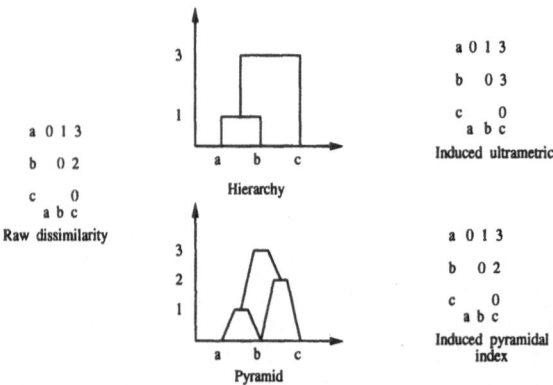

Figure 7.18: An example of pyramid and hierarchy corresponding to the same distance data.

In this section, an alternating optimization method developed by Hubert and Arabie 1994 in the least-squares fitting framework will be presented.

The presentation will be given in terms of similarity data. Let $a = (a_{ij})$ denote an original similarity matrix and $s = (s_{ij})$ a sought Robinson similarity matrix with respect to a sought ordering $P = (i_1, i_2, ..., i_N)$ of I. The problem of minimizing the least-squares criterion $L(s, P) = \sum_{i,j \in I}(a_{ij} - s_{ij})^2$ is treated with the following alternating minimization strategy: (1) matrix s given, minimize $L(s, P)$ with respect to ordering P, then (2) ordering P specified, minimize $L(s, P)$ with respect to s subject to the following Robinson constraints:

$$s_{i_k i_l} \geq s_{i_k i_{l-1}} \text{ and } s_{i_k i_l} \geq s_{i_{k-1} i_l}$$

for any k, l such that $1 \leq k < l \leq N$. As usual, the process stops when no criterion improvement is observed; the convergence of the procedure follows from the fact that there are only a finite number of orderings of I (though the authors believe the procedure may meet some computational anomalies, see Hubert and Arabie 1994, p. 12).

To initialize the procedure, an ordering P is selected somehow (Hubert and Arabie 1994 recommend choosing randomly; however, some expert-given or preprocessing-driven choice might be preferred, such as, for example, the seriation based on criterion LL, p. 181), and the initial estimates of s_{ij} are taken as $s_{i_k i_l} = N - |k - l|$, due to the ordering P, which provides an equally spaced Robinson structure.

The first step at an iteration, with s fixed, is performed maximizing the only varying part of the criterion, $g(P) = \sum_{k,l \in I} a_{kl} s_{i_k i_l}$, with a set of local operations on P. The authors employ the following three classes of operations: (i) all pairwise interchanges; (ii) all insertions of subintervals of length k (k varies from 1 to $N - 1$) between every pair of remaining objects; and (iii) all reversals of ordering in intervals of lengths from 2 to $N - 1$.

The second step, adjusting s while P is fixed, can be performed with any kind of algorithm for the linear-inequality constrained least-squares task. In particular, an iterative method by Dykstra 1983, in the present context, can be applied as follows (Hubert and Arabie 1994). The algorithm begins with $s = a$ and sequentially modifies the matrix based on considering all the order constraints in turn. If an order constraint fails to be satisfied by a pair of values, these are merely replaced by their average. The change made in the two values is labeled; to ensure convergence, a previous change must first be "undone" when the same constraint is reconsidered during the next pass through the set of constraints. Repeatedly cycling through the order constraints, a solution matrix s is found.

Obviously, the square scatter decomposition (2.21) in Section 2.3.4 holds for the result obtained:

$$\sum_{i,j \in I} a_{ij}^2 = \sum_{i,j \in I} s_{ij}^2 + L(s, P).$$

The procedure can be performed in the framework of the SEFIT strategy: repeatedly finding pairs s_t, P_t for the residual similarity matrix $a_t = a_{t-1} - s_t$, $t = 1, 2, ...$; a_0 is taken equal to a. This provides an extension of the square scatter decomposition to allow evaluation of the contribution for every Robinson matrix s_t found based just on the sum of its squared entries.

After employing this sequential fitting strategy, Hubert and Arabie 1994 recommend utilizing a follow-up sequential adjustment of the solutions s_t to the matrix $a - \sum_{u \neq t} s_u$, in turn; though, they observe that, actually, in their experiments, there is not a dramatic (if at all) difference between the initial SEFIT solutions and those after adjustment (which confirms the present author's experience with other kinds of clustering structures obtained with SEFIT).

Applied to the Kinship data summed up into a unique similarity matrix, then standardized, two Robinson matrices have been found in Hubert and Arabie 1994, accounting for 62% and 33% of the grand variance, respectively. The first matrix gives, basically, the solution already found in Section 4.6.2, Table 4.9: there are three major clusters, "Grand"ones, nuclear family, and collateral kinship; strong dyads within each of these reflect opposite-gender parallel terms (mother/father, son/sister, etc.) are underscored with a much smaller level of dissimilarity.

The second matrix points out division by sex (excluding "cousin" which is a gender-neutral kinship term). Within the two sex-based clusters, there are two parallel subgroupings according to generation: 1) son, nephew, grandson, brother; 2) father, grandfather, uncle; 3) daughter, niece, granddaughter, sister; and 4) grandmother, aunt, mother.

7.5.3   Discussion

In the concept of a Robinson matrix, two important mathematical structures, classification and order, are united. Two substantive lines of research, archaeology and genetics, contributed to the development of the corresponding mathematical concepts. Similar concepts have proven to be useful in some engineering problems. Mathematically, Robinson dissimilarity and pyramid is a structure between hierarchy and weak hierarchy; however, there is not much known about its properties.

The recognition and reconstruction problems are not difficult for this kind of structure. In contrast, the approximation problems seem quite hard. Hubert and Arabie 1994 provided a method for extracting the Robinson matrices from a dissimilarity/similarity matrix in the framework of a doubly local search strategy using square-error sequential fitting (see Section 2.3.4).

7.6   A Linear Theory for Binary Hierarchies

7.6.1   Binary Hierarchy Decomposition of a Data Matrix

The contents of this section is based on Mirkin 1995. Let us consider a binary hierarchy $S_W = \{S_w : w \in W\}$ and an $N \times n$ column-conditional data matrix $Y = (y_{ik})$ where rows $i \in I$ are the labels of the labeled singletons in the hierarchy (which means, basically, that $S_w \subseteq I$, for any $w \in W$). Let us suppose all the columns, $y_k = (y_{ik})$, $i \in I$, centered, $k = 1, ..., n$.

Let us define the nest indicator functions, ϕ_w, for every $w \in W$ and consider $N \times (N-1)$ matrix $\Phi = (\phi_{iw})$. Since the nest indicator functions form a basis of the $(N-1)$-dimensional space, every column-vector y_k, $k = 1, ..., n$, can be decomposed by the basis so that

$$Y = \Phi C \qquad (7.13)$$

where $C = (c_{wk})$ is the $(N-1) \times n$ matrix of the coefficients of the linear decompositions of y_ks (that are centered) by ϕ_ws, $y_{ik} = \sum_{w \in W} \phi_{iw} c_{wk}$.

Since $\Phi^T \Phi$ is the identity matrix, multiplying the equality in (7.13) by Φ^T leads to

$$C = \Phi^T Y \qquad (7.14)$$

which gives the value of every entry $c_{wk} = \sum_{i \in I} \phi_{iw} y_{ik}$ of matrix C expressed through the data as follows:

$$c_{wk} = a_w \sum_{i \in S_{w1}} y_{ik} - b_w \sum_{i \in S_{w2}} y_{ik} = \sqrt{\frac{N_{w1} N_{w2}}{N_w}} (\bar{y}_{w1k} - \bar{y}_{w2k}), \qquad (7.15)$$

where S_{w1} and S_{w2} are children of S_w (N_{w1}, N_{w2} and N_w are their respective cardinalities) and $\bar{y}_{w1k}$ and $\bar{y}_{w2k}$ are the means of k-th variable in S_{w1} and S_{w2}, respectively.

An entry in C, c_{wk}, can be referred to as *loading* of the variable k at cluster S_w, $w \in W$. After simple arithmetic transformations, we can arrive at yet another formula for c_{wk}:

$$c_{wk} = \sqrt{\frac{N_{w1} N_w}{N_{w2}}} (\bar{y}_{w1k} - \bar{y}_{wk}), \qquad (7.16)$$

where $\bar{y}_{wk}$ is the average of k-th variable in the whole $S_w = S_{w1} \cup S_{w2}$. An analogous formula holds with 2 and 1 exchanged (though, multiplied by -1).

Let us consider the m-dimensional vector y_w of the variable means within a cluster S_w, $w \in W$. The equation in (7.15) implies that Euclidean norm $\|c_w\| = \sqrt{(c_w, c_w)}$ of vector $c_w = (c_{wk})$ is equal to

$$\mu_w = \sqrt{\frac{N_{w1} N_{w2}}{N_w}} d(y_{w1}, y_{w2}) \qquad (7.17)$$

where $d(x, y)$ is the Euclidean distance between vectors x, y. The value μ_w is positive if $x \neq y$, and zero if $x = y$.

Let M be a diagonal $(N-1) \times (N-1)$ matrix with μ_w, $w \in W$, as its diagonal entries and let vectors c_w be normed. Then the equation in (7.13) becomes an analogue of the singular-value decomposition (SVD) of matrix Y since, in this case, $Y = \Phi M C$ where Φ is the matrix of an orthonormal vector set and M is a diagonal matrix with nonnegative diagonal entries (see Section 7.1.4). The weighted distances in (7.17) are analogues of the singular values; they will be referred to as the *cluster values*. Yet the decomposition is not a SVD since the vectors $c_w, w \in W$, in general, are not mutually orthogonal. For the sake of simplicity, in the rest, vectors c_w, $w \in W$, will be considered unnormed, thus holding all the formulas above as they are.

On the other hand, the decomposition in (7.13) has the nice property that the expression in (7.17) holds for any norm $\| \cdot \|$ as a function defined for the vectors $c_w, w \in W$, if the distance is defined accordingly as $d(y_{w1}, y_{w2}) = \|y_{w1} - y_{w2}\|$. Moreover, function $\| \cdot \|$ suffices to be any monotone function thus defining d as a dissimilarity measure which might fail to satisfy some of the metric properties. This obviously follows from (7.15).

Another useful property of the decomposition (7.13) is that

$$Y^T Y = C^T C \qquad (7.18)$$

which is proved multiplying (7.13) by its transposed version since $\Phi^T \Phi$ is the identity matrix.

Since the columns of Y are centered, the elements (y_k, y_l) of matrix $Y^T Y$ have the meaning of covariance coefficients between the variables k and l (differing from those only by the constant factor N, or by $N - 1$ in the estimation problems that are not considered here). When the columns of Y are also normed (in the Euclidean norm), these values are equal to the correlation coefficients (up to the same factor N or $N - 1$). Due to formula (7.15), equation (7.18) can be rewritten:

$$(y_k, y_l) = \sum_{w \in W} \frac{N_{w1} N_{w2}}{N_w} (y_{w1k} - y_{w2k})(y_{w1l} - y_{w2l}) \qquad (7.19)$$

This equality can be interpreted as a decomposition of the total covariation (correlation) coefficient by the hierarchy S_W clusters. When $k = l$, we have the variance of the variable k decomposed by the clusters:

$$(y_k, y_k) = \sum_{w \in W} c_{wk}^2 = \sum_{w \in W} \frac{N_{w1} N_{w2}}{N_w} (y_{w1k} - y_{w2k})^2 \qquad (7.20)$$

where

$$c_{wk}^2 = \frac{N_{w1} N_{w2}}{N_w} (y_{w1k} - y_{w2k})^2$$

presents the contribution of the cluster S_w, $w \in W$, to the variance of the variable k, $k = 1, .., M$.

Due to the orthonormality of the columns in Φ, yet another property of the SVD decomposition holds in the case considered: the equality (7.15) can be derived even when a part of the hierarchy is known only, as minimizing the least-squares difference between Y and corresponding part of the cluster structure. Let the part of all the nest indicator functions known consist of ϕ_w for $w \in W'$ where $W' \subset W$. Let us denote by Φ' the corresponding $N \times |W'|$ submatrix of Φ. The problem is to find a $|W'| \times M$ matrix $C' = (c'_{wk})$ minimizing the difference between Y and $\Phi'C'$ as measured by the ordinary least-squares criterion:

$$D_m(C') = Tr[(Y - \Phi'C')^T(Y - \Phi'C')] = \sum_{i=1}^{N}\sum_{k=1}^{M}(y_{ik} - \sum_{w \in W'} \phi_{iw}c_{wk})^2 \quad (7.21)$$

Statement 7.30. *The minimum value of $D_m(C')$ equals the sum of the squared cluster values by the clusters excluded, $D_m(C') = \sum_{w \notin W'} \mu_w^2$, while the optimal C' is determined by formula $C' = \Phi'^T Y$ which is analogous to (7.14).*

Proof: After differentiating criterion (7.21) by unknown C', the equality $C' = \Phi'^T Y$ is derived easily as the necessary condition for minimality of (7.21).

Putting this into (7.21), the equality $D_M(C') = Tr(Y^T Y - C'^T C')$ is derived. Since $Tr(C'^T C') = \sum_{w \in W'}\sum_k c_{wk}^2$ and $Tr(Y^T Y)$ equals the sum of expressions (7.20), the statement is proven. □

An important feature of the formula $C' = \Phi'^T Y$ is that it holds only when the least-square approximation is considered while the generic equality (7.14) holds always.

7.6.2 Cluster Value Strategy for Divisive Clustering

Amazingly, cluster value (7.17) squared is exactly the criterion

$$\mu_w^2 = \frac{N_{w1}N_{w2}}{N_w}d^2(y_{w1}, y_{w2}) \quad (7.22)$$

minimized at agglomerative steps in Ward's agglomerative clustering method (see p. 141). The same expression was employed by Edwards and Cavalli-Sforza 1965 for divisive clustering, to be maximized by splitting a cluster S_w into S_{w1} and S_{w2}. Equation

$$\sum_{i,k} y_{ik}^2 = Tr(Y^T Y) = Tr(C^T C) = \sum_{w \in W} \mu_w^2 \quad (7.23)$$

allows us to use the cluster values as the cluster contributions to the square data scatter (which is proportional to the sum of the variable variances, in this case). We can see that $\mu_w^2/Tr(Y^T Y)$ provides a measure of comparative salience of cluster S_w. With the cluster values renumbered in a decreasing order, the cumulative contribution $\sum_{w=1}^{m} \mu_w^2$ cannot be larger than the sum of the first m eigenvalues of $Y^T Y$ because vectors ϕ_w cannot fit better than the eigenvectors.

In these terms, error-square based divisive clustering looks quite analogous to the one-by-one SEFIT strategy of the principal component analysis. Indeed, finding the first cluster split (that is, vector ϕ_1 along with its cluster loadings), maximizing μ_1^2 in (7.22) to divide all the set I in two clusters is exactly what is to be done for finding the first singular triple except for the set of feasible solutions which is restricted here to consist of the nest indicator functions. Then, in SEFIT, the residual data matrix must be calculated; however, in this particular case, subtracting the matrix $\phi_1 c_1^T$ from Y may be skipped because the former matrix's rows coincide within the clusters; thus, the operation would not affect the within cluster distances since $d(x, y) = d(x - a, y - a) = \|x - y\|$ for every norm-driven metric, Euclidean distance included. Thus, the second cluster split can be sought by dividing one of the clusters found at the first step (by maximizing criterion (7.22) within the cluster). Reiterating the within-cluster splitting steps, we arrive, after $N - 1$ iterations, at the resulting binary hierarchy. Curiously, the order of splitting the clusters is not important since the operation of calculating the residuals is not performed and every nonsingleton cluster must be partitioned in two. Just the clusters can be enumerated in the order of the values of μ_w^2 in (7.22) decreased, after all of them are found.

On the other hand, if only a few upper clusters can satisfy the user, the order of the clusters to be partitioned becomes important: only the clusters having larger μ_w must be utilized. In this case, the general SEFIT stopping rule can be applied: the splitting stops when either the accumulated contribution to the data scatter becomes too large or a single cluster contribution becomes too small or just because the number of the clusters found has reached a pre-fixed limit (see p. 101).

Let us consider a cluster S_w splitting step in more detail. Depending on the formula for c_{wk}, (7.15) or (7.16), the value of the maximized criterion μ^2 can be expressed by formula (7.22) or

$$\mu_w^2 = \frac{N_w N_{w1}}{N_{w2}} d^2(y_{w1}, y_w) \qquad (7.24)$$

Each of these two formulas can be employed in a corresponding local search algorithm.

Formula (7.22) implies an algorithm which is just a version of the moving-center technique.

> **Maximizing (7.22)**
> Initially, the most distant points y_1 and y_2 in S_w are determined to be used as the initial centers of the clusters.
> Then, sequentially, the usual two steps are performed iteratively: (a) assigning the entities to the clusters (the nearest center wins) and (b) recomputing the centers (as the centers of gravity of the clusters obtained in (a)). The computation ends when the iteration (a) leaves the clusters unchanged.

Evidently, this version of the K-Means technique is nothing but the alternating minimization of the square error (WGSS) criterion by two groups of the variables, those related to membership of the entities to the clusters (a) and to the cluster centers (b). Simultaneously, it is an alternating maximization algorithm for the criterion (7.22).

The second algorithm, based on formula (7.24), is a seriation algorithm.

> **Splitting by Maximizing (7.24)**
> Initially, a point y_1 is found maximizing its distance to y_w, the center of S_w, setting $S_{w1} = \{y_1\}$. On a general step, a current S_{w1} along with its center y_{w1} is considered and an entity-point y_j, closest to y_{w1} by Euclidean distance, is sought. It is added to S_{w1} if the quotient $q = d^2(y_{w1}, y_w)/d^2$, where d is the distance between y_w and the center of $S_w \cup \{y_j\}$, satisfies the inequality
> $$q < \frac{N_1 N_2 + N_2}{N_1 N_2 - N_1},$$
> and the process stops if not.

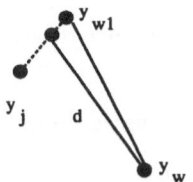

Figure 7.19: After point y_j is added, the cluster center y_{w1} moves toward y_j, which decreases the distance between y_w and the center.

The inequality involved is equivalent to the fact that value of μ_w^2 (7.24) increases when y_j is added to S_w. Basically, there is a trade-off between an increase of the coefficient N_{w1}/N_{w2} and a decrease of the distance $d^2(y_{w1}, y_w)$. The fact that the distance may only decrease in the adding process is well seen in Fig.7.19.

The latter algorithm is a seriation process based on linkage function $d(y, S) = d(y, c(S))$ where $c(S)$ is the gravity center of S.

Though the analogy of SVD one-by-one strategy and the square-error divisive clustering seems rather deep, there is a feature of the binary hierarchy which makes the two unlike: any binary hierarchy defines a SVD-like basis while there is only one genuine SVD basis consisting of the singular vectors. There are two issues following from the feature we wish to underscore. First, there is nothing to say about normalizing the variables. The bilinear model, with its residuals to be aggregated somehow, suggests a principle for that (see Section 6.1), but there are no residuals in decomposition (7.15), which means that there is no norming preferred. Second, the SVD-like one-by-one extracting strategy can be extended to any kind of dissimilarity function d in formulas (7.22) and (7.24).

This leads to a family of divisive clustering algorithms (both local splitting algorithms above remain valid with any dissimilarity function d) involving strictly defined cluster gravity centers and weighting coefficients with the arbitrary dissimilarity measure. The family is not included in the LW-algorithms family since there exist dissimilarity functions (such as that generated by Chebyshev norm, $l_\infty(x - y) = max_i|x_i - y_i|$) that do not allow expression of the merged cluster distances through the original distances (because, with the Chebyshev metric, the merged cluster center may have completely different components maximally deviated).

A computational strategy for divisive clustering, based on the theory above, can be set as follows:

1. Standardize the entity-to-variable data by shifting the origin into the point of the variable averages and norming the variables by a norm chosen.

2. Choose a dissimilarity function (it may be different from the distance driven by the norm chosen for standardizing).

3. Choose a clustering strategy (only the divisive one has been discussed above) and create a cluster hierarchy S_W with the strategy.

4. Draw a tree hierarchy representation reflecting the cluster values μ_w by the heights of the corresponding division nodes.

5. Interpret the hierarchy designed using:

   1) the drawn pattern of clustering;

2) contributions of the clusters and cluster–variable pairs to the square scatter of the data as reflected in values of $\sum_{k=1}^{n} c_{wk}^2$ and c_{wk}^2 (7.15), respectively $(w \in W, k = 1, ..., n)$;

3) the cluster variable-to-variable covariance/correlations, $N_{w1}N_{w2}(y_{w1k} - y_{w2k})(y_{w1l} - y_{w2l})/N_w$, as items in the additive decomposition of the overall covariance (7.19);

4) decompositions (7.13) of the entries y_{ik} by the clusters.

The hierarchical classification found with the divisive clustering algorithm maximizing the contribution to the total variance at each partition step (the first algorithm) is presented in Fig.7.20 as indexed by the corresponding cluster values (reflected in the heights of the vertical edges). The squared cluster values μ_w^2, which are equal to contributions of the divisions to the total variance, are presented (per cent) for the most contributing clusters. The general pattern of variable-to-variable correlation is pair-wise negative as

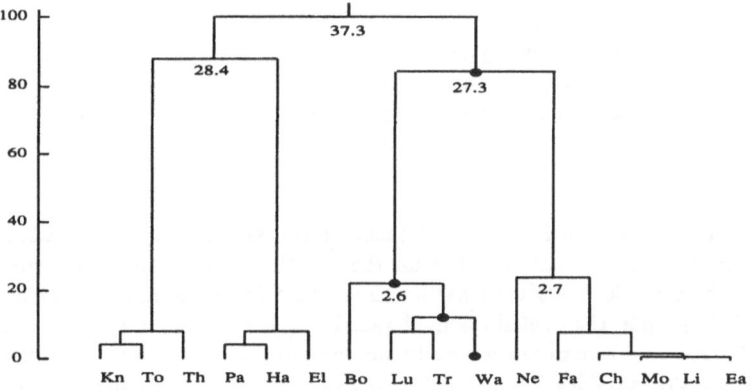

Figure 7.20: Binary hierarchy found for square scatter standardized Body data with the minimum-of-the-error criterion; the numbers show contributions of the major splits to the data variance.

is seen in the correlation matrix (the variances are presented in the principal diagonal):

$$
\begin{array}{llllll}
1 & 1.00 & & & \\
2 & -0.48 & 1.00 & & \\
3 & -0.24 & -0.27 & 1.00 & \\
4 & -0.45 & -0.15 & -0.24 & 1.00 \\
  & 1 & 2 & 3 & 4
\end{array}
$$

Its decomposition by the first three separations, due to formula (7.19), is presented by the following respective matrix terms:

$$
\begin{array}{llllll}
1 & 0.44 & & & \\
2 & -0.43 & 0.42 & & \\
3 & 0.32 & -0.31 & 0.23 & \\
4 & -0.42 & 0.41 & -0.30 & 0.40 \\
  & 1 & 2 & 3 & 4
\end{array}
$$

(first division),

$$
\begin{array}{llllll}
1 & 0.00 & & & \\
2 & -0.01 & 0.56 & & \\
3 & 0.00 & -0.01 & 0.00 & \\
4 & 0.01 & -0.57 & 0.01 & 0.58 \\
  & 1 & 2 & 3 & 4
\end{array}
$$

(second division),

$$
\begin{array}{llllll}
1 & 0.43 & & & \\
2 & -0.02 & 0.00 & & \\
3 & -0.54 & 0.02 & 0.60 & \\
4 & -0.01 & 0.00 & 0.02 & 0.00 \\
  & 1 & 2 & 3 & 4
\end{array}
$$

(third division).

These three matrix items take into account most of the variance and correlation. It can be seen from the diagonal entries, that all the variables contribute to the first separation, though the variable 3 is some-what less important (with its only 23% of the variance accounted for) while the contribution of variable 1 is some-what higher (44% of the variance). The second separation is due to the variables 2 and 4 while the third separation is made by the variables 1 and 3 (since the other variables in either case do not contribute to the variance at all).

Decomposition of the correlation coefficients confirms and details this conclusion. In particular, the negative correlations between the variables 1 and 3, as well as between 2 and 4, become positive at the first separation while sharper at the third and second separations, respectively. All the other correlations disappear in the clusters. The variance of variable 3 is not exhausted by the three first separations: this variable contributes to the separation of the smaller Head cluster.

The last interpretation aid concerns decomposition of all the standardized data entries y_{ik} by the clusters due to equation (7.13). Let us demonstrate the decomposition for the

16-th entity, Waist, belonging to the four clusters nested shown by the bold nodes in
Fig.7.20. The cluster decompositions of all the four variables, at Waist, are as follows:

$$
\begin{array}{rrrrrr}
1 & 0.52 = & -0.52+ & 1.09- & 0.01- & 0.05 \\
2 & 0.28 = & 0.50- & 0.04- & 0.06- & 0.12 \\
3 & -1.46 = & -0.37- & 1.20- & 0.36+ & 0.47 \\
4 & 0.36 = & 0.49- & 0.03- & 0.05- & 0.04
\end{array}
$$

Every single column of the decomposition relates to its cluster (as the weighted difference
between the centers of its split parts) reflecting the features of the cluster: the smaller
values of the variables 1 and 3 in the first cluster correspond to its Head–Chest nature
while the next cluster shows the split between these variables: enlarged 1 and reduced
3 correspond to the Chest membership of the entity. The last column represents the
individual traits of the entity.

Another tree (Fig.7.21) is generated with the divisive strategy when the criterion has
been changed for Chebyshev norm-driven metric. Every variable had been standardized
with the same norm: the maximum absolute deviation from the average became unity
after norming was completed. There are two major differences between the two trees

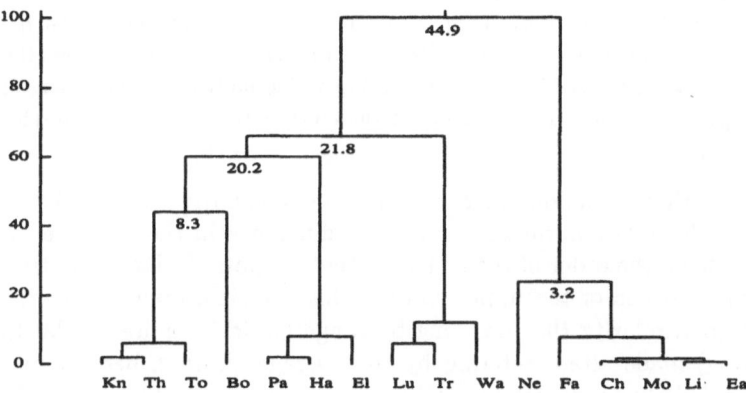

Figure 7.21: Binary hierarchy found for Body data with the Chebyshev norm; the numbers show the relative contributions of the major splits to the data variance.

presented in Fig.7.20 and 7.21: one is substantive (the "head" cluster is separated first in Fig.7.21 rather than the "extremity" cluster in Fig.7.20), the second is technical (the variance contribution of the first split in Fig.7.21 (44.9 %) is much higher than that in Fig.7.20 (37.3%)). The technical feature seems, at the first glance, really amazing. How it could occur that the maximized contribution (in Fig.7.20) turned out less than the contribution achieved when another (Chebyshev) criterion was optimized (Fig.7.21)? To address the issue, let us consider decomposition of the variances of the variables by the clusters as expressed in (7.20):

$$
\begin{array}{llllll}
1 & 0.36 = & 0.33+ & 0.00+ & 0.00 + \dots \\
2 & 0.18 = & 0.03+ & 0.02+ & 0.12 + \dots \\
3 & 0.20 = & 0.02+ & 0.15+ & 0.00 + \dots \\
4 & 0.19 = & 0.03+ & 0.03+ & 0.07 + \dots
\end{array}
$$

Again, only three major splits are represented in the decomposition. The variances (and, thus, the contributions to the square data scatter) of the variables now are different from the very beginning, which seems to determine the order they are involved in the division process: the most contributing variable 1 turns out to be the principal base of the first division; variable 3 having the second-place variance contributes mostly to the second division; the less contributing variables 3 and 4 are serving at the following divisions. Such a sequential involvement of the variables may generate a more complete account of the variance in splitting, which is reflected in the higher level of the variance extracted in the upper splits in Fig.7.21 as compared to those in Fig.7.20. This conclusion is supported by the results of the Euclidean-norm-based divisive clustering applied to the data standardized with Chebyshev norm (Fig.7.22). The variance contributions in the upper splits there are even greater (since the criterion is proper, at this time); evidently, it is the left four-element cluster in Fig.7.21 disappearing which makes that increasing of the variances in Fig.7.22 possible. The contents of the clusters in the latter figure also seem quite satisfactory.

In the present author's opinion, there is a general regularity (manifested in the example) that Chebyshev norming generates a difference in the variances of the variables influencing the order of their involvement in splits (fusions) and thus increasing the contributions of the upper splits. This principle might cause the empirically observed regularity that norming by range (which is quite similar to Chebyshev norming) made after centering by the average allows a best fit into Monte-Carlo generated cluster structures (Milligan and Cooper 1986).

7.6.3   Approximation of Square Tables

Let us briefly touch the problem of representing the square data matrix via linear combinations of the mutually orthogonal projection matrices $P_w = \phi_w \phi_w^T$ based on nest indicator functions ϕ_w of a binary hierarchy S_W. The projection matrices provide a basis only for the double-centered ultrasimilarity matrices related to S_W as defined in Section 7.1.4. This means that, in contrast to the case of the

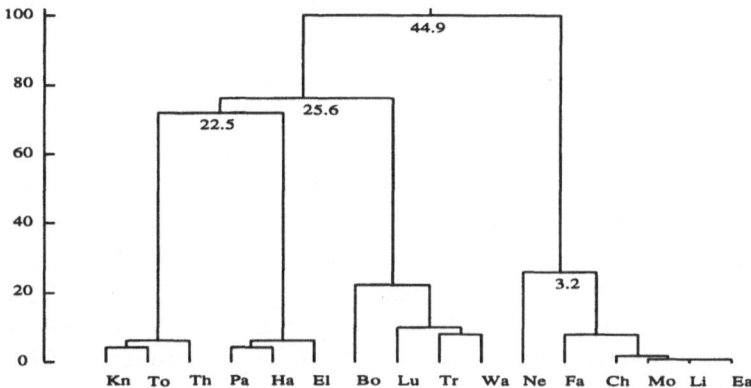

Figure 7.22: Binary hierarchy found with the square-error criterion for the Body data normed by Chebyshev norm.

rectangular data, the square data matrix, in general, cannot be entirely decomposed by the set of matrices P_w, $w \in W$, corresponding to a hierarchy S_W. This leads us to accept the following bilinear model decomposing a given square similarity matrix $b = (b_{ij})$ into a linear combination of the projection matrices P_w and residuals to be minimized.

$$b_{ij} = \sum_{w=1}^{N-1} \lambda_w \phi_{iw} \phi_{jw} + e_{ij} \qquad (7.25)$$

Matrix $b = (b_{ij})$ must be preliminarily double-centered (for instance, with the transformation (2.5) in Section 2.2.2).

The standardizing principles and options devised in Section 2.1.2 can be easily extended to this bilinear model. We consider here only the least-squares criterion.

Due to orthonormality of the set P_w, $w \in W$, for any binary hierarchy S_W, the square data scatter can be decomposed as follows:

$$\sum_{i,j \in I} b_{ij}^2 = \sum_{w=1}^{N-1} \lambda_w^2 + \sum_{i,j \in I} e_{ij}^2 \qquad (7.26)$$

whenever there is orthogonality of the residual matrix $E = (e_{ij})$ to the rest, as it is when $b = (b_{ij})$ has been projected in the subspace $L(\{P_w : w \in S_W\})$ in the parallel or sequential manner. Moreover, it can be found from (7.25) that the least-squares optimal coefficients are the scalar products (b, P_w) satisfying equations $\lambda_w = \sum_{i,j \in I} b_{ij} \phi_{iw} \phi_{jw}$.

Evidently, the entries of P_w are zeros outside S_w, and they are equal to $-1/N_w$ within S_w, with $1/N_{w1}$ or $1/N_{w2}$ added within its children, S_{w1} or S_{w2}, respectively. This implies that

$$\lambda_w = AL(S_{w1}) + AL(S_{w2}) - AL(S_w) \qquad (7.27)$$

where $AL(S) = \sum_{i,j \in S} b_{ij} / |S|$ (see Section 4.2.2 for discussion of the function). This gives an explicit combinatorial meaning to the least squares criterion as that one which is equivalent to the criterion of maximizing $\sum_{w=1}^{N-1} \lambda_w^2$ due to (7.26). We cannot maximize the criterion globally and suggest using the sequential fitting approach (one-by-one extracting P_ws starting from the larger clusters) which, however, is slightly easier here.

Let us consider a divisive clustering algorithm, at each splitting step maximizing the varying part of criterion (7.27),

$$g(S_{w1}, S_{w2}) = \sum_{i,j \in S_{w1}} b_{ij} / N_{w1} + \sum_{i,j \in S_{w2}} b_{ij} / N_{w2},$$

with regard to all biclass partitions $\{S_{w1}, S_{w2}\}$ of S_w $(w = 1, 2, ...)$. The criterion extends a criterion formula, (4.19), of the least-squares principal clustering for the entity-to-variable data to the arbitrary similarity data. That criterion is the semi-averaged within similarity criterion (B), p. 293, applied to the row-by-row scalar products.

Statement 7.31. *The sequential fitting procedure for the model (7.25) is equivalent to the algorithm of divisive clustering with criterion $g(S_{w1}, S_{w2})$.*

Proof: Since g is the varying part of (7.27) within S_w, optimizing g is equivalent to maximizing λ_w^2. We need to prove only that applying the criterion to the residual data is equivalent to applying it to the initial data. Indeed, extracting $\lambda_w \phi_{iw} \phi_{jw}$ from the data (after S_w has been found) means subtracting the same threshold value from all the within cluster similarities. However, criterion g remains invariant when a threshold, π, has been subtracted from all the similarities: $g(S_{w1}, S_{w2}, \pi) = \sum_{i,j \in S_{w1}} (b_{ij} - \pi) / N_{w1} + \sum_{i,j \in S_{w2}} (b_{ij} - \pi) / N_{w2} = g(S_{w1}, S_{w2}) - N_w \pi$, which proves the statement. □

Since the actual criterion maximized is g^2, the option of minimizing g, leading to anti-cluster rather than cluster splitting, should be considered as an available option. The user may control that due to the local character of the one-by-one splitting procedure.

When the model (7.25) is applied to the aggregable data, b_{ij} is assumed to be the result, q_{ij}, of the RCP transformation of the data, and the fitting criterion must be the weighted least squares (see Sections 2.2.3 and 4.5.4). In this case, the maximized splitting criterion becomes equal to $\lambda_w^2 = (q_{S_{w1}S_{w1}} + q_{S_{w2}S_{w2}} - 2q_{S_{w1}S_{w2}})^2$ (Lebart and Mirkin 1993).

7.6.4 Discussion

Decomposition of the data and the variances of variables by hierarchically structured factors has been made in the statistical discipline of analysis of variance for years. However, the analysis of variance addresses different issues, having nothing to do with the linear-wise theory developed here for binary hierarchies, due to particular nature of the orthonormal nest indicator functions having, in the binary hierarchy case, their entries completely defined by the standardizing conditions. Two features of the theory developed are:

1. Interpreting aids about the results of hierarchical clustering, whichever method had been utilized to get it, regarding decompositions of the data entries, correlation/covariation coefficients, and the variances by the clusters. Only the variance decomposition has been utilized, so far.

2. The cluster analogues of the singular-/eigen-values, allowing to interpret the square-error divisive clustering as an analogue of the sequential principal component analysis strategy. However, the cluster-value-based clustering strategies can be employed in many other partitioning problems. The "most-to-the-first" strategy of divisive clustering can be changed for an oppositely directed strategy, "least-to-the-last", which is equivalent to agglomerative clustering, or even for entirely different "least-to-the-first" or "all-equal" strategies reaching for the increasing or equal cluster values, respectively. The latter strategy aims at producing clusters, however nested, which have their centers at equal distances from each other. This shows that the cluster values are a flexible heuristic tool for formalizing different clustering goals.

The emphasis by the theory with the square-error criterion seems to be an explanation of the empirically observed phenomenon regarding better clustering results reached with the standardizing by range, not by deviation: perhaps, it is the difference in the variable variances which is responsible for that. However, this hypothesis needs further investigation.

However complete decomposition of the rectangular data is made, it does not much affect the similarity data which, in general, never can be entirely decomposed with a binary hierarchy: there are at least $(N - 1)N/2$ independent data entries while only $N - 1$ basis elements. This makes us return to the sequential fitting strategy.

Bibliography

[1] L.A. Abbott, F.A. Bisby, and D.J. Rogers (1985) *Taxonomic Analysis in Biology*, New York: Columbia University Press.

[2] E.N. Adams III (1986) N-Trees as nestings: complexity, similarity and consensus, *Journal of Classification, 3*, 299-317.

[3] R. Agarwala, V. Bafna, M. Farach, B. Narayanan, M. Paterson, and M. Thorup (1995) On the approximability of numerical taxonomy, DIMACS Technical Report 95-46, 9 p.

[4] A. Agresti (1984) *Analysis of Ordinal Categorical Data*, New York: J.Wiley & Sons.

[5] D.W. Aha and R.L. Bankert (1995) A comparative evaluation of sequential feature selection algorithms. In: D. Fisher and H.-J. Lenz (Eds.) *Learning from Data: AI and Statistics V.*, Springer-Verlag, 1-6.

[6] S.A. Aivazian, V.M. Buchshtaber, I.S. Eniukov, and L.D. Meshalkin (1989) *Applied Statistics: Classification and Reduction of Dimensionality*, Moscow, Finansy i Statistika (in Russian).

[7] H.C. Andrews (1972) *Introduction to Mathematical Techniques in Pattern Recognition*, New York: Wiley-Interscience.

[8] M. Andrews (1993) *Visual C++ Object-Oriented Programming*, Indianapolis, IN: Sams Publishing.

[9] K. Appel and W. Haken (1977) The solution of the four-color map problem, *Scientific Amer., 237*, 108-121.

[10] Yu. Apresian (1966) An algorithm for finding clusters by a distance matrix, *Computer Translation and Applied Linguistics, 9*, 72-79 (in Russian).

[11] P. Arabie, S.A. Boorman, and P.R. Levitt (1978) Constructing block models: how and why, *Journal of Mathematical Psychology, 17*, 21-63.

399

[12] P. Arabie and J.D. Carroll (1980) MAPCLUS: a mathematical programming approach to fitting the ADCLUS model, *Psychometrika, 45*, 211-235.

[13] P. Arabie, J.D. Carroll, and W. De Sarbo (1987) *Three-Way Scaling and Clustering*, Newbury Park, Ca.: Sage.

[14] P. Arabie and L. Hubert (1992) Combinatorial data analysis, *Annu. Rev. Psychol., 43*, 169-203.

[15] P. Arabie, S. Schleutermann, J. Daws, and L.J. Hubert (1988) Marketing applications of sequencing and partitioning of nonsymmetric and/or two-mode matrices. In: W. Gaul, and M. Schader (Eds.) *Data, Expert Knowledge and Decisions*, Berlin: Springer-Verlag, 215-224.

[16] A.G. Arkad'ev and E.M. Braverman (1967) *Computers and Pattern Recognition*, Washington: Thompson Book Co. (original Russian version published in 1964; 2-nd edition, 1971).

[17] K. Arrow (1951) *Collective Choice and Social Values*, New York: J.Wiley & Sons.

[18] M. Atkinson, D. Kilby, and I. Roca (1988) *Foundations of General Linguistics*, London: Unwin Hyman Ltd.

[19] P.O. Aven, I.B. Muchnik, and A.A. Oslon (1988) *Functional Multidimensional Scaling*, Moscow: Nauka (in Russian).

[20] S. Baase (1991) *Computer Algorithms*, Reading, Ma: Addison–Wesley.

[21] R. Baire (1905) *Leçons sur les Fonctions Discontinues*, Paris.

[22] P.V. (S.) Balakrishnan, M.C. Cooper, V.S. Jacob, and P.A. Lewis (1994) A study of the classification capabilities of neural networks using unsupervised learning: a comparison with K-means clustering, *Psychometrika, 59*, 509-525.

[23] G.H. Ball and D.J. Hall (1967) A clustering technique for summarizing multivariate data, *Behavioral Science, 12*, 153-155.

[24] H.-J. Bandelt (1990) Recognition of tree metrics, *SIAM J. Discrete Mathematics, 3*, 1-6.

[25] H.-J. Bandelt and A. Dress (1986) Reconstructing the shape of a tree from observed dissimilarity data, *Advances in Applied Mathematics, 7*, 309-343.

[26] H.-J. Bandelt and A.W.M. Dress (1989) Weak hierarchies associated with similarity measures – an additive clustering technique, *Bulletin of Mathematical Biology, 51*, 133-166.

[27] H.-J. Bandelt and A.W.M. Dress (1992) A canonical decomposition theory for metrics on a finite set, *Advances of Mathematics*, *92*, 47-105.

[28] J.D. Banfield, A.E. Raftery (1993) Model-based Gaussian and non-Gaussian Clustering, *Biometrics*, *49*, 803-821.

[29] E.R. Barnes, A.Vanelli, J.Q.Walker (1988) A new heuristic for partitioning the nodes of a graph, *SIAM Journal Discr.Math.*, 1, 299-305.

[30] J.-P. Barthélemy and A. Guenoche (1991) *Trees and Proximity Representations*, Chichester: J. Wiley & Sons.

[31] J.-P. Barthélemy, B. Leclerc, and B. Monjardet (1986) On the use of ordered sets in problems of comparison and consensus of classifications, *Journal of Classification*, *3*, 187-224.

[32] J.-P. Barthélemy and X. Luong (1986) Représentations arborées de mesures de dissimilarité, *Statistique et Analyse des Données*, *11*, 20-41.

[33] O.A. Bashkirov, E.M. Braverman, and I.B. Muchnik (1964) Potential function algorithms for pattern recognition learning machines, *Automation and Remote Control*, *25*, 692-695.

[34] F.M. Bass, E.A. Pessemier, and D. R. Lehmann (1972), An experimental study of relationships between attitudes, brand preference, and choice, *Behavioral Science*, *17*, 532-541.

[35] V. Batageli (1981) Note on ultrametric hierarchical clustering algorithms, *Psychometrika*, *46*, 351-353.

[36] F.B. Baulieu (1989) A classification of presence/absence based dissimilarity coefficients, *Journal of Classification*, *6*, 233-246.

[37] J.-P. Benzécri (1973) *L'Analyse des Données*, Paris: Dunod.

[38] J.C. Bezdek (1974) Numerical taxonomy with fuzzy sets, *Journal of Mathematical Biology*, *1*, 57-71.

[39] J.C. Bezdek (1981) *Pattern Recognition with Fuzzy Objective Function Algorithms*, New York: Plenum.

[40] J.C. Bezdek (1992) Integration and generalization of LVQ and c-means clustering, In: *SPIE, Vol. 1826 Intelligent Robots and Computer Vision XI*, Wachington, D.C.: Society of Photo-Optical Instrumentation Engineers, p. 280-299.

[41] P.W. Birkeland and E.E. Larson (1978) *Putnam's Geology*, New York: Oxford University Press.

[42] R.K. Blashfield, M.S. Aldenderfer and L.C. Morey (1982) Cluster analysis software, in P.R.Krishnaiah and L.N. Kanal (Eds.) *Handbook of Statistics, 2*, Amsterdam: North-Holland, 245-266.

[43] P.M. Blekher and M.Y. Kelbert (1978) Convergence of the algorithm FOREL. In: S. Aivazian (Ed.) *Applied Statistical Analysis*, Moscow: Nauka, 358-361 (in Russian).

[44] H.H. Bock (1974) *Automatische Klassifikation*, Goettingen: Vandenhoeck and Ruprecht.

[45] H.H. Bock (1989) Probabilistic aspects in cluster analysis. In: O.Opitz (Ed.) *Conceptual and Numerical Analysis of Data*, Berlin: Springer Verlag, 12-44.

[46] J.A. Bondy and U.S.R. Murty (1976) *Graph Theory with Applications*, New York: North-Holland.

[47] S.A. Boorman and D.C. Olivier (1973) Metrics on spaces of finite trees, *Journal of Mathematical Psychology, 10*, 26-59.

[48] K. Booth and G. Lueker (1976) Testing for the consecutive ones property, interval graphs, and graph planarity using PQ-tree algorithms, *J. Comput. Syst. Sci., 13*, 335-379.

[49] S. Bosanko (Ed.) (1983) *Predicting Your Future*, New York: The Diagram Visual Information Ld.

[50] E.M. Braverman (1966) Method of potential functions in problems of unsupervised learning, *Automation and Remote Control, 27*, 10, 1748-1770.

[51] E.M. Braverman (1970) Methods for extremal grouping of the variables and the problem of finding important factors, *Automation and Remote Control, 31*, 1, 123-132.

[52] E.M. Braverman, N.E. Kiseleva, I.B. Muchnik, and S.G. Novikov (1974), Linguistic approach to processing large bodies of data, *Automation and Remote Control, 35*, 11, 1768-1788.

[53] E.M. Braverman, B.M. Litvakov, I.B. Muchnik, and S.G. Novikov (1975), Stratified sampling for empirical data collecting, *Automation and Remote Control, 36*, 10, 1629-1641.

[54] E.M. Braverman and I.B. Muchnik (1983) *Structural Methods for Processing Empirical Data*, Moscow: Nauka (in Russian).

[55] J.N. Breckenridge (1989) Replicating cluster analysis: method, consistency, and validity, *Multivariate Behavioral Research, 24*, 147-161.

[56] R.L. Breiger (1981) The social class structure of occupational mobility, *American Journal of Sociology, 87*, 578-611.

[57] L. Breiman, J.H. Friedman, R.A. Olshen, C.J. Stone (1984) *Classification and Regression Trees*, Belmont, Ca: Wadsworth.

[58] S. Brew (1987) *Career Development Guide for Use with the Strong Interest Inventory*, Consulting Psychologist Press.

[59] P.G. Briant (1991) Large-sample results for optimization-based clustering methods, *Journal of Classification, 8*, 31-44.

[60] P.G. Briant (1994) Selecting models using the minimum description length principle, *UCD-CBA Working Paper 1994-14*, University of Colorado at Denver.

[61] L. Brillouin (1962) *Science and Information Theory*, New York: Academic Press.

[62] P. Buneman (1971) The recovery of trees from measures of dissimilarity, In: F. Hodson, D. Kendall, and P. Tautu (Eds.) *Mathematics in Archeological and Historical Sciences*, Edinburg: Edinburg University Press, 387-395.

[63] W. Buntine and T. Niblett (1992) A further comparison of splitting rules for decision-tree induction, *Machine Learning, 8*, 75-85.

[64] G.A. Carpenter and S. Grossberg (1992) Self-organizing neural networks for supervised and unsupervised learning and prediction. In: V. Cherkassky, J.H. Friedman, and H. Wechsler (Eds.) *From Statistics to Neural Networks*, New York: Springer-Verlag, 319-348.

[65] J.D. Carroll (1976) Spatial, non-spatial and hybrid models for scaling, *Psychometrika, 41*, 439-463.

[66] J.D. Carroll and P. Arabie (1983) INDCLUS: an individual differences generalization of the ADCLUS model and the MAPCLUS algorithm, *Psychometrika, 48*, 157-169.

[67] L. Cavalli-Sforza and A. Edwards (1967) Phylogenetic analysis models and estimation procedures, *Amer. J. Human Genetics, 19*, 233-257.

[68] L.L. Cavalli-Sforza, A. Piazza, P. Menozzi, and J. Mountain (1988) Reconstruction of human evolution: bringing together genetic, archaelogical, and linguistic data, *Proc. Natl Acad. Sci USA, 85*, 6002-6006.

[69] G. Celeux and G. Govaert (1991) Clustering criteria for discrete data and latent class models, *Journal of Classification, 8*, 157-196.

[70] G. Celeux and G. Govaert (1992) A classification EM algorithm for clustering and two stochastic versions, *Computational Statistics and Data Analysis, 14,* 315-332.

[71] J.M. Chambers and B. Kleiner (1982) Graphical techniques for multivariate data and for clustering. In: L.R. Krishnaiah and L.N. Kanal (Eds.) *Handbook of Statistics, 2,* Amsterdam: North-Holland, 209-244.

[72] I. Charon and O. Hudry (1993) The noising method: a new method for combinatorial optimization, *Operations Research Letters, 14,* 133-137.

[73] A. Chaturvedi and J.D. Carroll (1994) An alternating optimization approach to fitting INDCLUS and generalized INDCLUS models, *Journal of Classification, 11,* 155-170.

[74] Y.Q. Chen, D.W. Thomas, and M.S. Nixon (1994) Generating-shrinking algorithm for learning arbitrary classification, *Neural Networks, 7,* 1477-1489.

[75] Z. Chen and J. Van Ness (1994) Metric admissibility and agglomerative clustering, *Communications in Statistics: Simulation and Computation, 23,* 3.

[76] Z. Chen and J. Van Ness (1995) Space-conserving agglomerative algorithms, *Journal of Classification,* to appear.

[77] N. Chinchor, L. Hirshman, and D.D. Lewis (1993) Evaluating message understanding systems: An analysis of the Third Message Understanding Conference, *Computational Linguistics, 19,* 3, 409-450.

[78] W.J. Clancey (1985) Heuristic classification, *Artificial Intelligence, 27,* 289-351.

[79] N. Cliff, D.J. McCormick, J.L. Zlatkin, R.A. Cudeck, and L.M. Collins (1986) BINCLUS: Nonhierarchical clustering of binary data, *Multivariate Behavioral Research, 21,* 201-227.

[80] H.T. Clifford and W. Stephenson (1975) *An Introduction to Numerical Classification,* New York: Academic Press.

[81] R.A. Colombo and D.G. Morrison (1989) A Brand Switching Model with Implications for Marketing strategies, *Marketing Science, 8,* 89-99.

[82] H. Colonius and H.H. Schulze (1981) Tree structures for proximity data, *British J. Math. Stat. Psychol., 34,* 167-180.

[83] J.E. Corter and A. Tversky (1986) Extended similarity trees, *Psychometrika, 51,* 429-451.

[84] M. Csikszentmihalyi and E. Rochberg-Halton (1981) *The Meaning of Things: Symbols in the Development of the Self.* Cambridge, Ma.: Cambridge University Press.

[85] G. Cybenko (1989) Approximation by superpositions of a sigmoidal function, *Math. Control, Signals, Syst., 2,* 303-314.

[86] M. Damashek (1995) Gauging similarity with n-grams: language independent categorization of text, *Science, 267,* 843.

[87] H.E. Daniels (1944) The relation between measures of correlation in the universe of sample permutations, *Biometrika,* 33, 129-135.

[88] B.P. Dawkins (1995) Investigating the geometry of a p-dimensional data set, *Journal of the American Statistical Association, 90, 429,* 350-359.

[89] W.H.E. Day (1985) Optimal algorithms for comparing trees with labeled leaves, *Journal of Classification, 2,* 7-28.

[90] W.H.E. Day (1987) Computational complexity of inferring phylogenies from dissimilarity matrices, *Bulletin of Mathematical Biology, 49,* 461-467.

[91] W.H.E. Day (1993) *Classification Literature Automated Search Service, 1992, 21,* Hanover, Pa.: Classification Society of North America.

[92] P. De Boeck and S. Rosenberg (1988) Hierarchical classes: model and data analysis, *Psychometrika, 53,* 361-381.

[93] M. Delattre and P. Hansen (1980) Bicriterion cluster analysis, *IEEE Transactions on Pattern Analysis and Machine Intelligence (PAMI), 4,* 277-291.

[94] W.S. De Sarbo (1982) GENNCLUS: new models for general nonhierarchical clustering analysis, *Psychometrika, 47,* 446-449.

[95] G. De Soete, W.S. De Sarbo, G.W. Furnas, and J.D. Carroll (1984) The estimation of ultrametric and path length trees from rectangular proximity data, *Psychometrika, 49,* 289-310.

[96] L. Devroye and T.J. Wagner (1977) The strong uniform consistency of nearest neighbor density estimates, *The Annals of Statistics, 5,* 536-540.

[97] P. Diamond and P. Kloeden (1994) *Metric Spaces of Fuzzy Sets,* Singapore: World Scientific Publishing Co.

[98] J. Diatta and B. Fichet (1994) From Apresian hierarchies and Bandelt-Dress weak hierarchies to quasi-hierarchies. In: E. Diday, Y. Lechevallier, M. Schader, P. Bertrand, and B. Burtschy (Eds.) *New Approaches in Classification and Data Analysis,* Springer–Verlag, 111-118.

[99] E. Diday (1986) Orders and overlapping clusters by pyramids. In: J. de Leeuw, W. Heiser, J. Meulman, and F. Critchley (Eds.) *Multidimensional Data Analysis*, Leiden: DSWO Press, 201-234.

[100] E. Diday and P. Bertrand (1986) An extension of hierarchical clustering: the pyramidal presentation. In: E.S. Gelsema and L.N. Kanal (Eds.) *Pattern Recognition in Practice II*, Amsterdam: North-Holland, 411-423.

[101] E. Diday et al. (1979) *Optimisation en Classification Automatique*, Roquencourt: INRIA.

[102] E. Diday, J.V. Moreaux (1984) Learning hierarchical clustering from examples, *Pattern Recognition Letters, 2*, 365-378.

[103] J. Dopazo, A. Dress, and A. von Haeseler (1993) Split decomposition: A technique to analyze viral evolution, *Proc. Natl. Acad. Sci. USA, 90*, 10320-10324.

[104] A.A. Dorofeyuk (1966) Potential function based algorithms for unsupervised learning, *Automation and Remote Control, 27*, 10, 1728-1736.

[105] A.A. Dorofeyuk (1971) Methods for Automatic Classification: A Review, *Automation and Remote Control, 32*, 12, 1928-1958.

[106] J.L. Dubien and W.D. Warde (1979) A mathematical comparison of the members of an infinite family of agglomerative clustering algorithms, *Canadian Journal of Statistics, 7*, 27-33.

[107] R.O. Duda and P.E. Hart (1973) *Pattern Classification and Scene Analysis*, New York: J.Wiley & Sons.

[108] A.F. Dutka and H.H. Hanson (1989) *Fundamentals of Data Normalization*, Reading, Ma.: Addison-Wesley.

[109] B.S. Duran and P.L. Odell (1974) *Cluster Analysis: A Survey*, Springer-Verlag: Berlin.

[110] C. Durand, B. Fichet (1988) One-to-one correspondences in pyramidal representation: a unified approach. In: H.-H. Bock (Ed.) *Classification and Related Methods of Data Analysis*, Amsterdam: Elsevier, 85-90.

[111] E. Durkheim and M. Mauss (1958) *Primitive Classification*, Chicago: The University of Chicago Press.

[112] R.L. Dykstra (1983) An algorithm for restricted least square regression, *Journal of the American Statistical Association, 78*, 837-842.

[113] T. Eckes and P. Orlik (1993) An error variance approach to two-mode hierarchical clustering, *Journal of Classification, 10*, 51-74.

[114] C. Edelbrock (1979) Comparing the accuracy of hierarchical clustering algorithms: The problem of classifying everybody, *Multivariate Behavioral Research, 14*, 367-384.

[115] S. Edgell (1993) *Class*, London: Routledge.

[116] A.W.F. Edwards and L.L. Cavalli-Sforza (1965) A method for cluster analysis, *Biometrics, 21*, 362-375.

[117] V.N. Elkina and N.G. Zagoruiko (1966) An alphabet for recognized objects, *Computing Systems, 12*, Novosibirsk: Institute of Mathematics Press (in Russian).

[118] G.F. Estabrook, C. Johnson, and F. McMorris (1975) An idealized concept of the true cladistic character, *Math. Biosci., 23*, 263-272.

[119] G.F. Estabrook, F.R. McMorris, and A. Meacham (1985) Comparison of undirected phylogenetic trees based on subtrees of four evolutionary units, *Syst. Zool., 34*, 193-200.

[120] W.K. Estes (1994) *Classification and Cognition*, New York: Oxford University Press.

[121] B.S. Everitt (1974) *Cluster Analysis*, New York: J.Wiley & Sons.

[122] B.S. Everitt, G. Dunn (1992) *Applied Multivariate Data Analysis*, New York: Oxford University Press.

[123] J.S. Farris, A.G. Kluge, M.J. Eckart (1970) A numerical approach to phylogenetic systematics, *Systematic Zoology, 19*, 172-189.

[124] D.L. Featherman and R.M. Hauser (1978) *Opportunity and Change*, New York: Academic Press.

[125] D.W. Fisher (1987) Knowledge acquisition via incremental conceptual clustering, *Machine Learning, 2*, 139-172.

[126] D. Fisher, L. Xu, J.R. Carnes, Y. Reich, S.J. Fenves, J. Chen, R. Shiavi, G. Biswas, and J. Weinberg (1993) Applying AI clustering to engineering tasks. *IEEE Expert, December*, 51-60.

[127] L. Fisher and J.W. Van Ness (1971) Admissible clustering procedures, *Biometrika, 58*, 91-104.

[128] R.A. Fisher (1936) The use of multiple measurements in taxonomic problems, *Annals of Eugenics, VII, p. II*, 179-188.

[129] W.M. Fitch (1981) A non-sequential method for constructing trees and hierarchical classifications, *Journal of Molecular Evolution, 18*, 30-37.

[130] C. Flament (1976) *L'Analyse Bouleenne des Questionnaires*, Paris: Masson.

[131] K. Florek, J. Lukaszewicz, H. Perkal, H. Steinhaus, and S. Zubrzycki (1951) Sur la liason et la division des points d'un ensemble fini, *Colloquium Mathematicum, 2*, 282-285.

[132] L.R. Ford and D.R. Fulkerson (1962) *Flows in Networks*, Princeton: Princeton University Press.

[133] E.B. Fowlkes, and C.L. Mallows (1983) A method for comparing two hierarchical clusterings, *Journal of the American Statistical Association, 78*, 553-584.

[134] W.J. Frawley, G. Piatetsky-Shapiro, and C.J. Matheus (1992) Knowledge discovery in databases: an overview, *Artificial Intelligence Magazine, 13*, 3, 57-70.

[135] H.P. Friedman and J. Rubin (1967) On some invariant criteria for grouping data, *Journal of the American Statistical Association, 62*, 1159-1178.

[136] D.R. Fulkerson and O.A. Gross (1965) Incidence matrices and interval graphs, *Pacific Journal of Mathematics, 15*, 835-855.

[137] G. Gallo, M.D. Grigoriadis, and R.E. Tarjan (1989) A fast parametric maximum flow algorithm and applications. *SIAM Journal on Computing, 18*, 30-55.

[138] O. Gascuel (1994) A note on Sattah and Tversky's, Saitou and Nei's, and Studier and Keppler's algorithms for inferring phylogenies from evolutionary distances, *Molecular Biology and Evolution, 11*, 961-963.

[139] W. Gaul and M. Schader (1994) Pyramidal classification based on incomplete dissimilarity data, *Journal of Classification, 11*, 171-194.

[140] F. Gebhardt (1994) Discovering interesting statements from a database, *Applied Stochastic Models and Data Analysis, 10*, 1-14.

[141] A. Genkin and I. Muchnik (1993) Fixed point approach to clustering, *Journal of Classification, 10*, 219-240.

[142] J.H. Gennari (1989) Focused concept formation. In: *6-th Machine Learning Workshop*, p. 379-382.

[143] A. Gifi (1990) *Nonlinear Multivariate Analysis*, Chichester: Wiley.

[144] M.A. Gluck and J.E. Corter (1985) Information, uncertainty, and the utility of categories. *Proceedings of the Seventh Annual Conference of the Cognitive Science Society*. Irvine, Ca: L. Erlbaum Associates, 283-287.

[145] E. Godehardt (1990)*Graphs as Structural Models*, Wiesbaden: Vieweg.

[146] G.H. Golub and C.F. Van Loan (1989) *Matrix Computations*, Baltimore: J. Hopkins University Press.

[147] L.A. Goodman (1981) Criteria for determinimg whether certain categories in a cross-classification table should be combined, with special reference to occupational categories in an occupational mobility table, *American Journal of Sociology*, *87*, 612-650.

[148] L.A. Goodman and W. Kruskal (1979) *Measures of Association for Cross Classifications*, New York: Springer-Verlag.

[149] A.D. Gordon (1996) Hierarchical classification. In: P. Arabie, L. Hubert, and G. De Soete (Eds.) *Classification and Clustering*, Singapore: World Scientific.

[150] D. Gorenstein (1994) *Classification of the Finite Simple Groups*, Providence, R.I. : American Mathematical Society.

[151] G. Govaert (1980) Classification croisée de tableaux de contingence. In: *"Premiéres Journées Internationales Analyse des Données et Informatique (Versailles 1977)"*. Paris: CNRS.

[152] G. Govaert (1989) La classification croisée, *La Revue de MODULAD*, 4, 9-36.

[153] J.C. Gower (1966) Some distance properties of latent root and vector methods used in multivariate analysis, *Biometrika*, *53*, 325-338.

[154] J.C. Gower (1967) A comparison of some methods of cluster analysis, *Biometrics*, *23*, 623-637.

[155] J.C. Gower (1985) Measures of similarity, dissimilarity and distance. In: S. Kotz, N.L. Johnson, and C.B.Read (Eds.) *Encyclopedia of Statistical Sciences*, *5*, 397-405.

[156] J.C. Gower and G.J.S. Ross (1969) Minimum spanning tree and single linkage cluster analysis, *Applied Statistics*, *18*, 54-64.

[157] R.L. Graham and P. Hell (1985) On the history of the minimum spanning tree problem, *Annals of the History of Computing*, *7*, 43-57.

[158] M.J. Greenacre (1988) Clustering the rows and columns of a contingency table, *Journal of Classification*, *5*, 39-51.

[159] M.J. Greenacre (1993)*Correspondence Analysis in Practice*, Academic Press: San Diego, Ca.

[160] A. Guénoche, P. Hansen, and B. Jaumard (1991) Efficient algorithms for divisive hierarchical clustering with the diameter criterion, *Journal of Classification*, *8*, 5-30.

[161] D. Gusfield (1991) Efficient algorithms for inferring evolutionary trees, *Networks, 21*, 19-28.

[162] P. Hansen, B. Jaumard, and E. Da Silva (1993) Average-linkage divisive hierarchical clustering, *Les Cahiers du GERAD, G-91-55*, Montréal.

[163] P. Hansen and B. Jaumard (1993) Maximum split single cluster clustering, *Les Cahiers du GERAD*, 1-7.

[164] P. Hansen, B. Jaumard, and N. Mladenovic (1995) How to choose K entities among N. In: I.J. Cox, P. Hansen, and B. Julesz (Eds.)*Partitioning Data Sets*. DIMACS Series in Discrete Mathematics and Theoretical Computer Science, American Mathematical Society, 105-116.

[165] P. Hansen and N. Mladenovic (1992) A comparison of algorithms for the maximum clique problem, *Les Cahiers du GERAD*, G-92-28, 1-20.

[166] J.A. Hartigan (1967) Representation of similarity matrices by trees, *J. Amer. Stat. Assoc., 62*, 1140-1158.

[167] J.A. Hartigan (1972) Direct Clustering of a Data Matrix, *Journal of American Statistical Association, 67*, 123-129.

[168] J.A. Hartigan (1975) *Clustering Algorithms*, New York: J.Wiley & Sons.

[169] J.A. Hartigan (1976) Modal blocks in dentition of west coast mammals, *Systematic Zoology, 25*, 149-160.

[170] F. Hausdorff (1957) *Set Theory*, New York: Chelsea Pub. Co.

[171] P. Helman, B.M.E. Moret, H.D. Shapiro (1993) An exact characterization of greedy structures, *SIAM J. Disc. Math., 6*, 274-283.

[172] M.O. Hill (1979) TWINSPAN: a FORTRAN program for arranging multivariate data in an ordered two-way table by classification of the individuals and attributes. *Ecology and Systematics*, Ithaca, NY: Cornell University.

[173] K.J. Holzinger and H.H. Harman (1941) *Factor Analysis*, Chicago: University of Chicago Press.

[174] M. Hout (1986) *Mobility Tables*, Beverly Hills: Sage Publications.

[175] G. Howard, G.W. Evans, K. Pearce, V.J. Howard, R.A. Bell, E.J. Mayer, and G.L. Burke (1995) Is the Stroke Belt disappearing? *Stroke, 26*, 1153-1157.

[176] W.-L. Hsu and G.L. Nemhauser (1979) Easy and hard bottleneck location problems, *Discrete Applied Mathematics, 1*, 209-215.

[177] Z. Hubalek (1982) Coefficients of association and similarity, based on binary data: an evaluation. *Biological Review, 57*, 669-689.

[178] L.J. Hubert (1987) *Assignment Methods in Combinatorial Data Analysis*, New York: M. Dekker.

[179] L. Hubert and P. Arabie (1985) Comparing partitions, *Journal of Classification*, *2*, 193-218.

[180] L. Hubert and P. Arabie (1994) The analysis of proximity matrices through sums of matrices having (anti)-Robinson forms, *British Journal of Mathematical and Statistical Psychology*, *47*, 1-40.

[181] L. Hubert and J. Schultz (1976) Quadratic assignment as a general data-analysis strategy, *British Journal of Mathematical and Statistical Psychology*, *29*, 190-241.

[182] J.G. Hutchinson (1967) *Organizations: Theory and Classical Concepts*, New York: Holt, Rinehart and Winston.

[183] W.F. Hyde (1981) *Improving Productivity by Classification, Coding, and Data Base Standardisation*, New York: Marcel Dekker.

[184] A.G. Ivakhnenko, Y.V. Koppa, S.A. Petukhova, and M.A. Ivakhnenko (1985) Self-organization in partitioning of a data set in unknown number of clusters, *Automation*, 9-16 (in Russian).

[185] A.K. Jain and R.C. Dubes (1988) *Algorithms for Clustering Data*, Englewood Cliffs, NJ: Prentice Hall.

[186] M. Jambu (1978) *Classification Automatique pour l'Analyse des Données,I-Méthodes et Algorithms*, Paris: Dunod.

[187] K. Janich (1994) *Linear Algebra*, New York: Springer-Verlag.

[188] M.F. Janowitz (1978) An order theoretic model for cluster analysis, *SIAM J. Appl. Math.*, *34*, 55-72.

[189] M.F. Janowitz and R. Stinebrickner (1993) Compatibility in a graph-theoretic setting, *Math. Social Sciences*, *25*, 251-279.

[190] M.F. Janowitz and R. Wille (1995) On the classification of monotone-equivariant cluster methods. In: I.J. Cox, P. Hansen, and B. Julesz (Eds.)*Partitioning Data Sets*. DIMACS Series in Discrete Mathematics and Theoretical Computer Science, American Mathematical Society, 117-142.

[191] C.J. Jardine, N. Jardine, and R. Sibson (1967) The structure and construction of taxonomic hierarchies, *Math. Bioscience, 1*, 173-179.

[192] N. Jardine and R. Sibson (1971) *Mathematical Taxonomy*, New York: J.Wiley & Sons.

[193] C.V. Jawahar, P.K. Biswas, and A.K. Ray (1995) Detection of clusters of distinct geometry: A step toward generalized fuzzy clustering, *Pattern Recognition Letters, 16*, 1119-1123.

[194] W.S. Jevons (1958) *The Principles of Science*, New York: Dover Publications.

[195] I.T. Jolliffe (1986) *Principal Component Analysis*. New York: Springer-Verlag.

[196] G. John, R. Kohavi, and K. Pfleger (1994) Irrelevant features and the subset selection problem. In *Proceedings of the Eleventh International Machine Learning Conference* (pp. 121-129). New Brunswick, NJ: Morgan Kaufmann.

[197] S.C. Johnson (1967) Hierarchical clustering schemes, *Psychometrika, 32*, 241-245.

[198] B. Kamgar-Parsi, J.A. Gualtieri, J.E. Devaney, and B. Kamgar-Parsi (1990) Clustering with neural networks, *Biological Cybernetic, 63*, 201-208.

[199] Y. Kempner, B. Mirkin, and I. Muchnik (1995) Monotone linkage clustering and quasi-convex set functions, DIMACS Technical Report 95-44, 12 p.

[200] G. Keren and S. Baggen (1981) Recognition models of alphanumeric characters, *Perception and Psychophysics*, 234-246.

[201] B. Kernighan and S. Lin (1972) An efficient heuristic procedure for partitioning graphs, *Bell Systems Journal, 49*, 291-307.

[202] T. Kohonen (1989) *Self-Organization and Associative Memory*, Berlin: Springer-Verlag.

[203] T. Kohonen (1995) *Self-Organizing Maps*, Berlin: Springer-Verlag.

[204] A.P. Kovalenko (1993) Algorithm for k nearest neighbor unimodal clustering, *Automation and Remote Control, 54*, 5, 100-105.

[205] W.J. Krzanowski, F.H.C. Marriott (1994) *Multivariate Analysis*, London: Edward Arnold.

[206] M. Kubat, G. Pfurtscheller, and D. Flotzinger (1994) AI-based approach to automatic sleep classification, *Biological Cybernetic, 70*, 443-448.

[207] V.L. Kupershtoh and B.G. Mirkin (1971) Ordering of interrelated objects I, II, *Automation and Remote Control, 32*, 6, 924-929, 7, 1093-1098.

[208] V. Kupershtoh, B. Mirkin, and V. Trofimov (1976) Sum of within partition similarities as a clustering criterion, *Automation and Remote Control, 37*, 2, 548-553.

[209] V. Kupershtoh and V. Trofimov (1975) An algorithm for analysis of the structure in a proximity matrix, *Automation and Remote Control, 36*, 11, 1906-1916.

[210] P.J.M. van Laarhoven and E.H.L. Aarts (1987) *Simulated Annealing: Theory and Applications*, Dordrecht: Kluwer.

[211] G.N. Lance, and W.T. Williams (1967) A general theory of classificatory sorting strategies: 1. Hierarchical Systems, *Comp. Journal, 9*, 373-380.

[212] D.W. Langridge (1992) *Classification: Its Kinds, Systems, Elements and Applications*, London: Bowker-Saur.

[213] F.-J. Lapointe and P. Legendre (1992) Statistical significance of the matrix correlation coefficient for comparing independent phylogenetic trees, *Systematic Biology, 41*, 378-384.

[214] G.S. Lbov (1981) *Methods for Mixed Data Processing*, Novosibirsk: Nauka (in Russian).

[215] L. Lebart, B. Mirkin (1993) Correspondence analysis and classification. In: C.M. Quadras and C.R. Rao (Eds.) *Multivariate Analysis: Future Directions 2*, Amsterdam: North-Holland, 341-357.

[216] L. Lebart, A. Morineau, K. Warwick (1984)*Multivariate Descriptive Statistical Analysis*, New York: J.Wiley & Sons.

[217] B. Leclerc (1981) Description combinatoire des ultrametriques, *Mathematiques et Sciences Humaines, 73*, 5-37.

[218] B. Leclerc (1985) La comparaison des hierarchies: indices and metriques, *Mathematiques et Sciences Humaines, 92*, 5-40.

[219] B. Leclerc (1995) Minimum spanning trees for tree metrics: abridgments and adjustments, *Journal of Classification, 12*, 207-242.

[220] J. de Leeuw (1994) Block-relaxation algorithms in statistics. In: H.-H. Bock, W. Lenski, and M.M. Richter (Eds.) *Information Systems and Data Analysis*, Berlin: Springer-Verlag, 308-324.

[221] G. Lenski (1994) Societal taxonomies: mapping the social universe, *Annual Review of Sociology, 20*, 1-26.

[222] V. Levit (1988) An algorithm for finding a maximum perimeter submatrix containing only unity, in a zero/one matrix, In: V.S. Pereverzev-Orlov (Ed.) *Systems for Transmission and Processing of Data*, Moscow: Institute of Information Transmission Science Press, 42-45 (in Russian).

[223] W.-H. Li, and D. Graur (1991) *Fundamentals of Molecular Evolution*, Sunderland, Ma: Sinauer Associates.

[224] R.F. Ling (1973) Probability theory of cluster analysis, *Journal of American Statistical Association, 68*, 159-164.

[225] R.D. Luce, P.R. Bush and E. Galanter (Eds.) (1963-1965) *Handbook of Mathematical Psychology*, New York: J. Wiley & Sons.

[226] G.F. Luger and W.A. Stubblefield (1993) *Artificial Intelligence: Structures and Strategies for Complex Problem Solving*, Redwood city, Ca.: The Benjamin/Cummings.

[227] J.B. MacQueen (1967) Some methods for classification and analysis of multivariate observations, *Proceedings of 5th Berkeley Symposium, 2*, 281-297.

[228] I.D. Mandel (1988) *Cluster Analysis*, Moscow: Finansy i Statistika (in Russian).

[229] N. Mantel (1967) The detection of desease clustering and a generalized regression approach, *Cancer Research, 27*, 209-220.

[230] F. Marcotorchino (1987) Block seriation problems: a unified approach, *Journal of Applied Stochastical Models and Data Analysis, 3*, no.3, 73-93.

[231] T. Margush and F.R. McMorris (1981) Consensus n-trees, *Bulletin of Mathematical Biology, 43*, 239-244.

[232] S. McGuinness (1994) The greedy clique decomposition of a graph, *Journal of Graph Theory*, 18, 427-430.

[233] G. McLachlan and K. Basford (1988) *Mixture Models: Inference and Applications to Clustering*, New York: Marcel Dekker.

[234] J.E. Mezzich and H. Solomon (1980) *Taxonomy and Behavioral Science*, London: Academic Press.

[235] R.S. Michalski (1992). Concept learning. In: S.C. Shapiro (Ed.) *Encyclopedia of Artificial Intelligence*, New York: J. Wiley & Sons, 249-259.

[236] R.S. Michalski and R.E. Stepp (1983) Learning from observation: Conceptual clustering. In: R.S. Michalski, J. Carbonell, and T.M. Mitchell (Eds.)*Machine Learning: An Artificial Intelligence Approach*, Palo Alto, Ca.: Tioga, 331-363.

[237] G.W. Milligan (1979) Ultrametric hierarchical clustering algorithms, *Psychometrika, 44*, 343-346.

[238] G.W. Milligan (1981) A Monte Carlo study of thirty internal criterion measures for cluster analysis, *Psychometrika, 46*, 187-199.

[239] G.W. Milligan and M.C. Cooper (1986) A study of the comparability of external criteria for hierarchical cluster analysis, *Multivariate Behavioral Research, 21*, 441-458.

[240] G.W. Milligan and M.C. Cooper (1988) A study of standardization of the variables in cluster analysis, *Journal of Classification, 5*, 181-204.

[241] B. Mirkin (1975) On the problem of reconciling partitions, In: H. Blalock, A. Aganbegian, F. Borodkin, R. Boudon, and V. Capecchi (Eds.) *Quantitative Sociology: International Perspectives on Mathematical and Statistical Modeling*, New York: Academic Press, 441-449.

[242] B.G. Mirkin (1985) *Grouping in Socio-Economical Studies: Methods for Constructing and Analyzing*, Moscow: Finansy i Statistika (in Russian).

[243] B. Mirkin (1987a) Method of principal cluster analysis, *Automation and Remote Control, 48*, 1379-1388.

[244] B. Mirkin (1987b) Additive clustering and qualitative factor analysis methods for similarity matrices, *Journal of Classification, 4*, 7-31; Erratum (1989), *6*, 271-272.

[245] B. Mirkin (1990) A sequential fitting procedure for linear data analysis models, *Journal of Classification, 7*, 167-195.

[246] B. Mirkin (1994) Approximation of association data by structures and clusters, In: P.M. Pardalos and H. Wolkowicz (Eds.) *Quadratic Assignment and Related Problems*, DIMACS Series, Providence: American Mathematical Society, 293-316.

[247] B. Mirkin (1995a) Linear embedding the binary hierarchies, and its applications in clustering and querying, in J. Albus, A. Meystel, D. Pospelov and T. Reader (Eds.) *Proceedings of 1995 ISIC Workshop (10th IEEE International Symposium on Intelligent Control)*, Bala Cynvyd, Pa.: Adrem, 259-269.

[248] B. Mirkin (1995b) Clustering for contingency data: boxes and partitions, *Statistics and Computing* (to appear).

[249] B. Mirkin (1995c) Concept learning and feature selecting based on square-error clustering, *in press*.

[250] B. Mirkin, P. Arabie and L Hubert (1995) Additive two-mode clustering: the error-variance approach revisited, *Journal of Classification, 12*, 243-263.

[251] B.G. Mirkin and L.B. Cherny (1970) On a distance measure between partitions of a finite set, *Automation and Remote Control, 31*, 5, 786-792.

[252] B. Mirkin and I. Muchnik (1996) Clustering and multidimensional scaling in Russia (1960-1990): a review. In: P. Arabie, L. Hubert, G. De Soete (Eds.) *Classification and Clustering*, River Edge, NJ: World Scientific Publishers, 295-339.

[253] B. Mirkin, I. Muchnik, and T. Smith (1995) A biologically consistent model for comparing molecular phylogenies, *J. of Computational Biology*, *2*, 4, 493-507. forthcoming.

[254] B.G. Mirkin and V.V. Panfilova (1991) Analysis of socio-economical status of territorial units via multivariate clustering, *Economics and Mathematical Methods*, n.1, 212-221, Moscow (in Russian).

[255] B. Mirkin and S. Rodin (1984) *Graphs and Genes*, Bonn: Springer-Verlag (original work published in 1977, Moscow, in Russian).

[256] B.G. Mirkin and P.S. Rostovtsev (1978) Method to reveal associated subsets of the variables, In: B.G. Mirkin (Ed.) *Models for Socioeconomic Data Aggregation*, Novosibirsk: Institute of Economics Press, 107-112 (in Russian).

[257] B. Mirkin and G. Satarov (1990) Fuzzy additive types method in data analysis: I, II, *Automation and Rem. Control, 51*, 5, 683-688, 6,817-821.

[258] B. Mirkin and M. Yeremin (1991) *ClassMaster: A User's Guide*, Moscow: Stat-Dialogue (in Russian).

[259] B. Monjardet (1981) Metrics on partially ordered sets: A survey, *Discrete Mathematics, 35*, 173-184.

[260] J. Mullat (1976) Extremal subsystems of monotone systems: I, II, III; *Automation and Remote Control, 37*, 758-766, *37*, 1286-1294; *38*, 89-96.

[261] F. Murtagh (1996) Neural networks for clustering, In: P. Arabie, L. Hubert, G. De Soete (Eds.) *Combinatorial Data Analysis*, River Edge, NJ: World Scientific.

[262] S.K. Murti, S. Kasif, and S. Salzberg (1994) A system for induction of oblique decision trees, *Journal of Artificial Intelligence Research, 2*, 1-32.

[263] V.D. Nebylitsyn (1972) *Fundamental Properties of the Human Nervous System*, New York: Plenum Press.

[264] N.J. Nilsson (1965) *Learning Machines*, New York: McGraw-Hill.

[265] S. Nishisato (1994) *Elements of Dual Scaling: An Introduction to Practical Data Analysis*, Hillsdale, NJ: L.Erlbaum Associates.

[266] R. Ornstein (1985) *Psychology: The Study of Human Experience*, Orlando, Fla.: Hartcourt Brace Jovanovich.

[267] S. Osawa, and T. Honjo (Eds.) (1991) *Evolution of Life: Fossils, Molecules, and Culture*, Tokyo: Springer–Verlag.

[268] N. Oshumi, and N. Nakamura (1989) Space-distorting properties in agglomerative hierarchical clustering algorithms and a simplified method for combinatorial method. In: E. Diday (Ed.) *Data Analysis, Learning Symbolic and Numerical Knowledge*, New York: Nova Science Publishers, 103-108.

[269] C.V. Packer (1989) Applying row-column permutation to matrix representations of large citation networks, *Information Processing and Management, 25*, 307-314.

[270] G. Pagallo and D. Haussler (1990) Boolean feature discovery in empirical learning, *Machine Learning, 5*, 71-99.

[271] C.H. Papadimitriou and K. Steiglitz (1982) *Combinatorial Optimization: Algorithms and Complexity*, Englewood Cliffs, NJ: Prentice-Hall.

[272] P.M. Pardalos, F. Rendl, and H. Wolkowicz (1994) The quadratic assignment problem: a survey and recent developments. In: P. Pardalos and H. Wolkowicz (Eds.) *Quadratic Assignment and Related Problems*. DIMACS Series in Discrete Mathematics and Theoretical Computer Science, v. 16. American Mathematical Society.

[273] J.A. Paulos (1980) *Mathematics and Humor*, Chicago: The University of Chicago Press.

[274] D. Penny and M.D. Hendy (1985) The use of tree comparison metrics, *Syst. Zool., 34*, 75-82.

[275] D.T. Pham and E.J. Bayro-Corrochano (1994) Self-organizing neural-network-based pattern clustering method with fuzzy outputs, *Pattern Recognition, 27*, 1103-1110.

[276] B.T. Polak (1983) *Introduction to Optimization*, Moscow: Nauka (in Russian).

[277] J.R. Quinlan (1986) Induction of decision trees, *Machine Learning, 1*, 81-106.

[278] H. Ralambondrainy (1995) A conceptual version of the K-means algorithm, *Pattern Recognition Letters, 16*, 1147-1157.

[279] W.M. Rand (1971) Objective criteria for the evaluation of clustering methods, *Journal of the American Statistical Association, 66*, 846-850.

[280] H.T. Reynolds (1977) *The Analysis of Cross-Classifications*, New York: The Free Press.

[281] J. Rissanen (1989) *Stochastic Complexity in Statistical Inquiry*, Singapore: World Scientific Publishing Co.

[282] F.S. Roberts (1979) Indifference and seriation, *Annals New York Acad. Sci.*, 173-182.

[283] F.S. Roberts (1979) *Measurement Theory with Applications to Decision-Making, Utility, and the Social Sciences*, Reading, Mass. : Addison-Wesley.

[284] D.F. Robinson (1971) Comparison of labeled trees with valency three, *Journal of Combinatorial Theory*, *11*, 105-119.

[285] W.S. Robinson (1951) A method for chronologically ordering archaelogical deposites, *Amer. Antiquity*, *19*, 293-301.

[286] F.J. Rohlf (1970) Adaptive hierarchical clustering schemes, *Systematic Zoology*, *18*, 58-82.

[287] F.J. Rohlf (1974) Methods of comparing classifications, *Annual Review of Ecology and Systematics*, *5*, 101-113.

[288] H.C. Romesburg (1984) *Cluster Analysis for Researchers*, Belmont, Ca.: LLP.

[289] S. Rosenberg (1982) The method of sorting in multivariate research with applications selected from cognitive psychology and person perception. In: N. Hirchberg and L.G. Humphreys (Eds.) *Multivariate Applications in the Social Sciences*, University of Illinois at Urbana–Champaign: L. Erlbaum Assoc., 117 - 142.

[290] S. Rosenberg, I. Van Mechelen, and P. De Boeck (1996) A hierarchical class model: theory and method with applications in psychology and psychopathology. In: P. Arabie, L. Hubert, and G. De Soete (Eds.) *Classification and Clustering*, River Edge, NJ: World Scientific.

[291] P.S. Rostovtsev and B.G. Mirkin (1985) Methods for relative hierarchical grouping. In: B.G. Mirkin *Grouping in Socio-Economical Studies: Methods for Constructing and Analyzing*, Moscow: Finansy i Statistika, 126-134 (in Russian).

[292] D.E. Rumelhart, G.E. Hilton and R.J. Wilson (1986) Learning internal representation by error propagation. In: D.E. Rumelhart and J.L. McClelland (Eds.) *Parallel Distributed Processing: Explorations in the Microstructures of Cognition*, *1*, Cambridge: MIT Press, 318-362.

[293] H.J. Ryser (1973) Intersection properties of finite sets, *Journal of Combinatorial Theory*, *A14*, 79-92.

[294] N. Saitou and M. Nei (1987) The neighbor-joining method: a new method for reconstructing phylogenetic trees, *Molecular Biology and Evolution, 4*, 406-425

[295] G. Saporta (1988) About maximal association criteria in linear analysis and in cluster analysis, In: H.-H. Bock (Ed.) *Classification and Related Methods of Data Analysis*, Amsterdam: Elsevier, 541-550.

[296] H. Saran and V.V. Vazirani (1995) Finding k cuts within twice the optimal, *SIAM Journal on Computing, 24*, 1, 101-108.

[297] *SAS User's Guide: Statistics* (1982), SAS Institute, Cary N.C.

[298] G.A. Satarov (1991) Conceptual control: a new method for testing knowledge (Personal communication).

[299] S. Sattah and A. Tversky (1977) Additive similarity trees, *Psychometrika, 42*, 319-345.

[300] W.C. B. Sayers (1955) *An Introduction to Library Classification (ninth edition)*, London: Grafton & Co.

[301] Y.A. Schreider and A.A. Sharov (1982) *Systems and Models*, Moscow: Radio i Sviaz' (in Russian).

[302] M.R. Segal (1995) Extending the elements of tree-structured regression, *Statistical Methods in Medical Research, 4*, 219-236.

[303] M. Senechal (1990) *Crystalline Symmetries: An Informal Mathematical Introduction*, Bristol: Adam Hilger.

[304] R.N. Shepard (1966) Metric structures in ordinal data, *Journal of Mathematical Psychology, 3*, 287-315.

[305] R.N. Shepard (1988) George Miller's data and the development of methods for representing cognitive structures. In: W. Hirst (Ed.) *The Making of Cognitive Science: Essays in Honor of George A. Miller*, Cambridge: Cambridge University Press, 45-70.

[306] R.N. Shepard and P. Arabie (1979) Additive clustering: representation of similarities as combinations of overlapping properties, *Psychological Review, 86*, 87-123.

[307] J. Sherzer (1976) *An Areal-Typological Study of American Indian Languages North of Mexico*, Amsterdam: North-Holland.

[308] M.I. Shlezinger (1965) On unsupervised pattern recognition, In: V.M. Glushkov (Ed.) *Reading Automata*, Kiev: Naukova Dumka, 62-70 (in Russian).

[309] H.-Y. Shum, K. Ikeuchi, and R. Reddy (1995) PCA with missing data and its application to polyhedral object modeling, *IEEE Transactions on Pattern Analysis and Machine Intelligence, 17*, 9, 854-867.

[310] H.H. Sisler (1963) *Electronic Structure, Properties, and Periodic Law*, New York: Reinhold Publ.

[311] R.F. Smith and T.F. Smith (1990) Automatic generation of diagnostic sequence patterns from sets of related protein sequences, *Proc. Natl. Acad. Sci. USA, 87*, 118-122.

[312] Y.A. Smolensky (1969) A method for linear recording of graphs, *USSR Comput. Math. Phys., 2*, 396-397.

[313] P.H.A. Sneath (1995) Thirty years of numerical taxonomy, *Systematic Biology, 44*, 281-298.

[314] P.H.A. Sneath and R.R. Sokal (1973) *Numerical Taxonomy*, San Francisco: W.H. Freeman.

[315] R.R. Sokal and F.J. Rohlf (1962) The comparison of dendrograms by objective methods, *Taxon, 9*, 33-40.

[316] R.H. Somers (1962) A new asymmetric measure of association for ordinal variables, *American Sociological Review, 27*, 799-811.

[317] J.A. Sonquist, E.L. Baker, and J.N. Morgan (1973) *Searching for Structure*, Ann Arbor: Institute for Social Research, University of Michigan.

[318] E. Sontag (1989) Sigmoids distinguish more efficiently than Heavisides, *Neural Computation, 1*, 470-472.

[319] H. Spaeth (1985) *Cluster Dissection and Analysis*, Chichester: Ellis Horwood.

[320] H. Spaeth (1988) Homogeneous and heterogeneous clusters for distance matrices. In: H.-H. Bock (Ed.) *Classification and Related Methods of Data Analysis*, Amsterdam: North Holland, 157-164.

[321] M.A. Steel and D. Penny (1993) Distributions of tree comparison metrics - some new results, *Systematic Biology, 42*, 126-141.

[322] J.A. Studier and K.J. Keppler (1988) A note on neighbor-joining algorithm of Saitou and Nei, *Molecular Biology and Evolution, 5*, 729-731.

[323] C.J.F. Ter Braak (1986) Interpreting a hierarchical classification with simple discriminant functions: an ecological example. In: E.Diday et al. (Eds.) *Data Analysis and Informatics, IV*, Amsterdam: North-Holland, 11-21.

[324] R.C. Tryon (1939) *Cluster Analysis*, Edwards Bros.: Ann Arbor.

[325] L.R. Tucker (1964) The extension of factor analysis to three-dimensional matrices. In: N. Frederiksen, H. Gulliksen (Eds.) *Contributions to Mathematical Psychology*, New York: Holt, Rinehart, and Winston, p. 109-127.

[326] A. Tversky (1977) Features of similarity, *Psychological Review, 84*, 327-352.

[327] S. Van Buuren and W.J. Heiser (1989) Clustering N objects into K groups under optimal scaling of variables, *Psychometrika, 54*, 699-706.

[328] B. Van Cutsem (Ed.) (1994) *Classification and Dissimilarity Analysis*, Lecture Notes in Statisctics, 93, New York: Springer-Verlag.

[329] J. Van Ryzin (Ed.) (1977) *Classification and Clustering*, Academic Press: New York.

[330] V.N. Vapnik (1982) *Estimation of Dependencies Based on Empirical Data*, New York: Springer-Verlag.

[331] N. Vyssotskaya (1980) Method of chain classification in analysis of the life styles. In: B.G. Mirkin (Ed.) *Models for data analyzing and decision making*, Novosibirsk: Institute of Economics Press, 93-113 (in Russian)

[332] J.H. Ward, Jr (1963) Hierarchical grouping to optimize an objective function, *Journal of American Statist. Assoc., 58*, 236-244.

[333] S. Wasserman and K. Faust (1992) *Social Network Analysis: Methods and Applications*, New York: Cambridge University Press.

[334] M.S. Waterman, T.F. Smith, M. Singh, and W.A. Beyer (1977) Additive evolutionary trees, *Journal of Theoretical Biology, 64*, 199-213.

[335] M.S. Waterman, and T.F. Smith (1978) On the similarity of dendrograms, *Journal of Theor. Biology, 73*, 789-800.

[336] L. Wilkinson (1989) *SYSTAT: The System for Statistics*, Evanston, Il.: SYSTAT.

[337] R. Wille (1989) Knowledge acquisition by methods of formal concept analysis. In: E. Diday (Ed.) *Data Analysis, Learning Symbolic and Numerical Knowledge*, New York: Nova Science Publishers, 365-380.

[338] R. Wille (1991) Local completeness of conceptual knowledge systems. In: E. Diday and Y. Lechevallier (Eds.) *Symbolic-Numeric Data Analysis and Learning.* New York: Nova Science Publishers, 347-356.

[339] W.T. Williams and J.M. Lambert (1959) Multivariate methods in plant ecology. 1. Association analysis in plant communities, *Journal of Ecology, 47*, 83-101.

[340] M.P. Windham (1986) A unification of optimization-based numerical classification algorithms. In: W. Gaul and M. Shader (Eds.) *Classification as a Tool of Research*, Amsterdam: North-Holland, 447-452.

[341] D. Wishart (1969) Mode analysis: A generalization of nearest neighbor which reduces chaining effects. In: A.J. Cole (Ed.) *Numerical Taxonomy*, London: Academic Press, 282-319.

[342] J. Wnek and R.S. Michalski (1994) Hypothesis-driven constructive induction in AQ17-HCI: A method and experiments. *Machine Learning, 14*, 139-168.

[343] C.R. Woese (1971) Archaebacteria, *Scientific American, 244*, 6, 98-122.

[344] M.A. Wong, and T. Lane (1981) A kth nearest neighbour clustering procedure. In: *Comput. Sci. and Statistics*, Proceedings of 13th Symposium, Pittsburgh: Interface, 308-311.

[345] J. Xue (1994) Edge-maximal triangulated subgraphs and heuristics for the maximum clique problem, *Networks*, 24, 109-120.

[346] F.W. Young, R.A. Faldowski, and M.M. McFarlane (1993) Multivariate statistical visualization. In: C.R. Rao (Ed.) *Handbook of Statistics, 9*, Amsterdam: Elsevier, 959-998.

[347] M. Zacklad and D. Fontaine (1993) Systematic building of conceptual classification systems with C-KAT. In: N. Aussenac, G. Boy, B. Gaines, M. Linster, J.-G. Ganascia, and Y. Kodratoff (Eds.) *Knowledge Acquisition for Knowledge-Based Systems*. Lecture Notes in Artificial Intelligence, 723, Springer-Verlag, 79-102.

[348] N.G. Zagoruyko (1972) *Recognition Methods and Their Applications*, Moscow: Sovetskoye Radio (in Russian).

[349] Yu. Zaks, and I. Muchnik (1989) Incomplete Classifications of a Finite Set of Objects Using Monotone Systems, *Automation and Remote Control, 50*, 553-560.

[350] K.A. Zaretsky (1965) Reconstruction of a tree from the distances between its pendant vertices, *Uspekhi Math. Nauk (Russian Mathematical Surveys)*, *20*, 90-92 (in Russian).

[351] Q. Zhang and R.D. Boyle (1992) A new clustering algorithm with multiple runs of iterative procedures, *Pattern Recognition, 25*, 835-849.

[352] H.-J. Zimmerman (1991) *Fuzzy Set Theory and Its Applications*, Kluwer: Dordrecht.

Index

Nonconvex Optimization and Its Applications

KLUWER ACADEMIC PUBLISHERS – DORDRECHT / BOSTON / LONDON